Plastids

Annual Plant Reviews

A series for researchers and postgraduates in the plant sciences. Each volume in this series focuses on a theme of topical importance and emphasis is placed on rapid publication.

Editorial Board:

Professor Jeremy A. Roberts (Editor-in-Chief), Plant Science Division, School of Biosciences, University of Nottingham, Sutton Bonington Campus, Loughborough, Leics, LE12 5RD, UK. **Professor Hidemasa Imaseki**, Obata-Minami 2 4 19, Moriyama-ku, Nagoya 463, Japan. **Dr Michael McManus**, Department of Plant Biology and Biotechnology, Massey University, Palmerston North, New Zealand. **Professor David G. Robinson**, Heidelberg Institute for Plant Sciences, University of Heidelberg, Im Neuenheimer Feld 230, D-69120 Heidelberg, Germany. **Dr Jocelyn Rose**, Department of Plant Biology, Cornell University, Ithaca, New York 14853, USA.

Titles in the series:

Plastids

Edited by

SIMON GEIR MØLLER
Department of Biology
University of Leicester
UK

Blackwell
Publishing

CRC Press

© 2005 by Blackwell Publishing Ltd

Editorial offices:
Blackwell Publishing Ltd, 9600 Garsington
Road, Oxford OX4 2DQ, UK
    Tel: +44 (0)1865 776868
Blackwell Publishing Asia Pty Ltd, 550
Swanston Street, Carlton, Victoria 3053,
Australia
    Tel: +61 (0)3 8359 1011

ISBN 1-4051-1882-2
ISSN 1460-1494

Published in the USA and Canada (only) by
CRC Press LLC, 2000 Corporate Blvd., N.W.,
Boca Raton, FL 33431, USA Orders from the
USA and Canada (only) to CRC Press LLC

USA and Canada only:
ISBN 0-8493-2352-5
ISSN 1097-7570

This book contains information obtained from
authentic and highly regarded sources.
Reprinted material is quoted with permission,
and sources are indicated. Reasonable efforts
have been made to publish reliable data and
information, but the author and the publisher
cannot assume responsibility for the validity
of all materials or for the consequences of
their use.

Trademark notice: Product or corporate
names may be trademarks or registered
trademarks, and are used only for
identification and explanation, without intent
to infringe.

First published 2005

Library of Congress
Cataloging-in-Publication Data:
A catalog record for this title is available from
the Library of Congress

British Library Cataloguing-in-Publication
Data:
A catalogue record for this title is available
from the British Library

Set in 10/12 pt Times
by TechBooks
Printed and bound in Great Britain
by MPG Ltd, Bodmin, Cornwall.

The publisher's policy is to use permanent
paper from mills that operate a sustainable
forestry policy, and which has been
manufactured from pulp processed using
acid-free and elementary chlorine-free
practices. Furthermore, the publisher ensures
that the text paper and cover board used have
met acceptable environmental accreditation
standards.

For further information on Blackwell
Publishing, visit our website:
www.blackwellpublishing.com

Contents

3   Plastid metabolic pathways   60

IAN J. TETLOW, STEPHEN RAWSTHORNE, CHRISTINE
RAINES and MICHAEL J. EMES

List of contributors

Dr Zach Adam

The Robert H. Smith Institute of Plant Sciences and Genetics in Agriculture, The Hebrew University of Jerusalem, Rehovot 76100, Israel

Professor Henry Daniell

Department of Molecular Biology and Microbiology, University of Central Florida, Biomolecular Science, Bldg #20, Room 336, Orlando, FL 32816-2364, USA

Dr Andrew Devine

Department of Molecular Biology and Microbiology, University of Central Florida, Biomolecular Science, Bldg #20, Room 336, Orlando, FL 32816-2364, USA

Professor Michael J. Emes

College of Biological Sciences, University of Guelph, Guelph, Ontario N1G 2W1, Canada

Professor John C. Gray

Department of Plant Sciences, University of Cambridge, Downing Street, Cambridge CB2 3EA, UK

Dr Masahiro Kasahara

Gene Research Center, Tokyo University of Agriculture and Technology, Fuchu, Tokyo 183-8509, Japan

Dr Dario Leister

Abteilung für Pflanzenzüchtung
und Ertragsphysiologie,
Max-Planck-Institut für
Züchtungsforschung,
Carl-von-Linné Weg 10,
D-50829 Köln, Germany

Dr Alexandra Mant

Plant Biochemistry Laboratory,
Department of Plant Biology, Royal
Veterinary and Agricultural
University, Thorvaldsensvej 40,
DK-1871 Frederiksberg C, Denmark

Dr Simon Geir Møller

Department of Biology,
University of Leicester, University
Road, Leicester LE1 7RH, UK

Dr Paolo Pesaresi

Abteilung für Pflanzenzüchtung
und Ertragsphysiologie,
Max-Planck-Institut für
Züchtungsforschung,
Carl-von-Linné Weg 10,
D-50829 Köln, Germany

Dr Kevin Pyke

Plant Sciences Division, School
of Biosciences,University of
Nottingham, Sutton Bonington
Campus, Loughborough,
Leicestershire LE12 5RD, UK

Dr Christine Raines

Department of Biological Sciences,
John Tabor Laboratories, University
of Essex, Wivenhoe Park, Colchester,
CO4 3SQ, UK

Dr Stephen Rawsthorne

Department of Metabolic Biology,
John Innes Centre, Norwich
Research Park, Colney, Norwich
NR4 7UH, UK

Professor Colin Robinson

Department of Biological Sciences,
University of Warwick, Coventry
CV4 7AL, UK

Professor Dr Jürgen Soll

Department of Biology I,
Ludwig-Maximilians University,
Menzinger Str. 67, D-80638
München, Germany

Dr Ian J. Tetlow

College of Biological Sciences,
University of Guelph, Guelph,
Ontario N1G 2W1, Canada

Dr Ute C. Vothknecht

Department of Biology I, LMU
München, Menzinger Str. 67,
D-80638 München, Germany

Professor Masamitsu Wada

Department of Biological Sciences,
Graduate School of Science, Tokyo
Metropolitan University,
Minami-osawa, Hachioji, Tokyo
192-0397, Japan; and Division of
Biological Regulation and
Photobiology, National Institute
for Basic Biology, Okazaki,
Aichi 444-8585, Japan

Dr Mark Waters

Plant Sciences Division, School
of Biosciences, University of
Nottingham, Sutton Bonington
Campus, Loughborough,
Leicestershire LE12 5RD, UK

Preface

Plastids are essential plant organelles, vital for life on earth. They are important not just as photosynthetic organelles (chloroplasts) but also as sites involved in many fundamental intermediary metabolic pathways. Over the last decade, plastid research has seen tremendous advances and an exciting new picture is emerging of how plastids develop and function inside plant cells. The recent and rapid progress in the field has been due largely to reverse genetic approaches and forward genetic screening programs, which have resulted in the dissection of numerous chloroplast protein–function relationships.

This volume provides an up-to-date overview of our understanding of plastid biology. The initial chapter provides an insight into the genomic era of plastid research, describing recent genomics and proteomics approaches and setting the scene for later chapters. This is followed by two chapters on plastid development/differentiation and the integrated biochemistry of plastids within plant cells. There are chapters devoted to plastid division, chloroplast protein import, thylakoid membrane biogenesis and the regulation of chloroplast processes by proteolysis. The complex nature of plastid to nucleus signalling is then addressed, as is the ability of chloroplasts to relocate in response to various stimuli. The final chapter considers chloroplast genetic engineering and the use of plastids as biofactories, as viewed from a biotechnological perspective.

To my knowledge this is the first book combining plant physiology, cell biology, genetics, molecular biology and biochemistry to shed light on recent advances made in the field. Each chapter is designed to provide a detailed insight into the current state of research and future prospects, and an attempt has been made to integrate the coverage, providing the reader with an overall appreciation of this exciting era in plastid research.

The next challenge will be to dissect the cellular mode of action of the different plastid proteins and to understand how they act together to make a functional plastid. This will undoubtedly require interdisciplinary research efforts and collaborations within the international plant science community. I hope that this volume will serve as a platform towards reaching this goal.

I thank all authors for their participation in this project, and for providing such clear and informative chapters.

Simon Geir Møller

1   The genomic era of chloroplast research

Dario Leister and Paolo Pesaresi

1.1   Introduction

Plastids are essential organelles found in all living cells of plants, except pollen. As endosymbiotic remnants of a free-living cyanobacterial progenitor, plastids have, over evolutionary time, lost the vast majority of their genes. Indeed, depending on the organism, contemporary plastid genomes (plastomes) contain only 60–200 open reading frames. Most chloroplast proteins are nucleus-encoded and must be imported as precursor proteins from the cytoplasm. The plastids of a plant contain identical copies of the plastome. Nevertheless, plastids vary widely in their morphology and function, and can be divided into a number of types, based on colour, structure and developmental stage. All plastids originate from proplastids, which are colourless and lack an inner membrane system. In the absence of light, proplastids develop into yellow etioplasts, which contain a characteristic prolamellar body. Alternatively, proplastids can develop into chromoplasts or leucoplasts, which serve to store pigments or other molecules. Chromoplasts are carotenoid-rich plastids found in flowers, fruits, roots and senescing leaves, whereas leucoplasts are characterised by a lack of coloration. The leucoplasts can be further classified into amyloplasts (for starch storage), proteoplasts (for protein storage) or elaioplasts (for oil storage). Several plastid differentiations are reversible. Thus, chloroplasts or amyloplasts can evolve into chromoplasts and vice versa. The final stage in a plastid's life is the gerontoplast. These are plastids that have reached an irreversible state of senescence.

With one exception, the plastid forms mentioned above all derive their energy from imported compounds, such as hexosephosphates and ATP (Neuhaus and Emes, 2000). The only plastid type that is able to produce energy is the chloroplast, where all photosynthesis takes place. Mature chloroplasts are characterised by a complex and intricately folded membrane system, the thylakoids, which comprise two major domains: the grana and stroma lamellae enclosing the thylakoid lumen (Figure 1.1). The photosynthetic apparatus of higher plants is located in the thylakoids, and its various components tend to distribute unequally between grana and stroma. The grana are rich in photosystem II (PSII) complexes, whereas photosystem I (PSI) accumulates preferentially in the stroma lamellae. Besides photosynthesis, chloroplasts carry out many other essential functions, such as synthesis of amino acids, fatty acids and lipids, plant hormones, nucleotides, vitamins and secondary metabolites. Thus, photosynthesis takes place within a compartment

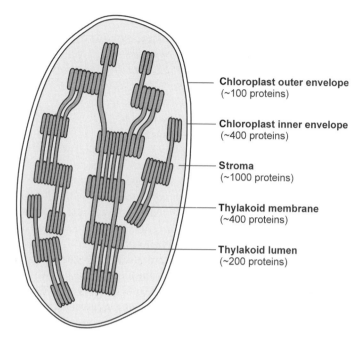

Figure 1.1 Compartments of chloroplasts and their tentative proteome sizes in *Arabidopsis*. Numbers are based on extrapolations and experimental analysis of (sub-)proteomes of the chloroplast, and are mostly derived from van Wijk (in press). The total number of cTP proteins is thought to be around 2000 (Richly and Leister, 2004).

that hosts many interdependent metabolic processes, which are subject to complex regulation in response to environmental fluctuations and changes in the developmental state of the organelle. These processes also have to be coordinated with the activities of the other compartments of the cell, including the nucleus and the mitochondria.

Technological developments in the genetics and genomics of *Arabidopsis thaliana*, including the complete sequencing of its nuclear genome (The *Arabidopsis* Genome Initiative, 2000), the large-scale generation and indexing of insertional mutants and advances in the large-scale analysis of proteins have stimulated a systematic approach to the functional characterisation of chloroplast proteins and elucidation of the regulation of the genes that encode them. As a result, a new picture is emerging of how the chloroplast operates in flowering plants, to what extent its functions differ from those of its cyanobacterial progenitor and how chloroplast functions depend upon, and/or interact with, other cellular processes. This chapter highlights new findings in chloroplast research, mostly in the model plant *A. thaliana*, focusing on proteomics, mutant identification and analysis, systematic mRNA expression profiling (transcriptomics) and genome comparisons (comparative genomics). Where possible, similar studies on other plastid types are also considered.

1.2 Chloroplast proteomics

For a comprehensive understanding of the biological functions of an organelle, its proteome has to be systematically characterised. The aim of proteomics is the definition of the function of every protein encoded by a given genome, and the analysis of how that function changes in different environmental or developmental conditions, with different modification states of the protein, and in interactions with different partners (Roberts, 2002). Although algorithms for the prediction of chloroplast proteins, as well as mass spectrometric techniques for their experimental identification, have improved significantly over the last few years, the analysis of complete proteomes, even from the simplest organism, still represents a formidable challenge. This is due to the limited throughput capacity of current proteomics technologies, to the fact that abundant proteins tend to mask other proteins present in low amounts, and to difficulties in developing a protein extraction strategy that is equally efficient for all proteins. As a consequence, before approaching whole-cell proteomics in plants, it is more realistic to characterise the proteomes of easily isolated compartments, such as chloroplasts or mitochondria. Both organelles, in fact, have been targeted by proteomics approaches in plants (van Wijk, 2000; Kruft *et al.*, 2001; Millar and Heazlewood, 2003). For chloroplasts, which contain an additional compartment (the thylakoids) relative to mitochondria, it is clear that proteome analysis must distinguish between the protein sets in each (sub)-compartment (van Wijk, 2001).

1.2.1 Predictions of chloroplast transit peptides

The vast majority of the plastid proteome is encoded by the nuclear genome. These proteins are generally synthesised as precursor proteins with cleavable, N-terminal, chloroplast transit peptides (cTPs) (Bruce, 2000). The availability of the complete genome sequence of *A. thaliana* (The *Arabidopsis* Genome Initiative, 2000), together with the development of algorithms for the computational identification of cTPs, has made large-scale prediction of cTP-containing proteins possible.

The first prediction of the number of cTPs encoded in the nuclear genome of *A. thaliana* was presented by Abdallah *et al.* (2000). These authors analysed the (incomplete) genomic sequence data then available for *A. thaliana*, employing the program ChloroP (Emanuelsson *et al.*, 1999), which is based on a neural-network approach. By extrapolation, they came up with a total number of around 2200 cTP proteins. Emanuelsson *et al.* (2000) went on to develop the TargetP program, which is based on the ChloroP algorithm. TargetP is able to discriminate among proteins destined for the mitochondrion, the chloroplast, the secretory pathway and 'other' localisations. By using TargetP, it was estimated that more than 3000 genes in the nuclear genome of *A. thaliana* code for proteins that have a cTP (Emanuelsson *et al.*, 2000).

More recently, the accuracy of the four most widely used cTP predictors – iPSORT (Bannai *et al.*, 2002), TargetP (Emanuelsson *et al.*, 2000), PCLR (Schein *et al.*, 2001) and Predotar (http://www.inra.fr/predotar/) – was found to be substantially lower than previously reported, using a test set of 2450 proteins whose subcellular locations are known (Richly and Leister, 2004). A combination of cTP predictors proved to be superior to any one of the predictors alone, and this was employed to estimate that around 2000 different cTP-bearing proteins should exist in *A. thaliana*.

In addition to proteins that carry a cTP, there are three other types of chloroplast proteins:

1. Nucleus-encoded plastid proteins without any obvious N-terminal transit peptide; these include stromal isoforms of 14-3-3 proteins (Sehnke *et al.*, 2000) and the inner envelope protein ceQORH (Miras *et al.*, 2002), and it is not yet clear how they are directed into plastids.
2. Proteins of the outer chloroplast envelope (see next section).
3. Proteins encoded by the chloroplast genome itself.

In *Arabidopsis*, a total of 87 genes, including 79 unique ones, are encoded by the plastid chromosome (Sato *et al.*, 1999).

1.2.2   Prediction of the proteome of the chloroplast outer envelope

Most outer envelope plastid proteins do not possess cTPs. Therefore, prediction of proteins of the outer membrane has to depend on features other than the presence of an N-terminal pre-sequence. Based on its evolutionary relationship to the outer membrane of Gram-negative bacteria, the outer envelope of the chloroplast should contain a large number of β-barrel proteins. Schleiff *et al.* (2003) have calculated the probability of the presence of β-sheet, β-barrel and hairpin structures for all proteins encoded by the *A. thaliana* genome, and selected a number of candidates for the outer envelope membrane. This protein pool was then analysed by TargetP to eliminate sequences with signals that would direct the proteins to organelles other than chloroplasts. The pool was further screened for the presence of proteins known to function outside of the chloroplast envelope. In total, a set of 891 potential outer membrane proteins were predicted, representing about 4.5% of all nuclear gene products. Among these were several that are known to be localised in the outer membrane, whereas others were good candidates for outer membrane proteins based on their putative sequence-based function.

1.2.3   Prediction of the proteome of the chloroplast inner envelope

No β-barrel proteins have been reported for the inner envelope or the thylakoid membrane. This indicates that integral membrane proteins in these compartments are of an α-helical nature, which in turn provides a starting point for genome-wide surveys for candidate inner membrane proteins. Ferro *et al.* (2002) used the following parameters to search the *Arabidopsis* genome for genes that encode candidate transporters that might be located in the inner envelope: (i) presence of a cTP, (ii) basic isoelectric point (pI), (iii) at least 4 transmembrane α-helices (TMs) and (iv) more than 1 TM per 100 amino acid residues. A set of 136 proteins was identified, and 35% of these belonged to plant transporter families or were homologous to transport systems present in other species. A few were known to be involved in lipid or pigment metabolism, whereas the remaining ones had unknown functions.

Koo and Ohlrogge (2002) performed a similar type of analysis to predict α-helical integral membrane proteins in the inner chloroplast envelope. In this case, proteins that (i) possess a cTP, (ii) contain membrane-spanning domains and (iii) are known not to be located in the thylakoids were selected, resulting in 541 putative inner envelope proteins. Putative functions, based on sequence, could be assigned to only 34% (or 183) of the candidates. Of the 183 candidates with assigned functions, 40% were classified in the category of 'transport facilitation'. This indicates that the proteome of the inner envelope is highly enriched in membrane transporters.

1.2.4   Prediction of the proteome of the thylakoid lumen

cTP-containing polypeptides without transmembrane domains either exist as soluble proteins in the stroma or in the thylakoid lumen, or are peripherally associated with the thylakoid or inner envelope membranes. Nucleus-encoded proteins of the thylakoid lumen can be predicted on the basis of the presence of an N-terminal lumenal transit peptide (lTP). The lTPs exhibit no obvious conserved sequence motif, but show a bias in amino acid content, rather similar to bacterial signal peptides used for the translocation of proteins from the cytosol to the periplast (Robinson et al., 2001). Peltier et al. (2000) identified among the proteins encoded in the nuclear genome of A. thaliana a set of 1224 proteins with potential lTPs, by selecting first all cTP proteins using TargetP (Emanuelsson et al., 2000) and then searching for proteins that had a signal peptide proximal to the cTP using the SignalP 2.0 HMM algorithm (Nielsen et al., 1997, 1999). A further constraint was imposed by specifying the amino acid motif present at the cleavage site. Furthermore, the total length of the predicted cTP + lTP was set to between 60 and 150 residues, and all sequences that were predicted to contain a TM region – either overlapping with the lTP cleavage site or downstream of it – were discarded. Finally, a set of 200 potential lumenal proteins was identified.

1.3   Experimental identification of the chloroplast proteome

The reliability of the predictions listed above largely depends on the accuracy of the algorithms employed. However, the large discrepancies among predictors highlight the need for continuous improvement of such programs. This can be achieved through the systematic experimental identification and characterisation of organelle proteomes, which in turn results in increasing numbers of proteins that can be included in the sets used to train the algorithms. Furthermore, proteins that lack conventional targeting sequences cannot be predicted. Such proteins have to be experimentally identified and analysed for the presence of novel consensus sequences that enable them to be targeted to subcellular compartments, and this, again, can be the starting point for novel prediction algorithms. In the following section, we summarise the results of recent advances in the experimental identification of proteins found in the sub-compartments of the chloroplast (for an overview, see van Wijk, in press).

1.3.1   Experimental identification of the proteomes of the chloroplast envelope and the thylakoid membrane

Two different groups have approached the identification of the proteome of the inner and/or the outer envelope of the chloroplast. Schleiff *et al.* (2003) reported on the characterisation of proteins from highly purified outer envelope membranes of chloroplasts from *Pisum sativum*. Four new proteins of the outer envelope membranes, in addition to the known components, were identified in this study.

Norbert Rolland, Jacques Joyard and colleagues have analysed a mixture of inner and outer envelope proteins of chloroplasts from spinach and *A. thaliana* (Seigneurin-Berny *et al.*, 1999; Ferro *et al.*, 2002). Several known, as well as novel, membrane proteins were identified. Envelope localisation of some of the new proteins was confirmed by transient expression of GFP (green fluorescent protein) fusions. In their latest, more extensive, study with mixed *A. thaliana* chloroplast envelope membranes, more than 100 proteins were identified (Ferro *et al.*, 2003). The envelope localisation of two phosphate transporters was verified by transient expression of GFP fusions. Almost one third of the identified proteins have as yet unknown functions, whereas more than 50% were very likely to be associated with the chloroplast envelope, based on their putative functions. These proteins were involved in either ion and metabolite transport or chloroplast lipid metabolism, or were components of the protein import machinery. Some soluble proteins, such as proteases and proteins involved in carbon metabolism or in responses to oxidative stress, were associated with envelope membranes.

Julian Whitelegge and colleagues reported on the identification of proteins in PSII-enriched thylakoid membranes from pea and spinach (Gomez *et al.*, 2002). Around 90 intact mass tags were detected, corresponding to approximately 40 gene products with variable post-translational modifications. A provisional identification of 30 of these gene products was proposed based upon coincidence of the measured mass with that calculated from the genomic sequence.

In the green alga *Chlamydomonas reinhardtii*, Michael Hippler's group analysed photosynthetic thylakoid membranes isolated from wild-type (WT) and mutant strains by high-resolution two-dimensional gel electrophoresis (Hippler *et al.*, 2001). More than 30 different LHCI and LHCII (light-harvesting complexes I and II) protein spots were detected. When isolated PSI complexes were separated, 18 LHCI protein spots were obtained, some of which were interpreted to arise by differential processing and post-translational modifications.

1.3.2 Experimental identification of the chloroplast lumenal proteome

By analysing soluble and peripheral proteins of pea thylakoids, Peltier *et al.* (2000) estimated that at least 200–230 different proteins are located in this compartment. Sixty-one proteins were identified, and for 33 of these proteins, a clear function or functional domain could be described. For 18 proteins, no expressed sequence tag or full-length gene was present in databases, despite experimental determination of a significant amount of amino acid sequence. Nine previously unidentified proteins with lTPs were found, of which seven possess the twin-arginine motif that is characteristic for substrates of the twin-arginine translocation (Tat) pathway.

In a subsequent study, the identity of 81 *Arabidopsis* proteins was established, and N-termini were sequenced to validate the predicted localisation (Peltier *et al.*, 2002). Expression of a surprising number of paralogous proteins was detected. Five isomerases of different classes, including FKBP isomerase-like proteins and TLP40, were identified. A function for these isomerases in the folding of thylakoid proteins or in signalling (such as TLP40) was suggested. Alternatively, these isomerases could be connected to a network of peripheral and lumenal proteins involved in antioxidative responses, including peroxiredoxins, m-type thioredoxins and a lumenal ascorbate peroxidase.

Wolfgang Schröder, Thomas Kieselbach and their colleagues also analysed the lumenal proteome of *A. thaliana* and spinach (Kieselbach *et al.*, 1998; Schubert *et al.*, 2002). Thirty-six proteins were identified, including a large group of proteases, peptidyl-prolyl *cis–trans* isomerases, a family of novel PsbP domain proteins, violaxanthin de-epoxidase, polyphenol oxidase and a novel peroxidase.

1.3.3 Experimental identification of stromal proteins or of proteins from other plastid types

Studies providing an exhaustive overview of the chloroplast stromal proteome or of the proteomes of other plastids have not been reported so far. However, an initial characterisation of the wheat amyloplast proteome led to the identification of 171 proteins (Andon *et al.*, 2002). In particular, 108 proteins from whole amyloplasts and 63 proteins from purified amyloplast membranes were identified. The majority of protein identities were derived from protein sequences from cereal crops other than wheat, as relatively little gene sequence data is available for the latter.

1.3.4 Identification of post-translational modifications in the chloroplast proteome

Stable or transient post-translational modifications can help to anchor proteins to membranes (in the case of lipid moieties), to regulate activity or protein interactions (phosphorylation, for example), to stabilise proteins (e.g. in the case of glycosylation or N-terminal formylation) or to target proteins for degradation (ubiquitination). Thus, the analysis of post-translational modifications is important, and mass spectrometry provides a suitable tool with which to investigate this issue.

In a recent analysis of the chloroplast granum proteome from pea and spinach, Gomez *et al.* (2002) identified several post-translational modifications. In particular, a minor fraction of the PSII protein D1 was isolated that was apparently palmitoylated. Based upon observed +80-Da adducts, the PSII proteins D1, D2, CP43 and PSII-H, as well as two proteins of LHCII, were shown to be phosphorylated, and a new phosphoprotein was proposed to be the product of the plastome *psbT* gene. The appearance of a second +80-Da adduct for PSII-H provided direct evidence for a second phosphorylation site. Adducts of +32 Da, which arise during illumination presumably owing to oxidative modification (such as the oxidative addition of dioxygen via sulphone or endoperoxide formation), were associated with more highly phosphorylated forms of PSII-H, implying a relationship between phosphorylation and oxidative modification.

Alexander Vener and colleagues have used the so-called 'parent ion scanning' technique to characterise phosphorylated thylakoid proteins (Vener *et al.*, 2001). From the analysis of tryptic peptides released from the surface of *Arabidopsis* thylakoids, phosphoproteins were identified by MALDI-TOF MS and ESI-MS/MS using a triple quadrupole instrument. This showed that the D1, D2 and CP43 proteins of the PSII core were phosphorylated at their N-terminal threonine residues (Thr), the PSII-H protein was phosphorylated at Thr-2 and LHCII proteins were phosphorylated at Thr-3. In addition, a doubly phosphorylated form of PSII-H, modified at both Thr-2 and Thr-4, was detected. By comparing the levels of phosphorylated and non-phosphorylated peptides, the *in vivo* phosphorylation state of these proteins was analysed under different physiological conditions. None of these thylakoid proteins were completely phosphorylated under continuous light, or completely dephosphorylated after long dark adaptation. However, rapid and reversible hyperphosphorylation of PSII-H at Thr-4 was detected in response to growth in the presence of light/dark transitions, and pronounced and specific dephosphorylation of the D1, D2 and CP43 proteins was observed during heat shock.

Additional protein modifications were reported previously for subsets of envelope proteins and for plastome-encoded proteins. Ferro *et al.* (2003) isolated a number of envelope proteins that were acetylated at their N-termini, whereas Giglione *et al.* (2003) reported on the removal of N-terminal methionine from plastome-encoded proteins by peptide deformylases.

1.3.5   Outlook and perspectives

During the past few years, much progress has been made in identifying proteins and their modifications, using a battery of newly developed, sophisticated genome-wide approaches. Nevertheless, there is still a need both for additional high-throughput technologies and for computational methods with which to analyse large data sets and integrate complex and disparate kinds of protein information. A breakthrough might be expected from high-throughput approaches for the identification of the subcellular locations of proteins based on the viral expression of cDNA-GFP fusions, as described recently by the group of Karl Oparka (Escobar *et al.*, 2003). The challenge for the proteomics community will be to proceed hand in hand with groups focused on biological problems, in order to convert the broad but shallow proteomic data into a deeper understanding. We may expect to have a reasonably complete picture of the proteome of a simple model organism, such as yeast, and of cellular sub-compartments of more complex organisms, such as the chloroplast, within the next decade.

1.4 Comparative genome analyses and chloroplast evolution

Chloroplasts arose through endosymbiosis from cyanobacteria, and therefore, numerous chloroplast proteins show significant homology to cyanobacterial proteins. Previous phylogenetic analyses of the cyanobacterial heritage of plant genomes were based on the cross-species comparison of relatively few genes, such as rRNA genes. The availability of complete genomic sequences for *Arabidopsis* and several cyanobacterial species, as well as the plastomes of a number of algal and plant species, has made novel types of phylogenetic analysis possible. Thus, Martin *et al.* (2002) compared 24,990 proteins encoded in the *Arabidopsis* genome to the proteins specified by three cyanobacterial genomes, 16 other prokaryotic reference genomes and yeast. Of the 9368 *Arabidopsis* proteins that were sufficiently conserved to permit primary sequence comparison, 866 detected homologues only in cyanobacteria and 834 others clustered with cyanobacterial homologues in phylogenetic trees. Extrapolation of these data to the whole genome suggested that approximately 4500 *Arabidopsis* protein-coding genes were acquired from the cyanobacterial ancestor of plastids.

Comparative analysis of plastome sequences *inter se* allows one to reconstruct the phylogeny of plastomes. In one of the first of such studies, Martin *et al.* (1998) compared the plastomes of a glaucocystophyte, a rhodophyte, a diatom, a euglenophyte and five land plants. In total, 210 different protein-coding genes were detected, of which 45 were common to all these species and to the cyanobacterium *Synechocystis*. A phylogenetic tree of the nine plastomes based on the 11,039 amino acid positions of the 45 common proteins allowed the authors to discern the pattern of gene loss from chloroplast genomes, revealing that independent parallel losses in multiple lineages outnumbered unique losses. Moreover, for 44 different plastid-encoded proteins, functional nuclear genes of chloroplast origin were identified.

This type of comparative analysis has since been extended to additional plastome sequences. Lemieux *et al.* (2000) compared the plastome sequence of the flagellate *Mesostigma* with those of three land plants and three chlorophyte algae, with the red alga *Porphyra purpurea* and *Synechocystis* as outgroups. They concluded that *Mesostigma* represents a lineage that emerged before the divergence of the *Streptophyta* (land plants and their closest green algal relatives, the charophytes) from *Chlorophyta* (green algae other than charophytes).

The most comprehensive phylogenetic analysis of plastomes considered 15 sequenced plastid chromosomes, with a total of 274 different protein-coding genes, and identified 117 nucleus-encoded proteins that are still encoded in at least one chloroplast genome (Martin *et al.*, 2002). A phylogenetic tree of the 15 chloroplast genomes based on 8303 amino acid positions in 41 proteins provided support for independent secondary endosymbiotic events for *Euglena*, *Guillardia* and *Odontella*. In contrast to Lemieux *et al.* (2000), these authors concluded that *Mesostigma* branched off basal to land plants but later than the chlorophyte algae *Chlorella* and *Nephroselmis*.

Because approximately 40 plastid genes are common to all extant chloroplasts (Martin *et al.*, 2002), the question arises why plastids have retained a separate genome and an energetically expensive expression apparatus for the production of relatively few proteins. Conversely, what has prevented the transfer of these genes to the nucleus? Such questions have been addressed repeatedly (Douglas, 1998; Martin and Herrmann, 1998; McFadden, 1999; Race *et al.*, 1999), and it appears that for this set of genes, positive selection for transcription/translation within the organelle accounts for their failure to be successfully incorporated into the nuclear genome.

Which chloroplast functions trace back to the cyanobacterial endosymbiont? Contemporary plastids resemble their prokaryotic relatives in several respects: they possess thylakoid membranes (Vothknecht and Westhoff, 2001) and 70S-type ribosomes (Yamaguchi *et al.*, 2000; Yamaguchi and Subramanian, 2000), use similar cell division proteins (Osteryoung and McAndrew, 2001), have light-dependent chlorophyll biosynthesis (Suzuki and Bauer, 1995) and have the secretory (Sec), twin-arginine translocation (Tat) and signal recognition particle (SRP) types of protein targeting to thylakoids (Robinson *et al.*, 2001). However, novel photosynthetic (Scheller *et al.*, 2001) and ribosomal proteins (Yamaguchi *et al.*, 2000; Yamaguchi and Subramanian, 2000), without obvious counterparts in prokaryotes, are found in the chloroplasts of land plants, and novel domains have been added to otherwise cyanobacterially derived proteins (e.g. in photosynthetic proteins such as PSI-D and PSI-E; Scheller *et al.*, 2001), as well as in the higher plant cytochrome c_6 homologue (Weigel *et al.*, 2003a). In addition, well-studied plastid functions not derived from prokaryotes include the machinery responsible for importing proteins across the plastid envelope (Jarvis and Soll, 2001; Soll, 2002), the 'spontaneous' targeting of proteins to thylakoid membranes (Robinson *et al.*, 2001) and the light-harvesting antenna complexes (LHCs) that have replaced the prokaryotic phycobilisomes (Montane and Kloppstech, 2000).

1.4.1   Outlook and perspectives

As more and more genome sequences become available, whole-genome comparisons and genome trees are emerging (Wolf *et al.*, 2002). However, a variety of different approaches have been used to compare whole genomes, and almost all require a substantial amount of computation. The problem of how to treat comparisons within gene families in complex genomes is still not satisfactorily solved. Therefore, it is rather unlikely that whole-genome comparisons will soon become a standard procedure that can be performed in most laboratories. A major challenge for comparative genomics in the field of chloroplast evolution will be to analyse as many cyanobacterial genomes as possible, in order to identify the cyanobacterial lineage from which chloroplasts are descended. Moreover, cross-species comparisons of the entire complements of nuclear chloroplast genes will become possible as soon as high-quality genomic sequences from rice (Sasaki *et al.*, 2002) and other plant species become publicly available.

1.5 Mutants for chloroplast function

Intensive efforts have been dedicated to the systematic identification of the functions of chloroplast proteins. This has been stimulated by the elucidation of the complete sequence of the *Arabidopsis* nuclear genome, as well as the assembly of large collections of insertional or chemically mutagenised lines. *Arabidopsis* mutant populations have been used for a number of phenotypic screens ('forward genetics'), leading to the identification of diverse classes of mutants for chloroplast functions. These include mutants affected in photoprotection (Niyogi *et al.*, 1998; Shikanai *et al.*, 1999), photosynthetic performance (Varotto *et al.*, 2000a), state transitions (Allen and Race, 2002; O. Kruse and colleagues, unpublished results, 2003), thylakoid biogenesis (Vothknecht and Westhoff, 2001), carotenoid (Norris *et al.*, 1995; Pogson *et al.*, 1996) and chlorophyll (Meskauskiene *et al.*, 2001) biosyntheses, plastid-to-nucleus signalling (reviewed in Surpin *et al.*, 2002), plastid replication (summarised in Pyke, 1999), leaf coloration (Leister, 2003) and seedling viability (Budziszewski *et al.*, 2001) (Figure 1.2). In a complementary approach, mutant collections have been searched for mutations in specific genes of interest; the most advanced tools consist of sequence-indexed populations, in which insertions in genes of interest can simply be identified by database searches (e.g. the SALK collection; Alonso *et al.*, 2003). As an alternative to insertional mutagenesis, loss-of-function alleles induced by EMS mutagenesis can be identified by TILLING, a gel-based method for the identification of mismatched heteroduplexes (McCallum *et al.*, 2000; Colbert *et al.*, 2001). Moreover, the targeted inactivation of nuclear genes by antisense, co-suppression or RNAi strategies has also been widely adopted. Taken together, these genetic tools are providing a growing catalogue of protein–function relationships for the chloroplast, making this organelle one of the best understood compartments of the plant cell.

A recent survey of the results of diverse screens for mutations in chloroplast functions has been provided by Leister (2003). In this review, we summarise recent progress in the mutational dissection of (i) protein targeting to/within chloroplasts, (ii) the photosynthetic process, (iii) plastid-to-nucleus signalling and (iv) chloroplast biogenesis.

1.5.1 Mutants for the chloroplast protein-sorting machinery

To gain more insight into the function of the Toc (translocon on the outer envelope of chloroplasts) and Tic (translocon on the inner envelope membrane) complexes,

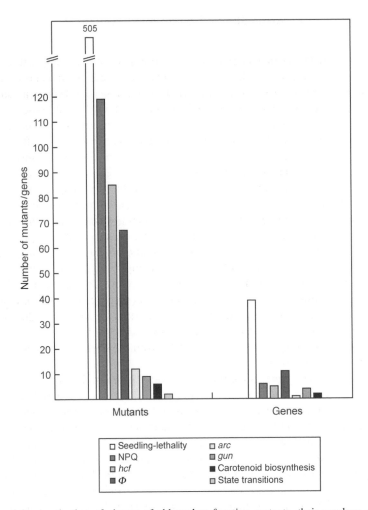

Figure 1.2 A selection of classes of chloroplast function mutants, their numbers and the number of identified mutated genes. Budziszewski *et al.* (2001) identified 505 seedling-lethal mutants, most of which were affected in chloroplast functions, and identified 39 of the mutated genes. The groups led by Kris Niyogi and Toshiharu Shikanai identified more than 100 mutants altered in chlorophyll fluorescence, a fraction of them affected in NPQ, using a video imaging system (Niyogi *et al.*, 1998; Shikanai *et al.*, 1999). For a number of these mutants the affected gene has been identified, including *PsbS* (Li *et al.*, 2000), *PetC* (Munekage *et al.*, 2001), *PGR5* (Munekage *et al.*, 2002) and *PsbO* (Murakami *et al.*, 2002). *High chlorophyll fluorescence* (*hcf*) mutants have been identified in several species; in *Arabidopsis*, 85 *hcf* mutants have so far been identified (P. Westhoff, personal communication, 2003). Cloned *HCF* genes include *HCF136* (Meurer *et al.*, 1998), *HCF164* (Lennartz *et al.*, 2001), *HCF107* (Felder *et al.*, 2001) and *HCF109* (Felder *et al.*, 2001). Screening for mutants altered in the effective quantum yield of PSII (Φ_{II}) and the isolation of corresponding genes has been reviewed recently (Leister, 2003; Leister and Schneider, 2003). Twelve *accumulation and replication of chloroplasts* (*arc*) mutants are known (Pyke, 1999). The *ARC6* gene has been cloned recently, and its product is homologous to the cyanobacterial cell division protein Ftn2 (Vitha *et al.*, 2003). Only relatively few mutants have been selected for defects in plastid signalling (*gun* mutants; Surpin *et al.*, 2002), carotenoid biosyntheses (*pds* and *lut* mutants; Norris *et al.*, 1995, Pogson *et al.*, 1996) or state transitions (Allen and Race, 2002; O. Kruse, unpublished results, 2003).

which facilitate the passage of precursors of chloroplast proteins across the chloroplast envelope (Jarvis and Soll, 2001; Soll, 2002), corresponding mutants have been identified by forward and reverse genetics. Gutensohn *et al.* (2000) generated antisense plants for the two *Arabidopsis* Toc subunits atToc34 and atToc33. While antisense plants for atToc33 had a pale yellowish coloration, antisense plants for atToc34 were more similar to WT, suggesting that the two proteins differ in their specificity for certain imported precursor proteins. An *Arabidopsis* mutant that lacks atToc33 was also isolated in a screen for mutants altered in leaf coloration (*ppi1*; Jarvis *et al.*, 1998). Bauer *et al.* (2000) identified the *Arabidopsis* mutant *ppi2*, which lacks atToc159. In *ppi2*, photosynthetic proteins that are abundant in WT are transcriptionally repressed. In the mutant, such proteins were found in much lower amounts in the plastids, although the mutation affected neither expression nor import of less abundant, non-photosynthetic, plastid proteins. These findings suggest that atToc159 is required for the quantitative import of photosynthetic proteins. Budziszewski *et al.* (2001) showed that disruption of the *Tic40* gene of *A. thaliana* resulted in seedling lethality. The role of atTic20 in chloroplast protein import was investigated in antisense lines, which exhibited pale leaf coloration, reduced accumulation of plastid proteins and significant growth defects (Chen *et al.*, 2002). The severity of the phenotypes correlated directly with the degree of reduction in the level of atTic20 expression.

Once chloroplast proteins are transferred into the chloroplast stroma, a fraction of them are targeted to the thylakoid membranes or to the thylakoid lumen, via one of four different pathways: for lumenal proteins, these are the (i) twin-arginine translocation (Tat) and (ii) secretory (Sec) pathways, while the (iii) signal recognition particle (SRP) and (iv) the 'spontaneous' pathways (reviewed in Robinson *et al.*, 2001) are used by thylakoid membrane proteins. Mutant plants altered in these pathways have been identified, revealing some of the functions involved. The *Arabidopsis alb3* mutant was disrupted in a gene encoding a protein homologous to the yeast OXA1 protein (Sundberg *et al.*, 1997), and ALB3 was shown to be required for the insertion of LHC proteins into thylakoids via the SRP pathway (Moore *et al.*, 2000). The *ffc* and *cao* mutants of *Arabidopsis* were disrupted in the genes coding for the 54- and 43-kDa subunits of the chloroplast signal recognition particle (cpSRP), respectively. Both mutants accumulated reduced amounts of LHC proteins, implying a crucial role for the cpSRP complex in targeting these proteins to the thylakoid membranes (Amin *et al.*, 1999; Klimyuk *et al.*, 1999; Hutin *et al.*, 2002). A seedling-lethal *Arabidopsis* mutation caused by disruption of the *TatC* gene, which codes for a component of the Tat pathway, has recently been identified by Budziszewski *et al.* (2001).

1.5.2 Mutants for the chloroplast photosynthetic apparatus

Mutagenesis or silencing of most of the nuclear genes encoding proteins of the photosynthetic machinery has markedly improved our understanding of their functions. The specific roles of CP29 and CP26, two minor antenna proteins of PSII, in light harvesting and energy dissipation were investigated in *A. thaliana* by using antisense lines (Andersson *et al.*, 2001). These lines had distinct chlorophyll fluorescence characteristics, indicating a change in the organisation of the light-harvesting antenna. However, the overall rate of photosynthesis in both lines was similar to that in WT, with a normal qE-type of non-photochemical fluorescence quenching (NPQ), indicating that CP29 and CP26 are unlikely to be the sites of NPQ.

In the course of a screen for *A. thaliana* mutants that were unable to dissipate excess light energy by NPQ, the line *npq4* was isolated (Li *et al.*, 2000). This mutant did not accumulate the PSII-S protein, and its characterisation showed that PSII-S was necessary for NPQ, but not for efficient light harvesting or photosynthesis. Subsequent studies showed that plants with a twofold increase in qE capacity could be produced by over-expressing PSII-S, demonstrating that the level of PSII-S limits the qE capacity in WT plants (Li *et al.*, 2002).

Arabidopsis antisense lines affected in proteins that form the light-harvesting complex of PSII (LHCII) (Andersson *et al.*, 2003; Ruban *et al.*, 2003) have PSII supercomplexes in almost identical abundance and with a similar structure to those found in WT plants. In these lines, however, LHCII itself was replaced by a trimeric form of CP26 (Ruban *et al.*, 2003). These results highlight the flexibility and importance of the PSII macrostructure: in the absence of one of its main components a different protein was recruited to allow it to assemble and function.

Extensive reverse genetics analyses have been performed to investigate the role of nucleus-encoded PSI subunits. In particular, Scheller and co-workers generated a collection of *A. thaliana* lines in which individual nucleus-encoded subunits of PSI were down-regulated by antisense or co-suppression strategies (Haldrup *et al.*, 1999, 2000, 2003; Naver *et al.*, 1999; Jensen *et al.*, 2000, 2002; Lunde *et al.*, 2000). This approach was effective even in cases where the same subunit is encoded by two functional genes. These studies have revealed that PSI-K plays a role in organising LHCI (Jensen *et al.*, 2000), PSI-N is necessary for the interaction of plastocyanin with PSI (Haldrup *et al.*, 1999), PSI-H appears to provide an attachment site for LHCII during state transitions (Lunde *et al.*, 2000), PSI-F seems to have a role in stabilising PSI complexes (Haldrup *et al.*, 2000) and PSI-D is essential for the accumulation of a functional PSI (Haldrup *et al.*, 2003).

Analyses of stable knockout PSI mutants generated by T-DNA or transposon insertions have also been performed. The *psae1-1* mutant of *Arabidopsis* was identified on the basis of its decreased photosynthetic performance, and the mutation responsible was localised to *PsaE1*, one of two *Arabidopsis* genes that encode subunit E of PSI (Varotto *et al.*, 2000b). The entire stromal side of PSI was affected by disruption of the *PsaE1* gene (Varotto *et al.*, 2000b), and furthermore, the interaction between PSI and LHCII was perturbed (Pesaresi *et al.*, 2002).

Plants that have a PSI complex that lacks PSI-G have been generated in two laboratories (Jensen *et al.*, 2002; Varotto *et al.*, 2002). The analyses showed that PSI-G plays a role in stabilizing the binding of the peripheral antenna. The increased activity of PSI in the absence of PSI-G suggests that this subunit could play an important role in the regulation of PSI (Jensen *et al.*, 2002).

The antisense strategy was also employed to dissect the roles of the four different proteins (Lhca1 to Lhca4) that make up the LHC of PSI (LHCI). Zhang *et al.* (1997) produced transgenic lines with reduced amounts of Lhca4. Low-temperature fluorescence analysis indicated that Lhca4-bound chlorophylls are responsible for emission of most of the long-wavelength fluorescence. In addition, some Lhca4 antisense lines showed a delay in flowering and an increase in seed weight. Antisense inhibition of either Lhca2 or Lhca3 resulted in a concomitant decrease in the levels of both proteins (Ganeteg *et al.*, 2001), suggesting that Lhca2 and Lhca3 can form heterodimers, although no evidence for their existence could be found by chemical cross-linking (Jansson *et al.*, 1996).

Besides the photosystems themselves, the subunits of the electron transport chain that connects them have also been investigated by genetic methods. Maiwald *et al.* (2003) reported on the characterisation of a T-DNA tagged mutant disrupted in the gene for the Rieske protein of cytochrome b_6/f (cyt b_6/f). The mutant was seedling-lethal, while heterotrophically grown plants displayed a high-chlorophyll-fluorescence phenotype. Lack of the Rieske protein destabilised cyt b_6/f and also affected the levels of other thylakoid proteins, particularly those of PSII. In addition, linear electron flow was completely blocked, clearly demonstrating the essential role of Rieske protein in electron transport.

In 2002, two groups independently identified a cytochrome c_6 (cyt c_6) like protein in higher plants (Gupta *et al.*, 2002; Wastl *et al.*, 2002). Prior to this, it was generally accepted that this protein had been lost during the evolution of angiosperms, and only algae and cyanobacteria were thought to use either plastocyanin or cyt c_6 as electron donors to PSI. From biochemical and genetic analyses, Luan and co-workers concluded that the cyt c_6 like protein is targeted to the thylakoid lumen, where it can replace plastocyanin in reducing PSI (Gupta *et al.*, 2002). Two more recent studies (Molina-Heredia, 2003; Weigel *et al.*, 2003b) have challenged the contention that the higher plant cyt c_6 homologue donates electrons to PSI and is capable of functionally replacing plastocyanin. Weigel *et al.* (2003b) showed that *Arabidopsis* plants mutated in both of the plastocyanin-coding genes, but retaining a functional cyt c_6, could not grow photoautotrophically because of a complete block in light-driven electron transport. Even increased dosage of the gene encoding the cyt c_6 like protein could not complement the double-mutant phenotype, demonstrating that in *Arabidopsis* only plastocyanin can donate electrons to PSI *in vivo*. Furthermore, structural and kinetic data showed that *Arabidopsis* cyt c_6 cannot carry out the same function as *Arabidopsis* plastocyanin or as cyt c_6 from the alga *Monoraphidium braunii* (Molina-Heredia *et al.*, 2003).

Only three subunits of the chloroplast ATPase (cpATPase) of *A. thaliana* are encoded by nuclear genes: the b′, the δ- and the γ-subunit. A knockout allele of the *Arabidopsis AtpD* gene encoding the δ-subunit of ATPase (cpATPase-δ) has been identified. Plants carrying this allele in the homozygous state are seedling-lethal when grown on soil (Budziszewski *et al.*, 2001). Indeed, the absence of the δ-subunit leads to destabilisation of the entire cpATPase complex, whereas residual accumulation of cyt b_6/f and of the photosystems still allows linear electron flow (Maiwald *et al.*, 2003).

Knockout of *AtpC1*, one of the two genes coding for the γ-subunit of the cpATPase in *A. thaliana*, also results in loss of cpATPase function and in seedling lethality (Bosco *et al.*, 2004).

1.5.3 Mutants for chloroplast-nucleus signalling

The distribution of the genes encoding plastid proteins between two genetic compartments has led to the evolution of mechanisms that serve to integrate nuclear and organellar gene expression. Hence, inter-organellar signalling, and the coordinated expression of sets of chloroplast nuclear genes, operate to control the metabolic and developmental status of the chloroplast. These mechanisms include both anterograde (nucleus-to-plastid) and retrograde (plastid-to-nucleus) controls. Anterograde mechanisms coordinate gene expression in the plastid with endogenous and environmental signals that are perceived by the nucleus (Goldschmidt-Clermont, 1998). This type of control depends upon nuclear proteins that regulate the transcription and translation of plastid genes. Retrograde signalling regulates the expression of nuclear chloroplast genes in response to the metabolic and/or developmental state of the plastid. Early evidence that nuclear genes are regulated by signals originating from the plastid came from studies of plants with photo-oxidised chloroplasts (Oelmüller, 1989; Mayfield, 1990). These plants bleach when exposed to high light levels, and show decreased expression of nuclear photosynthetic genes. Regulation occurs frequently at the transcriptional level, and the *Lhcb* genes are found to be down-regulated most.

Thomas Pfannschmidt and colleagues demonstrated that the redox state of the plastoquinone pool affects nuclear photosynthetic gene expression in higher plants (Pfannschmidt *et al.*, 2001). These authors measured the transcriptional response of selected nuclear photosynthetic genes to excitation pressure applied to the two photosystems, and also investigated the effects of inhibitors of photosynthetic electron transport. It emerged that the *PetH* promoter did not respond to redox signals, while the *PsaD* and *PsaF* promoters responded to redox signals originating from the plastoquinone pool and PSI, or reacted to the overall electron transport capacity. The *PetE* promoter was regulated specifically by the redox state of the plastoquinone pool.

The *Arabidopsis* chlorophyll *a/b* binding protein under-expression mutant *cue1* offered additional evidence for the involvement of the redox state of the plastoquinone pool in the regulation of nuclear photosynthetic genes (Streatfield *et al.*, 1999). *CUE1* codes for the phosphoenol/phosphate translocator in the inner chloroplast envelope. In *cue1* plants, biosynthesis of aromatic compounds was reduced, resulting in reduced flux through the shikimate pathway – the source of phenolic UV protectants, such as flavonoids, hydroxycinnamic acids and phenolics. Plastoquinone itself is also derived from the shikimate pathway, and the relative size of the plastoqinone pool in the *cue1* mutants was lower than in WT plants. Furthermore, measurements of photosynthetic electron transport and of photochemical and non-photochemical quenching showed that plastoquinone was more susceptible to reduction by brief illumination in the *cue1* mutant, and that an altered redox state might, in fact, affect *Lhcb* expression (Streatfield *et al.*, 1999). In addition, the reduced levels of phenolics in the mutant render *cue1* more susceptible to high-light-induced repression of *Lhcb* gene transcription.

In addition to redox signalling, the tetrapyrrole-dependent pathway seems to control the expression of nucleus-encoded photosynthetic proteins. Tetrapyrroles, which are synthesised in the plastids, are the intermediates and end products of heme, chlorophyll and phytochromobilin biosyntheses (Rodermel and Park, 2003). In *Chlamydomonas*, studies of protoporphyrin accumulation in appropriate mutants, and the results of feeding of inhibitors of the chlorophyll biosynthetic pathway to WT cells, suggested that intermediates in the chlorophyll biosynthetic pathway inhibit the expression of *Lhcb* genes and of the *RbcS* gene that codes for the small subunit of Rubisco (Johanningmeier and Howell, 1984; Johanningmeier, 1988).

Insight into tetrapyrrole signalling in higher plants has come from a mutational analysis involving a screen for *Arabidopsis* mutants that do not repress *Lhcb* transcription upon photo-oxidative damage (Susek *et al.*, 1993). Because none of the selected mutants, *genomes uncoupled 1–5* (*gun1–5*) (Susek *et al.*, 1993; Mochizuki *et al.*, 2001; Larkin *et al.*, 2003), affected the tissue- and cell-specific, light-dependent or circadian regulation of *Lhcb* genes, these genotypes appeared to be specifically impaired in the plastid-mediated regulation of nuclear transcription. In contrast to *GUN1*, the genes *GUN2–5* are essential for normal tetrapyrrole metabolism (Vinti *et al.*, 2000; Mochizuki *et al.*, 2001). The products of *GUN2* and *GUN3* form part of the 'iron branch' of tetrapyrrole biosynthesis, whereas *GUN5* encodes the ChlH subunit of the Mg-chelatase (Mochizuki *et al.*, 2001). A role for the ChlH subunit of Mg-chelatase as a tetrapyrrole sensor in chloroplast-to-nucleus signalling has been discussed (Mochizuki *et al.*, 2001; Surpin *et al.*, 2002). This idea was recently revised in favour of the tetrapyrrole intermediate Mg-protoporphyrin IX, which was suggested to act as a signalling molecule between chloroplast and nucleus (Strand *et al.*, 2003). The *GUN4* gene was recently cloned (Larkin *et al.*, 2003); its product binds the product and substrate of Mg-chelatase, and activates Mg-chelatase. Thus, it is thought that GUN4 participates in plastid-to-nucleus signalling by regulating Mg-protoporphyrin IX synthesis or trafficking.

The role of tetrapyrrole intermediates as regulators of nuclear gene expression has been supported by the isolation of the *Arabidopsis* mutant *long after far-red 6* (*laf6*), which exhibits reduced responsiveness to continuous far-red light (Møller *et al.*, 2001). *LAF6* encodes a chloroplast-targeted ATP-binding-cassette (atABC1) protein of 557 amino acids with high homology to ABC-like proteins from lower eukaryotes. atABC1 deficiency results in the accumulation of the chlorophyll precursor Mg-protoporphyrin IX and in attenuation of far-red regulated gene expression. In agreement with the notion that ABC proteins are involved in transport, these observations suggest that atABC1 is required for the transport and correct distribution of Mg-protoporphyrin IX.

1.5.4   Mutants affected in chloroplast development and division

Although several pathways that target proteins into and across the thylakoids have been described, little is known about the origin of this membrane system. Thylakoid membranes contain about 70–80% galactolipids, which are synthesised at the inner envelope of the chloroplast (Douce, 1974). It is concluded that an intra-organellar lipid transport system must exist that transfers lipids from their site of synthesis to the thylakoids. Mutational analysis has led to the identification of a T-DNA-tagged line that is altered in thylakoid membrane formation. The mutant was disrupted in the single-copy gene *VIPP1* (*vesicle-inducing protein in plastids 1*), which codes for a hydrophilic protein associated with both the inner envelope and the thylakoid membrane (Kroll *et al.*, 2001). In the mutant, the vesicle buds that are normally formed on the inner envelope of WT plastids are absent, indicating the essential role of VIPP1 in the formation and/or maintenance of thylakoid membranes by a vesicle transport pathway.

Genetic screens in *Arabidopsis* have also been extremely useful in dissecting the mechanism of plastid division in higher plants. Microscopy-based screens have led to the identification of a collection of *Arabidopsis* mutants with altered numbers of chloroplasts per cell (summarised in Pyke, 1999). The characterisation of these *accumulation and replication of chloroplasts* (*arc*) mutants has shown that some nuclear genes play specific roles both in the chloroplast division process itself and in the control of the size of the chloroplast population in a cell during its development. In total, 12 *arc* mutants, showing a variety of chloroplast division phenotypes, were identified. *arc6* and *arc12* contained an average of two enlarged chloroplasts per mesophyll cell instead of the usual >100 chloroplasts per cell; *arc3* and *arc5* had about 15 chloroplasts per leaf mesophyll cell, and the *arc1* and the *arc7* mutants had a larger number of smaller chloroplasts per cell than WT. Of particular interest is *arc10*; in mesophyll cells of this mutant, chloroplasts were highly heterogeneous in size within a single cell, most probably owing to the presence of a subpopulation of chloroplasts that did not divide, or to other forms of abnormal chloroplast division.

1.5.5   Outlook and perspectives

Genetic screens for mutations that affect chloroplast function have taken advantage of the advanced state of molecular genetics in *A. thaliana*. For all 26,000 genes of *Arabidopsis*, saturating phenotypic screens of a non-redundant set of loss-of-function mutants may become feasible within the next couple of years. A further conceivable improvement might involve the systematic generation of double mutants for segmentally duplicated genes which exhibit (partial) functional redundancy. Another promising approach involves the systematic collection of loss-of-function mutants for all predicted or known nuclear chloroplast genes and their systematic analysis by a battery of assays for chloroplast phenotypes.

1.6   Transcriptomics

As was mentioned above, the evolution of the endosymbiotic progenitor into the chloroplast organelle was associated with a massive transfer of chloroplast genes into the nucleus. This made it necessary to establish a coordinated regulatory system for the expression of the nuclear and plastome genes for chloroplast proteins, in order to ensure effective functioning of the chloroplast. Previous studies considered a limited set of target genes that respond to plastid signals, and the recent progress in the genomics of nuclear *Arabidopsis* genes, as well as the availability of numerous plastome sequences, now allows more advanced approaches to the genome-wide analysis of the transcriptional regulation of chloroplast function.

A macroarray representing all 118 genes and 11 open reading frames of the tobacco plastid chromosome has been constructed by spotting corresponding amplicons on nylon membranes (Legen *et al.*, 2002). This plastome array was used to investigate the transcription rates and transcript patterns of the entire plastid chromosome from WT leaves, as well as from tobacco plants lacking the plastome-encoded RNA polymerase (PEP). Hybridisation was performed using either labelled run-on transcripts, or total plastid RNA phosphorylated with ^{32}P at the 5'-end, as probes. The run-on transcription data show that all plastid genes were transcribed in the PEP-deficient mutant background, though the overall profile differed from that in WT plastids. In many cases, steady-state transcript levels correlated with the findings of the run-on analyses. The data clearly showed that the two chloroplast RNA polymerases, PEP and NEP, are not responsible for the transcription of specific classes of genes in the plastome, as previously proposed (Hajdukiewicz *et al.*, 1997).

The group led by Joanne Chory used a commercially available DNA array representing 8200 *Arabidopsis* genes to study nuclear mRNA expression in WT and *gun* mutants before and after treatment with norflurazon, a non-competitive inhibitor of carotenoid biosynthesis (Strand *et al.*, 2003). Three hundred and twenty-two genes were identified whose expression levels changed more than threefold upon treatment of WT seedlings with norflurazon. Of these 322 genes, 152 showed more than a threefold difference in one or more of the three *gun* mutants – *gun1*, *gun2* and *gun5*. Cluster analysis of those 152 genes showed that the expression profiles of the *gun2* and *gun5* mutants clustered together (Strand *et al.*, 2003), supporting the results of previous genetic analyses which had suggested that *gun1* is involved in a separate signalling pathway (Mochizuki *et al.*, 2001).

In our laboratory, a macroarray-based approach has been established (Kurth *et al.*, 2002; Richly *et al.*, 2003) to characterise specifically the expression patterns of those *Arabidopsis* nuclear genes that code for chloroplast proteins (the nuclear chloroplast transcriptome). Profiling of the nuclear chloroplast transcriptome under different genetic or environmental conditions may allow us to assign functions to genes that have not been characterised so far, as well as to correlate the transcriptional effects of different mutations or stimuli with the physiological perturbations observed under these conditions. A set of 1827 nuclear genes coding for chloroplast proteins were amplified from genomic DNA by PCR, and spotted on nylon membranes to generate arrays of gene-sequence tags (GSTs) (Varotto *et al.*, 2001; Kurth *et al.*, 2002). This 1827-GST array was employed to compare mRNA levels in dark- vs. light-grown seedlings, as well as in WT vs. *prpl11* mutant plants. In the *prpl11* mutant, the nuclear gene coding for the L11 subunit of the plastid ribosome is disrupted, severely affecting translation in plastids; this in turn causes a drop in the levels of components of the photosynthetic apparatus and of Rubisco, as monitored by PAGE and Western analyses (Pesaresi *et al.*, 2001). mRNA profiling of *prpl11* plants showed that transcription levels of nuclear genes coding for proteins of the plastid ribosome, of the photosynthetic apparatus and of the small subunit of Rubisco were up-regulated (Kurth *et al.*, 2002), indicating that the mutant plant is able to monitor the altered physiological state of the chloroplast and reacts by up-regulating appropriate nuclear genes. This supports the idea that regulatory networks operate in plant cells that can sense the levels of key proteins in the chloroplast and transmit a signal to the nucleus, which then acts to compensate for the relevant deficit. In the case of the photosystems and of Rubisco, which contain nucleus- and plastome-encoded subunits in a fixed stoichiometry, however, up-regulation of appropriate nuclear genes cannot repair the structural defect in *prpl11* plants, because the associated decrease in the level of plastome-encoded proteins also seems to limit the concentration of nucleus-encoded protein subunits in the chloroplast.

Recently, the 1827-GST array has been replaced by a 3300-GST array, which covers almost all of the ~2000 nuclear *Arabidopsis* genes predicted to encode chloroplast-targeted proteins, and has been employed to analyse mRNA expression under a variety of conditions, as well as to characterise mutants (Richly *et al.*, 2003). When gene expression profiles observed under 35 different genetic/environmental conditions were compared, three major types of transcriptome responses were identified: two of these were predominantly associated with either up-regulation or down-regulation of substantial fractions of the nuclear chloroplast transcriptome (Figure 1.3). A third type of response involved approximately equal numbers of up- and down-regulated genes (Richly *et al.*, 2003). Hierarchical clustering showed that sets consisting mostly of the same genes were up- or down-regulated coordinately depending on the condition analysed. The degree of covariation in the expression of a large set of genes has been interpreted as evidence for the existence of a major switch that regulates the response of the nuclear chloroplast transcriptome to changes in the metabolic state of plants. Such coordinate expression of nuclear genes in response to various treatments has been described for prokaryotes and eukaryotes. Examples include the SOS response in *Escherichia coli*, in which at least 30 genes exhibit a coordinate increase in expression level following treatments that lead to DNA damage (Sutton *et al.*, 2000; Khil and Camerini-Otero, 2002), and the so-called environmental stress response in yeast – in which a set of about 900 genes appear to be activated upon exposure to multiple stressful stimuli (Gasch *et al.*, 2000).

The 3300-GST array has been employed to interpret the specific effects of diverse mutations on the physiological state of chloroplasts. Pesaresi *et al.* (2003) found that lack of cytoplasmic N-terminal acetylation catalysed by the AtMAK3 protein resulted in a specific drop in the synthesis of the plastome-encoded D1 and CP47 proteins. In *atmak3-1*, 777 of the 3292 genes tested displayed significant differences in mRNA expression relative to WT plants. In the mutant, moreover, the fraction of differentially regulated genes coding for non-chloroplast proteins was higher (200 different transcripts, or 26%) than noted under any of the other 34 genetic or environmental conditions tested, suggesting that the function of

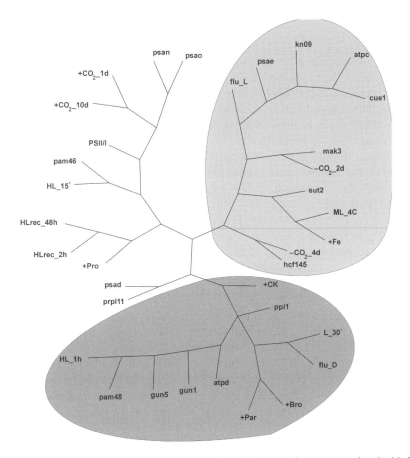

Figure 1.3 Relationships between the nuclear chloroplast transcriptomes associated with 35 different genetic/environmental conditions according to Richly *et al.* (2003). The displayed cladogram is based on the hierarchical clustering of the expression profiles of 1972 genes that showed differential expression under at least 33 of the 35 conditions tested. The two major classes of transcriptome change, which affect the expression of similar sets of genes in diametrically opposite directions, are indicated in dark grey (most genes up-regulated) and light grey (preferential down-regulation). Acronyms are as following: (i) Treatments: PSII/I, growth under PSII-specific light vs. PSI-specific light; L_30′, 30 min light vs. darkness; HL_15′, 15-min high-light stress vs. no stress; HL_1h, 1-h high-light stress vs. normal light; HLrec_2h, 2-h recovery after 1h high-light stress vs. before stress; HLrec_48h, 48-h recovery after 1-h high-light stress vs. before stress; ML_4C, 24-h medium light at 4°C vs. 20°C; +Par, treatment with benzoquinone herbicide paraquat vs. no treatment; +Bro, treatment with nitrile herbicide bromoxynil vs. untreated; $-CO_2$_2d, low-CO_2 stress: 2 days of 0.003% (v/v) CO_2 vs. normal CO_2 level; $-CO_2$_4d, low-CO_2 stress: 4 days of 0.003% (v/v) CO_2 vs. normal CO_2 level; $+CO_2$_1d, high-CO_2 stress: 1 day of 1% (v/v) CO_2 vs. normal CO_2 levels; $+CO_2$_10d, high-CO_2 stress: 10 days of 1% (v/v) CO_2 vs. normal CO_2 levels; +Fe, high-iron stress: spraying with iron solution vs. no treatment; +Pro, 48 h 100 mM proline vs. no treatment; +CK, cytokinin-treated (2 h) cell culture vs. untreated cell culture. (ii) Mutants: prpl11, *prpl11-1* vs. WT; psad, *psad1-1* vs. WT; psae, *psae1-1* vs. WT; psan, *psan-1* vs. WT; psao, *psao-1* vs. WT; atpc, *atpc1-1* vs. WT; atpd, *atpd-1* vs. WT; hcf145, *hcf145* vs. WT; gun1, *gun1* vs. WT; gun5, *gun5* vs. WT; cue1, *cue1-1* vs. WT; flu_D, *flu* (dark) vs. WT (dark); flu_L, *flu* (light) vs. WT (dark); ppi1, *ppi1* vs. WT; kn09, *kn09* vs. WT; sut2, *sut2* vs. WT; mak3, *atmak3-1* vs. WT, pam48, *pam48* vs. WT; pam46, *pam46* vs. WT.

AtMAK3 is not restricted to chloroplast processes. Among the 577 chloroplast-protein-coding genes that were differentially expressed in the mutant with respect to WT plants, 121 were up- and 456 down-regulated. Genes for transcription and protein synthesis/degradation were down-regulated less than others, indicating that the plant may be able to monitor the change in plastid protein synthesis/accumulation due to the *atmak3-1* mutation.

Kubis *et al.* (2003) employed the 3300-GST array to analyse the expression profile of the atToc33 knockout, *ppi1*, and found that photosynthetic genes were moderately, but specifically, down-regulated in the mutant.

Additional analyses of the nuclear chloroplast transcriptome, employing the 3300-GST array, have included investigations of the photosynthetic mutants *petc-2*, *atpd-1*, *pete1*, *pete2*, *pete1pete2* and *atcx* (Maiwald *et al.*, 2003; Weigel *et al.*, 2003a). Direct comparison of the differential expression profiles of the knockout of the Rieske protein *petc-2* and the chloroplast ATPase knockout mutant *atpd-1* revealed that among all genes differentially expressed in the two genotypes, 451 genes showed the same trend (Maiwald *et al.*, 2003). A further set of 346 genes showed opposite trends in transcriptional regulation in the two lines. In *petc-2*, a balanced response of the nuclear chloroplast transcriptome was observed, with about equal fractions of genes being up- or down-regulated. Relatively more genes for photosynthesis tended to be down-regulated, whereas, in this genotype, genes for stress responses represented the largest group of up-regulated genes. In contrast, in the *atpd-1* mutant, 88% of the differentially regulated genes were up-regulated. Most of the different functional gene classes followed this trend, again with the exception of genes coding for proteins involved in photosynthesis (only 27% of which were up-regulated). The data showed that, with the exception of photosynthetic genes – which are predominantly down-regulated in both genotypes, the mutations *petc-2* and *atpd-1* result in very different transcriptional responses of the nuclear chloroplast transcriptome. These different transcriptional responses were interpreted as manifestations of the effects of different types of plastid signalling pathways.

The expression profile of the *Arabidopsis atcx* mutant, in which the gene coding for the cyt c_6 homologue cyt c_x is disrupted, differed markedly from that of single or double plastocyanin mutants, suggesting that lack of plastocyanin or cyt c_x induces quite distinct physiological states (Weigel *et al.*, 2003a). Interestingly, transcript levels of genes coding for proteins of the photosynthetic machinery were similarly altered in the plastocyanin single mutants and *atcx*, which might indicate that cyt c_x participates in the regulation of photosynthetic electron flow (Weigel *et al.*, 2003a).

1.6.1 Outlook and perspectives

Over the past few years, significant progress has been made in dissecting the transcriptional regulation of chloroplast functions. Whereas transcript levels – in contrast to protein levels – are relatively easy to measure, differential expression of transcripts does not necessarily or faithfully reflect changes in the abundance of the corresponding polypeptides. This concerns, for example, compensatory transcriptional responses in which a drop in the abundance of certain proteins is compensated for by up-regulating corresponding nuclear genes – without achieving a net increase in the abundance of the protein. The analysis of polysome-bound mRNA instead of total RNA might increase the power of the transcriptomics approach, but transcriptomics will achieve its full potential only in combination with information on the abundance of other cellular compounds, such as proteins and/or metabolites.

Acknowledgements

We thank Francesco Salamini and Paul Hardy for critical reading of the manuscript. Peter Westhoff, Jörg Meurer and Olaf Kruse are acknowledged for making unpublished data available.

References

Abdallah, F., Salamini, F. and Leister, D. (2000) A prediction of the size and evolutionary origin of the proteome of chloroplasts of *Arabidopsis*. *Trends Plant Sci.*, 5, 141–142.

Allen, J.F. and Race, H.L. (2002) Will the real LHC II kinase please step forward? *Sci. STKE*, 2002, PE43.

Alonso, J.M., Stepanova, A.N., Leisse, T.J. *et al.* (2003) Genome-wide insertional mutagenesis of *Arabidopsis thaliana*. *Science*, 301, 653–657.

Amin, P., Sy, D.A., Pilgrim, M.L., Parry, D.H., Nussaume, L. and Hoffman, N.E. (1999) *Arabidopsis* mutants lacking the 43- and 54-kilodalton subunits of the chloroplast signal recognition particle have distinct phenotypes. *Plant Physiol.*, 121, 61–70.

Andersson, J., Walters, R.G., Horton, P. and Jansson, S. (2001) Antisense inhibition of the photosynthetic antenna proteins CP29 and CP26: implications for the mechanism of protective energy dissipation. *Plant Cell*, 13, 1193–1204.

Andersson, J., Wentworth, M., Walters, R.G. *et al.* (2003) Absence of the Lhcb1 and Lhcb2 proteins of the light-harvesting complex of photosystem II – effects on photosynthesis, grana stacking and fitness. *Plant J.*, 35, 350–361.

Andon, N.L., Hollingworth, S., Koller, A., Greenland, A.J., Yates, J.R., III and Haynes, P.A. (2002) Proteomic characterization of wheat amyloplasts using identification of proteins by tandem mass spectrometry. *Proteomics*, 2, 1156–1168.

Bannai, H., Tamada, Y., Maruyama, O., Nakai, K. and Miyano, S. (2002) Extensive feature detection of N-terminal protein sorting signals. *Bioinformatics*, 18, 298–305.

Bauer, J., Chen, K., Hiltbunner, A. *et al.* (2000) The major protein import receptor of plastids is essential for chloroplast biogenesis. *Nature*, 403, 203–207.

Bosco, C.D., Lezhneva, L., Biehl, A. *et al.* (2004) Inactivation of the chloroplast ATP synthase γ subunit results in high non-photochemical fluorescence quenching and altered nuclear gene expression in *Arabidopsis thaliana*. *J. Biol. Chem.*, 279, 1060–1069.

Bruce, B.D. (2000) Chloroplast transit peptides: structure, function and evolution. *Trends Cell Biol.*, 10, 440–447.

Budziszewski, G.J., Lewis, S.P., Glover, L.W. *et al.* (2001) *Arabidopsis* genes essential for seedling viability: isolation of insertional mutants and molecular cloning. *Genetics*, 159, 1765–1778.

Chen, X., Smith, M.D., Fitzpatrick, L. and Schnell, D.J. (2002) In vivo analysis of the role of atTic20 in protein import into chloroplasts. *Plant Cell*, 14, 641–654.

Colbert, T., Till, B.J., Tompa, R. *et al.* (2001) High-throughput screening for induced point mutations. *Plant Physiol.*, 126, 480–484.

Douce, R. (1974) Site of biosynthesis of galactolipids in spinach chloroplasts. *Science*, 183, 852–853.

Douglas, S.E. (1998) Plastid evolution: origins, diversity, trends. *Curr. Opin. Genet. Dev.*, 8, 655–661.

Emanuelsson, O., Nielsen, H., Brunak, S. and von Heijne, G. (2000) Predicting subcellular localization of proteins based on their N-terminal amino acid sequence. *J. Mol. Biol.*, 300, 1005–1016.

Emanuelsson, O., Nielsen, H. and von Heijne, G. (1999) ChloroP, a neural network-based method for predicting chloroplast transit peptides and their cleavage sites. *Protein Sci.*, 8, 978–984.

Escobar, N.M., Haupt, S., Thow, G., Boevink, P., Chapman, S. and Oparka, K. (2003) High-throughput viral expression of cDNA-green fluorescent protein fusions reveals novel sub-cellular addresses and identifies unique proteins that interact with plasmodesmata. *Plant Cell*, 15, 1507–1523.

Felder, S., Meierhoff, K., Sane, A.P. *et al.* (2001) The nucleus-encoded *HCF107* gene of *Arabidopsis* provides a link between intercistronic RNA processing and the accumulation of translation-competent *psbH* transcripts in chloroplasts. *Plant Cell*, 13, 2127–2141.

Ferro, M., Salvi, D., Brugiere, S. *et al.* (2003) Proteomics of the chloroplast envelope membranes from *Arabidopsis thaliana. Mol. Cell Proteomics*, 2, 325–245.

Ferro, M., Salvi, D., Riviere-Rolland, H. *et al.* (2002) Integral membrane proteins of the chloroplast envelope: identification and subcellular localization of new transporters. *Proc. Natl. Acad. Sci. U.S.A.*, 99, 11487–11492.

Ganeteg, U., Strand, A., Gustafsson, P. and Jansson, S. (2001) The properties of the chlorophyll *a/b*-binding proteins Lhca2 and Lhca3 studied *in vivo* using antisense inhibition. *Plant Physiol.*, 127, 150–158.

Gasch, A.P., Spellman, P.T., Kao, C.M. *et al.* (2000) Genomic expression programs in the response of yeast cells to environmental changes. *Mol. Biol. Cell*, 11, 4241–4257.

Giglione, C., Vallon, O. and Meinnel, T. (2003) Control of protein life-span by N-terminal methionine excision. *EMBO J.*, 22, 13–23.

Goldschmidt-Clermont, M. (1998) Coordination of nuclear and chloroplast gene expression in plant cells. *Int. Rev. Cytol.*, 177, 115–180.

Gomez, S.M., Nishio, J.N., Faull, K.F. and Whitelegge, J.P. (2002) The chloroplast grana proteome defined by intact mass measurements from liquid chromatography mass spectrometry. *Mol. Cell Proteomics*, 1, 46–59.

Gupta, R., He, Z. and Luan, S. (2002) Functional relationship of cytochrome c_6 and plastocyanin in *Arabidopsis. Nature*, 417, 567–571.

Gutensohn, M., Schulz, B., Nicolay, P. and Flugge, U.I. (2000) Functional analysis of the two *Arabidopsis* homologues of Toc34, a component of the chloroplast protein import apparatus. *Plant J.*, 23, 771–783.

Hajdukiewicz, P.T., Allison, L.A. and Maliga, P. (1997) The two RNA polymerases encoded by the nuclear and the plastid compartments transcribe distinct groups of genes in tobacco plastids. *EMBO J.*, 16, 4041–4048.

Haldrup, A., Lunde, C. and Scheller, H.V. (2003) *Arabidopsis thaliana* plants lacking the PSI-D subunit of photosystem I suffer severe photoinhibition, have unstable photosystem I complexes, and altered redox homeostasis in the chloroplast stroma. *J. Biol. Chem.*, 278, 33276–33283.

Haldrup, A., Naver, H. and Scheller, H.V. (1999) The interaction between plastocyanin and photosystem I is inefficient in transgenic *Arabidopsis* plants lacking the PSI-N subunit of photosystem I. *Plant J.*, 17, 689–698.

Haldrup, A., Simpson, D.J. and Scheller, H.V. (2000) Down-regulation of the PSI-F subunit of photosystem I (PSI) in *Arabidopsis thaliana*. The PSI-F subunit is essential for photoautotrophic growth and contributes to antenna function. *J. Biol. Chem.*, 275, 31211–31218.

Hippler, M., Klein, J., Fink, A., Allinger, T. and Hoerth, P. (2001) Towards functional proteomics of membrane protein complexes: analysis of thylakoid membranes from *Chlamydomonas reinhardtii*. *Plant J.*, 28, 595–606.

Hutin, C., Havaux, M., Carde, J.P. *et al.* (2002) Double mutation cpSRP43⁻/cpSRP54⁻ is necessary to abolish the cpSRP pathway required for thylakoid targeting of the light-harvesting chlorophyll proteins. *Plant J.*, 29, 531–543.

Jansson, S., Andersen, B. and Scheller, H.V. (1996) Nearest-neighbor analysis of higher-plant photosystem I holocomplex. *Plant Physiol.*, 112, 409–420.

Jarvis, P., Chen, L.J., Li, H., Peto, C.A., Fankhauser, C. and Chory, J. (1998) An *Arabidopsis* mutant defective in the plastid general protein import apparatus. *Science*, 282, 100–103.

Jarvis, P. and Soll, J. (2001) Toc, Tic, and chloroplast protein import. *Biochim. Biophys. Acta*, 1541, 64–79.

Jensen, P.E., Gilpin, M., Knoetzel, J. and Scheller, H.V. (2000) The PSI-K subunit of photosystem I is involved in the interaction between light-harvesting complex I and the photosystem I reaction center core. *J. Biol. Chem.*, 275, 24701–24708.

Jensen, P.E., Rosgaard, L., Knoetzel, J. and Scheller, H.V. (2002) Photosystem I activity is increased in the absence of the PSI-G subunit. *J. Biol. Chem.*, 277, 2798–2803.

Johanningmeier, U. (1988) Possible control of transcript levels by chlorophyll precursors in *Chlamydomonas*. *Eur. J. Biochem.*, 177, 417–424.

Johanningmeier, U. and Howell, S.H. (1984) Regulation of light-harvesting chlorophyll-binding protein mRNA accumulation in *Chlamydomonas reinhardtii*. Possible involvement of chlorophyll synthesis precursors. *J. Biol. Chem.*, 259, 13541–13549.

Khil, P.P. and Camerini-Otero, R.D. (2002) Over 1000 genes are involved in the DNA damage response of *Escherichia coli*. *Mol. Microbiol.*, 44, 89–105.

Kieselbach, T., Hagman, A., Andersson, B. and Schröder, W.P. (1998) The thylakoid lumen of chloroplasts. Isolation and characterization. *J. Biol. Chem.*, 273, 6710–6716.

Klimyuk, V.I., Persello-Cartieaux, F., Havaux, M. *et al.* (1999) A chromodomain protein encoded by the *Arabidopsis* CAO gene is a plant-specific component of the chloroplast signal recognition particle pathway that is involved in LHCP targeting. *Plant Cell*, 11, 87–99.

Koo, A.J. and Ohlrogge, J.B. (2002) The predicted candidates of *Arabidopsis* plastid inner envelope membrane proteins and their expression profiles. *Plant Physiol.*, 130, 823–836.

Kroll, D., Meierhoff, K., Bechtold, N. *et al.* (2001) *VIPP1*, a nuclear gene of *Arabidopsis thaliana* essential for thylakoid membrane formation. *Proc. Natl. Acad. Sci. U.S.A.*, 98, 4238–4242.

Kruft, V., Eubel, H., Jansch, L., Werhahn, W. and Braun, H.P. (2001) Proteomic approach to identify novel mitochondrial proteins in *Arabidopsis*. *Plant Physiol.*, 127, 1694–1710.

Kubis, S., Baldwin, A., Patel, R. *et al.* (2003) The *Arabidopsis ppi1* mutant is specifically defective in the expression, chloroplast import, and accumulation of photosynthetic proteins. *Plant Cell*, 15, 1859–1871.

Kurth, J., Varotto, C., Pesaresi, P. *et al.* (2002) Gene-sequence-tag expression analyses of 1,800 genes related to chloroplast functions. *Planta*, 215, 101–109.

Larkin, R.M., Alonso, J.M., Ecker, J.R. and Chory, J. (2003) GUN4, a regulator of chlorophyll synthesis and intracellular signaling. *Science*, 299, 902–906.

Legen, J., Kemp, S., Krause, K., Profanter, B., Herrmann, R.G. and Maier, R.M. (2002) Comparative analysis of plastid transcription profiles of entire plastid chromosomes from tobacco attributed to wild-type and PEP-deficient transcription machineries. *Plant J.*, 31, 171–188.

Leister, D. (2003) Chloroplast research in the genomic age. *Trends Genet.*, 19, 47–56.

Leister, D. and Schneider, A. (2003) From genes to photosynthesis in *Arabidopsis thaliana*. *Int. Rev. Cytol.*, 228, 31–83.

Lemieux, C., Otis, C. and Turmel, M. (2000) Ancestral chloroplast genome in *Mesostigma viride* reveals an early branch of green plant evolution. *Nature*, 403, 649–652.

Lennartz, K., Plucken, H., Seidler, A., Westhoff, P., Bechtold, N. and Meierhoff, K. (2001) *HCF164* encodes a thioredoxin-like protein involved in the biogenesis of the cytochrome $b_6 f$ complex in *Arabidopsis*. *Plant Cell*, 13, 2539–2551.

Li, X.P., Bjorkman, O., Shih, C. *et al.* (2000) A pigment-binding protein essential for regulation of photosynthetic light harvesting. *Nature*, 403, 391–395.

Li, X.P., Muller-Moule, P., Gilmore, A.M. and Niyogi, K.K. (2002) PsbS-dependent enhancement of feedback de-excitation protects photosystem II from photoinhibition. *Proc. Natl. Acad. Sci. U.S. A.*, 99, 15222–15227.

Lunde, C., Jensen, P.E., Haldrup, A., Knoetzel, J. and Scheller, H.V. (2000) The PSI-H subunit of photosystem I is essential for state transitions in plant photosynthesis. *Nature*, 408, 613–615.

Maiwald, D., Dietzmann, A., Jahns, P. *et al.* (2003) Knock-out of the genes coding for the Rieske protein and the ATP-synthase δ-subunit of *Arabidopsis*. Effects on photosynthesis, thylakoid protein composition, and nuclear chloroplast gene expression. *Plant Physiol.*, 133, 191–202.

Martin, W. and Herrmann, R.G. (1998) Gene transfer from organelles to the nucleus: how much, what happens, and why? *Plant Physiol.*, 118, 9–17.

Martin, W., Rujan, T., Richly, E. *et al.* (2002) Evolutionary analysis of *Arabidopsis*, cyanobacterial, and chloroplast genomes reveals plastid phylogeny and thousands of cyanobacterial genes in the nucleus. *Proc. Natl. Acad. Sci. U.S.A.*, 99, 12246–12251.

Martin, W., Stoebe, B., Goremykin, V., Hapsmann, S., Hasegawa, M. and Kowallik, K.V. (1998) Gene transfer to the nucleus and the evolution of chloroplasts. *Nature*, 393, 162–165.

Mayfield, S.P. (1990) Chloroplast gene regulation: interaction of the nuclear and chloroplast genomes in the expression of photosynthetic proteins. *Curr. Opin. Cell Biol.*, 2, 509–513.

McCallum, C.M., Comai, L., Greene, E.A. and Henikoff, S. (2000) Targeting induced local lesions in genomes (TILLING) for plant functional genomics. *Plant Physiol.*, 123, 439–442.

McFadden, G.I. (1999) Endosymbiosis and evolution of the plant cell. *Curr. Opin. Plant Biol.*, 2, 513– 519.

Meskauskiene, R., Nater, M., Goslings, D., Kessler, F., op den Camp, R. and Apel, K. (2001) FLU: a negative regulator of chlorophyll biosynthesis in *Arabidopsis thaliana*. *Proc. Natl. Acad. Sci. U.S.A.*, 98, 12826–12831.

Meurer, J., Plucken, H., Kowallik, K.V. and Westhoff, P. (1998) A nuclear-encoded protein of prokaryotic origin is essential for the stability of photosystem II in *Arabidopsis thaliana*. *EMBO J.*, 17, 5286–5297.

Millar, A.H. and Heazlewood, J.L. (2003) Genomic and proteomic analysis of mitochondrial carrier proteins in *Arabidopsis*. *Plant Physiol.*, 131, 443–453.

Miras, S., Salvi, D., Ferro, M. *et al.* (2002) Non-canonical transit peptide for import into the chloroplast. *J. Biol. Chem.*, 277, 47770–47778.

Mochizuki, N., Brusslan, J.A., Larkin, R., Nagatani, A. and Chory, J. (2001) *Arabidopsis genomes uncoupled 5* (*GUN5*) mutant reveals the involvement of Mg-chelatase H subunit in plastid-to-nucleus signal transduction. *Proc. Natl. Acad. Sci. U.S.A.*, 98, 2053–2058.

Molina-Heredia, F.P.E.A.N. (2003) Photosynthesis: a new function for an old cytochrome? *Nature*, 424, 33–34.

Møller, S.G., Kunkel, T. and Chua, N.H. (2001) A plastidic ABC protein involved in intercompartmental communication of light signaling. *Genes Dev.*, 15, 90–103.

Montane, M.H. and Kloppstech, K. (2000) The family of light-harvesting-related proteins (LHCs, ELIPs, HLIPs): was the harvesting of light their primary function? *Gene*, 258, 1–8.

Moore, M., Harrison, M.S., Peterson, E.C. and Henry, R. (2000) Chloroplast Oxa1p homolog Albino3 is required for post-translational integration of the light harvesting chlorophyll-binding protein into thylakoid membranes. *J. Biol. Chem.*, 275, 1529–1532.

Munekage, Y., Hojo, M., Meurer, J., Endo, T., Tasaka, M. and Shikanai, T. (2002) PGR5 is involved in cyclic electron flow around photosystem I and is essential for photoprotection in *Arabidopsis*. *Cell*, 110, 361–371.

Munekage, Y., Takeda, S., Endo, T., Jahns, P., Hashimoto, T. and Shikanai, T. (2001) Cytochrome b_6f mutation specifically affects thermal dissipation of absorbed light energy in *Arabidopsis*. *Plant J.*, 28, 351–359.

Murakami, R., Ifuku, K., Takabayashi, A., Shikanai, T., Endo, T. and Sato, F. (2002) Character-ization of an *Arabidopsis thaliana* mutant with impaired *psbO*, one of two genes encoding extrinsic 33-kDa proteins in photosystem II. *FEBS Lett.*, 523, 138–142.

Naver, H., Haldrup, A. and Scheller, H.V. (1999) Cosuppression of photosystem I subunit PSI-H in *Arabidopsis thaliana*. Efficient electron transfer and stability of photosystem I is dependent upon the PSI-H subunit. *J. Biol. Chem.*, 274, 10784–10789.

Neuhaus, H.E. and Emes, M.J. (2000) Nonphotosynthetic metabolism in plastids. *Annu. Rev. Plant Physiol. Plant Mol. Biol.*, 51, 111–140.

Nielsen, H., Brunak, S. and von Heijne, G. (1999) Machine learning approaches for the prediction of signal peptides and other protein sorting signals. *Protein Eng.*, 12, 3–9.

Nielsen, H., Engelbrecht, J., Brunak, S. and von Heijne, G. (1997) A neural network method for identification of prokaryotic and eukaryotic signal peptides and prediction of their cleavage sites. *Int. J. Neural Syst.*, 8, 581–599.

Niyogi, K.K., Grossman, A.R. and Bjorkman, O. (1998) *Arabidopsis* mutants define a central role for the xanthophyll cycle in the regulation of photosynthetic energy conversion. *Plant Cell*, 10, 1121–1134.

Norris, S.R., Barrette, T.R. and DellaPenna, D. (1995) Genetic dissection of carotenoid synthesis in *Arabidopsis* defines plastoquinone as an essential component of phytoene desaturation. *Plant Cell*, 7, 2139–2149.

Oelmüller, R. (1989) Photooxidative destruction of chloroplasts and its effect on nuclear gene expression and extraplastidic enzyme levels. *Photochem. Photobiol.*, 49, 229–239.

Osteryoung, K.W. and McAndrew, R.S. (2001) The plastid division machine. *Annu. Rev. Plant Physiol. Plant Mol. Biol.*, 52, 315–333.

Peltier, J.B., Emanuelsson, O., Kalume, D.E. *et al.* (2002) Central functions of the lumenal and peripheral thylakoid proteome of *Arabidopsis* determined by experimentation and genome-wide prediction. *Plant Cell*, 14, 211–236.

Peltier, J.B., Friso, G., Kalume, D.E. *et al.* (2000) Proteomics of the chloroplast: systematic identification and targeting analysis of lumenal and peripheral thylakoid proteins. *Plant Cell*, 12, 319–341.

Pesaresi, P., Gardner, N.A., Masiero, S. *et al.* (2003) Cytoplasmic N-terminal protein acetylation is required for efficient photosynthesis in *Arabidopsis*. *Plant Cell*, 15, 1817–1832.

Pesaresi, P., Lunde, C., Jahns, P. *et al.* (2002) A stable LHCII-PSI aggregate and suppression of photosynthetic state transitions in the *psae1-1* mutant of *Arabidopsis thaliana*. *Planta*, 215, 940–948.

Pesaresi, P., Varotto, C., Meurer, J., Jahns, P., Salamini, F. and Leister, D. (2001) Knock-out of the plastid ribosomal protein L11 in *Arabidopsis*: effects on mRNA translation and photosynthesis. *Plant J.*, 27, 179–189.

Pfannschmidt, T., Schutze, K., Brost, M. and Oelmuller, R. (2001) A novel mechanism of nuclear photosynthesis gene regulation by redox signals from the chloroplast during photosystem stoichiometry adjustment. *J. Biol. Chem.*, 276, 36125–36130.

Pogson, B., McDonald, K.A., Truong, M., Britton, G. and DellaPenna, D. (1996) *Arabidopsis* carotenoid mutants demonstrate that lutein is not essential for photosynthesis in higher plants. *Plant Cell*, 8, 1627–1639.

Pyke, K.A. (1999) Plastid division and development. *Plant Cell*, 11, 549–556.

Race, H.L., Herrmann, R.G. and Martin, W. (1999) Why have organelles retained genomes? *Trends Genet.*, 15, 364–370.

Richly, E., Dietzmann, A., Biehl, A. *et al.* (2003) Covariations in the nuclear chloroplast tran-scriptome reveal a regulatory master-switch. *EMBO Rep.*, 4, 491–498.

Richly, E. and Leister, D. (2004) An improved prediction of chloroplast proteins reveals diversities and commonalities in the chloroplast proteomes of *Arabidopsis* and rice. *Gene*, 329, 11–16.

Roberts, J.K. (2002) Proteomics and a future generation of plant molecular biologists. *Plant Mol. Biol.*, 48, 143–154.

Robinson, C., Thompson, S.J. and Woolhead, C. (2001) Multiple pathways used for the targeting of thylakoid proteins in chloroplasts. *Traffic*, 2, 245–251.

Rodermel, S. and Park, S. (2003) Pathways of intracellular communication: tetrapyrroles and plastid-to-nucleus signaling. *Bioessays*, 25, 631–636.

Ruban, A.V., Wentworth, M., Yakushevska, A.E. *et al.* (2003) Plants lacking the main light-harvesting complex retain photosystem II macro-organization. *Nature*, 421, 648–652.

Sasaki, T., Matsumoto, T., Yamamoto, K. *et al.* (2002) The genome sequence and structure of rice chromosome 1. *Nature*, 420, 312–316.

Sato, S., Nakamura, Y., Kaneko, T., Asamizu, E. and Tabata, S. (1999) Complete structure of the chloroplast genome of *Arabidopsis thaliana*. *DNA Res.*, 6, 283–290.

Schein, A.I., Kissinger, J.C. and Ungar, L.H. (2001) Chloroplast transit peptide prediction: a peek inside the black box. *Nucleic Acids Res.*, 29, E82.

Scheller, H.V., Jensen, P.E., Haldrup, A., Lunde, C. and Knoetzel, J. (2001) Role of subunits in eukaryotic photosystem I. *Biochim. Biophys. Acta*, 1507, 41–60.

Schleiff, E., Eichacker, L.A., Eckart, K. *et al.* (2003) Prediction of the plant β-barrel proteome: a case study of the chloroplast outer envelope. *Protein Sci.*, 12, 748–759.

Schubert, M., Petersson, U.A., Haas, B.J., Funk, C., Schroder, W.P. and Kieselbach, T. (2002) Proteome map of the chloroplast lumen of *Arabidopsis thaliana*. *J. Biol. Chem.*, 277, 8354–8365.

Sehnke, P.C., Henry, R., Cline, K. and Ferl, R.J. (2000) Interaction of a plant 14-3-3 protein with the signal peptide of a thylakoid-targeted chloroplast precursor protein and the presence of 14-3-3 isoforms in the chloroplast stroma. *Plant Physiol.*, 122, 235–242.

Seigneurin-Berny, D., Rolland, N., Garin, J. and Joyard, J. (1999) Technical Advance. Differential extraction of hydrophobic proteins from chloroplast envelope membranes: a subcellular-specific proteomic approach to identify rare intrinsic membrane proteins. *Plant J.*, 19, 217–228.

Shikanai, T., Munekage, Y., Shimizu, K., Endo, T. and Hashimoto, T. (1999) Identification and characterization of *Arabidopsis* mutants with reduced quenching of chlorophyll fluorescence. *Plant Cell Physiol.*, 40, 1134–1142.

Soll, J. (2002) Protein import into chloroplasts. *Curr. Opin. Plant. Biol.*, 5, 529–535.

Strand, A., Asami, T., Alonso, J., Ecker, J.R. and Chory, J. (2003) Chloroplast to nucleus communication triggered by accumulation of Mg-protoporphyrin IX. *Nature*, 421, 79–83.

Streatfield, S.J., Weber, A., Kinsman, E.A. *et al.* (1999) The phosphoenolpyruvate/phosphate translocator is required for phenolic metabolism, palisade cell development, and plastid-dependent nuclear gene expression. *Plant Cell*, 11, 1609–1622.

Sundberg, E., Slagter, J.G., Fridborg, I., Cleary, S.P., Robinson, C. and Coupland, G. (1997) *ALBINO3*, an *Arabidopsis* nuclear gene essential for chloroplast differentiation, encodes a chloroplast protein that shows homology to proteins present in bacterial membranes and yeast mitochondria. *Plant Cell*, 9, 717–730.

Surpin, M., Larkin, R.M. and Chory, J. (2002) Signal transduction between the chloroplast and the nucleus. *Plant Cell*, 14 (Suppl.), S327–S338.

Susek, R.E., Ausubel, F.M. and Chory, J. (1993) Signal transduction mutants of *Arabidopsis* uncouple nuclear *CAB* and *RBCS* gene expression from chloroplast development. *Cell*, 74, 787–799.

Sutton, M.D., Smith, B.T., Godoy, V.G. and Walker, G.C. (2000) The SOS response: recent insights into umuDC-dependent mutagenesis and DNA damage tolerance. *Annu. Rev. Genet.*, 34, 479–497.

Suzuki, J.Y. and Bauer, C.E. (1995) A prokaryotic origin for light-dependent chlorophyll biosynthesis of plants. *Proc. Natl. Acad. Sci. U.S.A.*, 92, 3749–3753.

The *Arabidopsis* Genome Initiative (2000) Analysis of the genome sequence of the flowering plant *Arabidopsis thaliana*. *Nature*, 408, 796–815.

van Wijk, K. (2000) Proteomics of the chloroplast: experimentation and prediction. *Trends Plant Sci.*, 5, 420–425.

van Wijk, K. (in press) Chloroplast proteomics. In *Plant Functional Genomics* (ed. D. Leister), The Haworth Press Inc., Binghamton.

van Wijk, K.J. (2001) Challenges and prospects of plant proteomics. *Plant Physiol.*, 126, 501–508.

Varotto, C., Pesaresi, P., Jahns, P. *et al.* (2002) Single and double knockouts of the genes for photosystem I subunits G, K, and H of *Arabidopsis*. Effects on photosystem I composition, photosynthetic electron flow, and state transitions. *Plant Physiol.*, 129, 616–624.

Varotto, C., Pesaresi, P., Maiwald, D., Kurth, J., Salamini, F. and Leister, D. (2000a) Identification of photosynthetic mutants of *Arabidopsis* by automatic screening for altered effective quantum yield of photosystem 2. *Photosynthetica*, 38, 497–504.

Varotto, C., Pesaresi, P., Meurer, J. *et al.* (2000b) Disruption of the *Arabidopsis* photosystem I gene *psaE1* affects photosynthesis and impairs growth. *Plant J.*, 22, 115–124.

Varotto, C., Richly, E., Salamini, F. and Leister, D. (2001) GST-PRIME: a genome-wide primer design software for the generation of gene sequence tags. *Nucleic Acids Res.*, 29, 4373–4377.

Vener, A.V., Harms, A., Sussman, M.R. and Vierstra, R.D. (2001) Mass spectrometric resolution of reversible protein phosphorylation in photosynthetic membranes of *Arabidopsis thaliana*. *J. Biol. Chem.*, 276, 6959–6966.

Vinti, G., Hills, A., Campbell, S. *et al.* (2000) Interactions between *hy1* and *gun* mutants of *Arabidopsis*, and their implications for plastid/nuclear signalling. *Plant J.*, 24, 883–894.

Vitha, S., Froehlich, J.E., Koksharova, O., Pyke, K.A., van Erp, H. and Osteryoung, K.W. (2003) ARC6 is a J-domain plastid division protein and an evolutionary descendant of the cyanobacterial cell division protein Ftn2. *Plant Cell*, 15, 1918–1933.

Vothknecht, U.C. and Westhoff, P. (2001) Biogenesis and origin of thylakoid membranes. *Biochim. Biophys. Acta*, 1541, 91–101.

Wastl, J., Bendall, D.S. and Howe, C.J. (2002) Higher plants contain a modified cytochrome c_6. *Trends Plant Sci.*, 7, 244–245.

Weigel, M., Pesaresi, P. and Leister, D. (2003a) Tracking the function of the cytochrome c_6-like protein in higher plants. *Trends Plant Sci.*, 8, 513–517.

Weigel, M., Varotto, C., Pesaresi, P. *et al.* (2003b) Plastocyanin is indispensable for photosynthetic electron flow in *Arabidopsis thaliana*. *J. Biol. Chem.*, 278, 31286–31289.

Wolf, Y.I., Rogozin, I.B., Grishin, N.V. and Koonin, E.V. (2002) Genome trees and the tree of life. *Trends Genet.*, 18, 472–479.

Yamaguchi, K. and Subramanian, A.R. (2000) The plastid ribosomal proteins. Identification of all the proteins in the 50 S subunit of an organelle ribosome (chloroplast). *J. Biol. Chem.*, 275, 28466–28482.

Yamaguchi, K., von Knoblauch, K. and Subramanian, A.R. (2000) The plastid ribosomal proteins. Identification of all the proteins in the 30 S subunit of an organelle ribosome (chloroplast). *J. Biol. Chem.*, 275, 28455–28465.

Zhang, H., Goodman, H.M. and Jansson, S. (1997) Antisense inhibition of the photosystem I antenna protein Lhca4 in *Arabidopsis thaliana*. *Plant Physiol.*, 115, 1525–1531.

2 Plastid development and differentiation

Mark Waters and Kevin Pyke

2.1 Introduction

The second law of thermodynamics dictates that all living organisms require an exogenous energy source for growth, development and reproduction. Whilst heterotrophic organisms obtain their energy and carbon from other organisms through nutrition, the ultimate source of this energy must be inorganic. Organisms that exploit such forms of energy and fix inorganic carbon are known as *autotrophs*. Although some chemoautotrophic bacteria fix inorganic carbon by using energy derived from the oxidation of chemical sources such as H_2S, the vast majority of autotrophic life, and subsequently heterotrophic life, is based on the harnessing of energy from the Sun. The ubiquitous process by which sunlight is converted into chemical energy, *photosynthesis*, arose approximately 3.6 billion years ago in a prokaryote (Niklas, 1997). The atmosphere of the early Earth was more reducing than that of the present, and oxygen did not reach high enough concentrations (1–2% of present-day levels) to support aerobic respiration until somewhere between 2.4 and 2.8 billion years ago (Knoll, 1992). Given in addition that most present-day photosynthetic eubacteria neither produce nor consume molecular oxygen, it seems probable that the first photosynthetic bacteria were anoxygenic. Eukaryotic photosynthesis liberates oxygen however, and it is now widely accepted that photosynthesis in eukaryotes arose as a result of an endosymbiotic event between an aerobic proto-eukaryote and an *oxygenic* photosynthetic prokaryote, most probably cyanobacterium-like in form (McFadden, 2001). Whilst the original photosynthetic prokaryote and its host are now inextricably associated, owing in no small part to the transfer of genetic information from endosymbiont to the host nucleus, this symbiosis is the defining feature of all extant photosynthetic eukaryotes; that is, the fundamental photosynthetic events (i.e. the net fixation of carbon dioxide) occur in the evolutionary remnant of this prokaryote, the plastid.

The evolution of photosynthetic eukaryotes has followed a trend of increasing genetic and developmental complexity. Embryophytes, the land plants, are thought to have evolved from a freshwater multicellular green alga of the order Coleochaetales about 450 million years ago, when various adaptations such as a waxy cuticle permitted survival in the desiccating terrestrial environment (Niklas, 1997). The earliest embryophyte probably resembled a liverwort, with a free-living gametophyte and an ephemeral sporophyte, and without vascular tracheids. From this ancestral land plant evolved the monophyletic group of embryophytes observable today. Apart from a limited number of parasitic angiosperms, all of these organisms derive their energy from photosynthesis, and all contain a plastid compartment within their cells. Additionally, in line with the increase in morphological complexity and diversity from the liverwort-like ancestor to the angiosperms, the plastid compartment itself has also attained a variety of forms and functions during evolution, and yet has conserved a sufficient suite of characters that alludes to its prokaryotic ancestry.

Plastids in lower plants (green algae, liverworts, mosses, hornworts) contrast with those in higher (vascular) plants in various ways. Unicellular green algae like *Chlamydomonas* possess only one plastid, which occupies a large proportion of the cell volume. Many multicellular algae also contain single, spiral plastids that span the entire length of a cell, e.g. *Spirogyra*, a common filamentous green alga of ponds and streams. In contrast, vascular plants possess from several to hundreds of plastids per cell, which is presumably an adaptation to coping with varying light conditions, because several, smaller chloroplasts can move within a cell to intercept or avoid light more efficiently than fewer, larger ones (Pyke, 1999; Jeong *et al.*, 2002). Secondly, the extent of plastid differentiation in the lower plants is restricted relative to higher plants, whose plastids perform a variety of functions and differentiate concomitantly with the cell type. Full chloroplast differentiation in angiosperms requires light, but most green algae synthesise chlorophyll in the dark. In *Chlamydomonas reinhardtii*, although transcript levels for chlorophyll biosynthetic genes and Rubisco are attenuated when it is grown in the dark (Cahoon and Timko, 2000), the plastid still accumulates some chlorophyll and is competent to carry out photosynthesis upon transfer to the light. The ability to synthesise chlorophyll in the dark is also retained in mosses and some Pteridiophytes such as *Selaginella* and *Isoetes*, but not in others such as the Equisitaceae (Kirk and Tilney-Bassett, 1978). Thirdly, the segregation of plastids between daughter cells during cell division varies amongst different taxa. In the moss *Anthoceros*, plastids are passed on to the daughter cell during mitosis in the form of chloroplasts, which contrasts with *Isoetes* and higher vascular plants that have either one or several colourless proplastids, respectively, in meristematic cells (Kirk and Tilney-Bassett, 1978). Such evidence upholds the established view that plastids are not created *de novo* but are part of a continuum of multiplying plastids transmitted from cell to cell. Given that plastids in these lower plants are generally chloroplastic in nature, even in the dark, it seems plausible that there has been little adaptation on the 'default' plastid form of the ancestral green algae, and that plastid differentiation is generally limited to the chloroplast. The primary plastid function in these plants, therefore, appears to be photosynthesis.

The vast majority of biochemical, ultrastructural and molecular-genetic studies on plastids have been performed on angiosperms. These plants are important agronomically, are easy to cultivate and offer a range of experimental advantages such as ease of genetic manipulation and availability of a wide range of mutants. In contrast with the non-vascular plants described earlier, plastids in angiosperms vary in size, shape, content and function. Plastids can perform several interrelated roles simultaneously, and the various types are dynamically interconvertible (Figure 2.1). Plastids are hence aptly named, the term originating from the Greek *plastikos*, meaning 'plastic, mouldable'. Traditionally, plastids have been classified according to the obvious function of the plastid in question, generally based on their morphological appearance: a green plastid in leaf cells, a chloroplast, a colourless one with starch grains, an amyloplast, etc. (Kirk and Tilney-Bassett, 1978). Such a classification is useful for describing the scope of plastid forms and how they are critical to plant development and reproductive success but is an arbitrary and overly simplistic one, since frequently a particular plastid expresses features of more than one type. A more flexible classification system might be based upon the

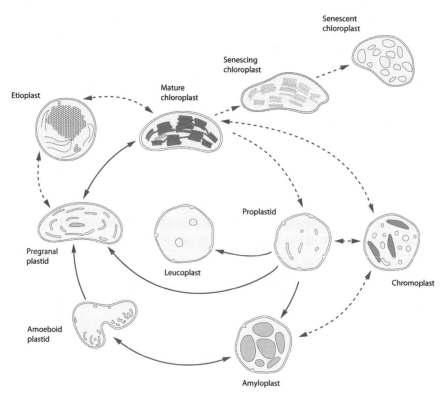

Figure 2.1   Interrelations between the different forms of plastids in higher plants. Solid lines indicate conversions that occur in most higher plants under normal conditions; dashed lines specify conversions that are possible under a limited set of conditions or that are unusual. Note that this diagram is not exhaustive, as some plastid types do not neatly fall into one of these categories. Redrawn with permission from Buchanan *et al.* (2000). Copyright (2000) The American Society of Plant Physiologists.

physiological and biochemical properties of the plastid, or some other way of reflecting the range of forms a plastid can take that are intermediate between those somewhat rigid classifications. Nevertheless, distinct states of plastid differentiation do exist, each with specific though not necessarily unique properties.

In this chapter we consider the major types of plastids found in cells of higher plants and what is known about the mechanisms that influence their differentiation and their developmental programmes.

2.2 Meristematic proplastids

The plastid type basic to most pathways of plastid development and differentiation in higher plants is the proplastid (Figure 2.2A). *Proplastids* are small, unpigmented plastids found in meristematic tissues and from which all plastids within a plant are ultimately derived. The original source of the plastid population within an entire plant is the plastids contained within the egg cell and the pollen prior to gamete fertilization and zygote formation. In the majority of angiosperms the pollen is devoid of functional plastids so that the plastid population is derived entirely from the few plastids within the egg cell (Corriveau and Coleman, 1998). Hence, in most species plastids are maternally inherited. During embryogenesis, proplastids derived from the egg cell divide as cell division occurs and the basic plan of the embryo is constructed. Upon establishment of clear meristematic regions at the shoot and root apices, proplastids proliferate in these regions as cells divide to ensure continuity of the plastid line within daughter cells (Chaley and Possingham, 1981). Although proplastids are essentially confined to meristematic regions in most of plant development, in the earliest stages of embryogenesis the plastids found throughout the body of the embryo are probably best described as proplastids. Although fundamental to plastid development, an understanding of proplastid biology is poor principally because of difficulties in observing, tracking and isolating such small, unpigmented organelles from dense tissues such as meristems. Much of the knowledge of proplastid behaviour during cell division and development within the meristem comes from electron microscopy studies, which clearly show populations of proplastids dispersed throughout the cell and in various states of division (Chaley and Possingham, 1981; Robertson *et al.*, 1995). Attempts to estimate proplastid population size in these cells in different species, such as maize root meristem (Juniper and Clowes, 1965), spinach shoot meristem (Possingham and Rose, 1976) and *Arabidopsis* (Pyke and Leech, 1992), provides a general consensus of 10–20 proplastids per cell, although this number may change according to the stage of the cell cycle and the cellular position within the meristem. During this division stage, proplastids replicate their DNA, complexed within up to 10 nucleoid structures within the organelle (Miyamura *et al.*, 1990). Interestingly, even at the earliest stages of cellular development in the meristem, there are differences in proplastid morphology and DNA content between proplastids from different layers of the shoot apical meristem – L1, L2 and

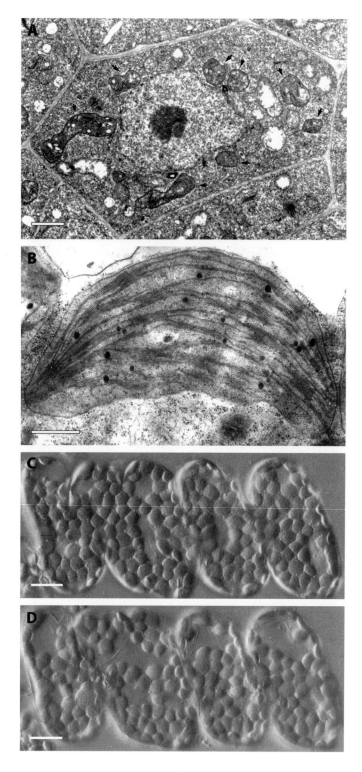

34

L3 (Fujie *et al.*, 1994), suggesting that even proplastids may show tissue-specific characteristics.

Although proplastid segregation at cell division is of crucial importance to future cell viability, a distinct mechanism ensuring correct segregation of proplastids into daughter cells has not been elucidated. Most likely the process is dependent upon a moderately even distribution of proplastids throughout the cell (Figure 2.2A), which ensures proplastids are present at either cell pole and hence in the daughter cells. Interestingly giant proplastids, which are reduced in number in plastid division mutants, apparently are still able to ensure continuity of proplastids through cell lineages in meristems and mature tissues (Robertson *et al.*, 1995).

In terms of internal structure, proplastids contain little definable structure other than traces of thylakoid-like membrane and sometimes starch grains. Several efforts have been made to assess levels of gene transcription by plastid DNA in proplastids and expression of nuclear genes for plastid-targeted proteins. None of these studies has been able to measure activities at the individual cellular or tissue level within the meristem, which is technically very demanding. However, studies using proplastids in spinach cotyledons (Harak *et al.*, 1995; Mache *et al.*, 1997), proplastids in cultured Bright-Yellow 2 (BY-2) cells (Sakai *et al.*, 1998) and meristematic tissues at the base of barley leaves (Baumgartner *et al.*, 1989) together show that the level of proplastid DNA transcription is very low and that the progressive development of the proplastid towards the chloroplast requires the expression of nuclear genes for ribosomal structures preceding those that are plastid encoded. The emphasis in these studies has been on the initiation of the plastid differentiation pathway and little is known of the essentially housekeeping metabolism that occurs during proplastid division and growth in cells in the central zone of the shoot apical meristem or in the root apical meristem.

2.3 Chloroplast biogenesis and cell differentiation

Upon germination, a seedling must establish an independent energy source before it depletes the storage reserves present in the seed. Attainment of this state is dependent on the formation of photosynthetically competent chloroplasts, triggered by the perception of light. The light signal is translated into an induction of novel gene expression and protein synthesis, marking the beginnings of a complex chain of events that require tight metabolic coordination between the nuclear and plastid

Figure 2.2 Proplastids and chloroplasts. (A) Transmission electron micrograph of a cell in the shoot apical meristem of *Arabidopsis*. Note the central nucleus surrounded by proplastids, some of which are arrowed. Bar: 1 μm. (B) Transmission electron micrograph of a mature mesophyll cell chloroplast from an *Arabidopsis* leaf. Bar: 1 μm. (C, D) Two fixed, isolated mesophyll cells from a wheat leaf viewed by Nomarski optics and showing the distribution of chloroplasts on the upper cell surface (C) and the lower cell surface (D). The left-hand cell has three lobes at the top and the right-hand cell has two lobes at the top. Bar: 10 μm.

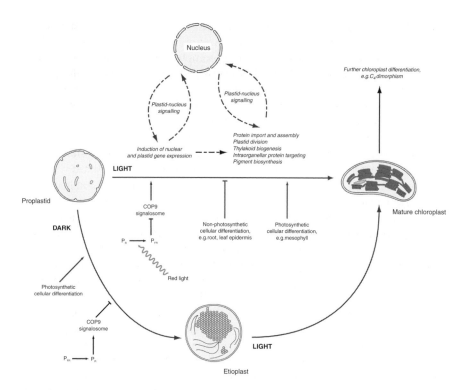

Figure 2.3 Overview of chloroplast biogenesis. Meristematic and cotyledonary proplastids differentiate into either chloroplasts or etioplasts, depending on the detection of red light by phytochrome, which relieves suppression of photomorphogenesis by the COP9 signalosome. Chloroplast development is modulated by continual feedback between the plastid and the nucleus, and also through complex interplay with cellular differentiation, such that functional chloroplasts form in a cell-type-specific manner. The etioplast therefore represents a partial chloroplast whose development has been promoted by cellular differentiation but prevented from reaching completion by the absence of light. Processes are in *italic* type; inhibitory/promoting factors are in roman type.

compartments and to ensure that chloroplast biogenesis proceeds in concert with cell differentiation. Much of the detail on the processes involved in chloroplast development is considered elsewhere in this work. Here we present a themed overview that outlines the conversion from the basal proplastid to a functional chloroplast (Figure 2.3).

2.3.1 Photomorphogenesis

Prior to light reaching the developing seedling, plastids in the aerial parts of the plant are present either as meristematic proplastids or in the form of partially developed chloroplasts, called *etioplasts*, generally found in the cotyledons of a typical epigeal seedling. We consider the etioplast to represent a blocked stage along the path to chloroplast development. Etioplasts are typified by a semi-crystalline prolamellar body and high levels of protochlorophyllide, such that the plastid is in a state primed for rapid thylakoid biogenesis and chlorophyll biosynthesis upon illumination. If cotyledons are thought of as functionally equivalent to leaves, then the plastids are forced to develop as far as possible without light in line with cellular differentiation, and thus attain the blocked etioplast state. However, meristematic proplastids destined to become leaf mesophyll chloroplasts probably never establish an etioplast-like state since leaf development and chloroplast differentiation occur rapidly and simultaneously once photomorphogenesis is initiated (Brutnell and Langdale, 1998). The etioplast has attracted a great deal of attention in the past for studying chloroplast development, but is probably best regarded as an unusual plastid type that does not accurately reflect normal chloroplast differentiation.

The perception of light by the chromoprotein phytochrome triggers chloroplast biogenesis as part of the general photomorphogenic response (Bauer *et al.*, 2001). Phytochrome absorbs red light, undergoes photoconversion to an active form and then translocates from the cytoplasm to the nucleus (Smith, 2000) where it interacts with at least one transcription factor (PIF3) in a light-reversible manner (Ni *et al.*, 1999). Plants antisense for *PIF3* show lower levels of *CAB3* (chlorophyll *a/b* binding protein) gene expression, indicating that photomorphogenesis is inhibited and thus chloroplast development is retarded (Ni *et al.*, 1998). PIF3 has since been shown to bind to the G-box sequence motif present in the promoters of a number of light-regulated nuclear genes, including the small subunit of Rubisco, *RBCS* (Martínez-García *et al.*, 2000). As a primary recipient of the phytochrome signal, PIF3 is one likely candidate for triggering the very initial stages of chloroplast biogenesis brought about by changes in nuclear gene expression, presumably upstream of the COP/DET/FUS group of photomorphogenesis repressors (Ni *et al.*, 1998; Kim *et al.*, 2002). It is likely that these repressors must also operate in non-green tissues such as roots in order to inhibit chloroplast biogenesis. In the *det1* mutant of *Arabidopsis*, which is partially photomorphogenic in the light, root plastids partially differentiate into chloroplasts as is evident from the accumulation of chlorophyll, together with *psbA*, *psaA*, *rbcL* and *rbcS* transcripts (Chory and Peto, 1990). The chloroplast does not develop under the sole control of the nucleus, however; the developing chloroplast produces a retrograde signal to the nucleus as an indication of its developmental status (Rodermel, 2001). Treatment of developing seedlings with norflurazon to induce photo-oxidative damage to the chloroplasts results in significant reduction in the expression of nuclear photosynthetic genes such as *RBCS* and *CAB* (Burgess and Taylor, 1987; La Rocca *et al.*, 2001). A number of mutants have been isolated in which this regulation is defective; that is, they exhibit normal nuclear photosynthetic gene expression despite photodamaged chloroplasts (Susek *et al.*, 1993). Analysis of the nuclear, recessive nature of these *gun* (*genomes uncoupled*) mutations suggests that one aspect of the 'chloroplast signal' is an intermediate in the tetrapyrrole biosynthetic pathway as the messenger (Strand *et al.*, 2003). Since the assembly of a functional chloroplast is strictly dependent on the integration of nuclear- and plastid-encoded gene products, the existence of such a regulatory signal is clearly of adaptive significance.

2.3.2   Specific processes

The mature chloroplast is a structurally complex entity with a large proteome consisting of contributions from two different genomes. Once a plastid has commenced the pathway towards chloroplast development, several interrelated activities take place during chloroplast biogenesis, each of which is essential for complete chloroplast functionality. Some of these processes are simply summarised here; they are discussed in greater detail throughout this book.

The developing chloroplast must import some 3000 proteins encoded by nuclear genes (Martin *et al.*, 2002; Leister, 2003). Proteins must possess an N-terminal transit peptide that is recognised by the chloroplast protein import apparatus, and are then transported in an unfolded state to the chloroplast envelope by cytosolic chaperones (Bauer *et al.*, 2001). Chloroplast protein import is an early, critical event that depends on a complex of several nuclear-encoded proteins (see Chapter 5). Various mutants disrupted in components of the import apparatus have been characterised, such as *ppi1*, which exhibits a pale green phenotype owing to improper chloroplast differentiation (Jarvis *et al.*, 1998). Furthermore, there is evidence that photosynthetic and non-photosynthetic proteins are imported through different protein import receptors, providing import specificity that may be important in directing plastid differentiation (Kubis *et al.*, 2003).

Imported proteins must subsequently be sorted among the various suborganellar locations available, such as either of the two envelope membranes, the stroma, the thylakoid membranes and the thylakoid lumen. This is achieved by several parallel targeting pathways (see Chapter 6). Imported proteins attain functionality only when correctly assembled together with other subunits in the correct stoichiometery. Rubisco, for example, is a hexadecameric holoenzyme of eight large and eight small subunits encoded by the plastidial *rbcL* and nuclear *rbcS* genes respectively. Stoichiometry is maintained through modulation of translation initiation of rbcL mRNA and one interpretation is that translation is negatively regulated by the presence of excess RbcL subunits or, alternatively, it is activated by excess RbcS subunits (Rodermel, 2001).

Since plastids do not arise *de novo*, cellular growth and expansion in developing leaf primordia must be accompanied by division of plastids, such that they are appropriately distributed between daughter cells and maintain a density suitable for efficient photosynthesis. The molecular mechanics of division are highly conserved between ancestral cyanobacteria and plastids, and have been extensively studied (Osteryoung and McAndrew, 2001; see Chapter 4). However, the genetic mechanisms that regulate plastid size and density – factors that vary substantially between cell type and species – have yet to be identified.

A chloroplast must simultaneously harvest sufficient light whilst coping with excess light through the biosynthesis of a suite of light-harvesting and accessory pigments. The biosynthesis of the chlorophylls is performed entirely in the chloroplast from the simple precursor glutamate. There is strong evidence that the presence of chlorophyll is necessary for stabilising the thylakoid membrane system and light-harvesting complex (León and Arroyo, 1998), reflecting a recurring theme in chloroplast biogenesis that plastid metabolism is constantly self-regulating and that no single process occurs in isolation from any other. On a morphological level, the formation of internal thylakoid membranes marks the process of chloroplast maturation. Thylakoid formation requires the reorganisation and biogenesis of internal membranes, together with the assembly of thylakoid-localised protein complexes. Thylakoid biogenesis is initiated by the development of long lamellae, which are later complemented by smaller, disc-shaped structures to form granal stacks. The mature chloroplast contains an interlocking network of granal thylakoid stacks connected by thylakoid lamellae, with a densely packed stroma containing all of the soluble proteins involved in photosynthesis and other metabolic processes. Thylakoid membranes are thought to be derived from invaginations of the inner membrane, as maturing chloroplasts sometimes exhibit a continuum between the inner membrane and internal membrane structures (Vothknecht and Westhoff, 2001), although this continuum is not present in mature chloroplasts. It has been suggested that vesicle trafficking from the inner membrane to the thylakoids allows maintenance and regeneration of these structures in the mature chloroplast (Vothknecht and Westhoff, 2001). Furthermore, an ATP-dependent factor involved in vesicle fusion within pepper chromoplasts has been isolated and the gene cloned (Hugueney et al., 1995b). Such a 'budding' mechanism of thylakoid biogenesis would explain how other hydrophobic membrane components (e.g. carotenoids, galactolipids), synthesised on the chloroplast envelope, are able to reach the thyklakoid membranes themselves.

2.3.3 Chloroplast development and cellular differentiation

Numerous lines of evidence indicate that cellular differentiation in the leaf is tightly dependent on proper chloroplast development. Mutants deficient in fundamental aspects of chloroplast biogenesis have a pleiotropic effect on mesophyll and palisade cell differentiation, probably as a result of impaired feedback signalling from the chloroplast to the nucleus that would otherwise orchestrate completion of leaf development (León and Arroyo, 1998). The CLA1 locus of Arabidopsis encodes the first enzyme of the plastid-based isoprenoid biosynthetic pathway, necessary for chlorophyll and carotenoid biosynthesis (Estévez et al., 2000). cla1-1 mutants are albino and show aberrant chloroplasts with reduced thylakoid stacking. This in turn leads to defective mesophyll tissue formation (Mandel et al., 1996; Estévez et al., 2000). Similar phenotypes have been described in the dcl-m mutant of tomato (Lycopersicon esculentum) (Keddie et al., 1996) and dag of Antirrhinum (Chatterjee et al., 1996). Both of these mutants possess white sectors with proplastid-like plastids and aborted palisade cell development. In the Arabidopsis immutans (im) mutant, which is defective in a chloroplast homologue of the mitochondrial alternative oxidase, palisade cells in white sectors fail to expand properly with generally perturbed tissue organisation following photodamage to carotenoid-deficient plastids (Aluru *et al.*, 2001). It is also possible to disrupt tissue development by preventing chloroplast formation through pharmacological means. When applied to *Brassica napus* explants, spectinomycin, an inhibitor of plastid protein synthesis, induces white sectors in which plastids lack ribosomes and in which palisade cell development is arrested (Pyke *et al.*, 2000).

Chloroplasts also differentiate in a manner appropriate to the cell type within the leaf. In C_4 plants such as maize, atmospheric CO_2 is initially fixed in the mesophyll (M) cells and shuttled across to the bundle sheath (BS) cells in the form of malate, which is decarboxylated in the BS cell chloroplasts to supply the Calvin cycle with CO_2. This CO_2-concentrating mechanism requires differential expression of both the C_4 cycle genes and Calvin cycle enzymes, especially Rubisco, across the two cell types (Sheen, 1999). Furthermore, the two chloroplast types are morphologically different. Mesophyll cell chloroplasts are starchless and possess numerous grana, whereas bundle sheath cell chloroplasts accumulate starch grains and thylakoid membranes are largely unstacked (Brutnell and Langdale, 1998). Whilst these differences are probably a result of differences in transcription of nuclear genes (Sheen, 1999), mutant analysis of maize leaf development has revealed that primary defects in BS cell chloroplast development can lead to pleiotropic aberrancies in BS cell differentiation as well (Brutnell *et al.*, 1999). Plastid dimorphism can even be achieved within a single cell. The leaf chlorenchyma cells of the succulent dicotyledon *Bienertia cycloptera* possess two forms of plastid morphologically similar to those observed in the M and BS cells of maize. The two chloroplast forms also appear to be functionally identical and the differential expression of Rubisco between the two chloroplast types, as well as the spatial separation of cytosolic C_4 cycle enzymes between different cytoplasmic compartments, allow efficient C_4 photosynthesis within the same cell (Voznesenskaya *et al.*, 2002). As such, both cell and chloroplast development are inextricably linked and rely on constant communication between the nucleus and plastid compartments; however, the molecular nature of this interplay, beyond the early role of the chloroplast signal described above, has so far proved elusive.

2.4   Stromules: an enigmatic feature of plastid development

Studies into the physical appearance of the plastid have a long history. The comprehensive work of Schimper (1885), describing an array of plastid types in angiosperms and organising them according to their colour, has formed the basis of their current classification. Besides pigmentation, different plastid types vary dramatically in size and shape, both within and between species. These differences result from long-term shifts in plastid function due to differentiation, but there is also plenty to be learnt from looking at changes in plastid morphology over a much shorter time-scale. Much of our knowledge regarding cellular dynamics results from meticulous observation through microscopes, allowing real-time visualisation of cellular behaviour in living tissue. A great deal has been found out about the plant cell in this way, especially with regards to organelle movement. It is well documented that organelles stream in plant cells (Williamson, 1993), and plastids in particular are able to move in response to changes in light intensity (see Chapter 10). However, a distinction ought to be made between this kind of organelle movement within cells and the autonomous, pleiomorphic movement of individual organelles, even though the molecular basis, at least in terms of motor proteins bringing about the locomotion, may be similar. It is this aspect of plastid motility that is of particular interest to understanding plastid development. One notable feature that has consistently emerged from watching plastids *in vivo* is that plastids are not static, independent organelles as is often assumed, but are instead highly dynamic entities that frequently connect with one another. Membranous conduits emanating from the plastid surface extend and retract into the cytoplasm, and sometimes join up with other plastids. These protrusions, now known as *stromules*, are tubular extensions of the plastid envelope that contain stroma, and permit the exchange of molecules between individual plastids. The use of green fluorescent protein (GFP) to highlight plastids has provided the means with which to see these structures readily and reliably, but references to plastid protrusions and dynamics can be found in the literature that spans the past one hundred years or so (Gray *et al.*, 2001; Figure 2.4).

Prior to the use of GFP, one of the most convincing testimonies of plastid motility is that of work performed by Wildman and colleagues at the University of California. Using a combination of phase contrast microscopy and cinephotomicrography, Wildman *et al.* (1962) describe how chloroplasts in living spinach palisade cells consist of two visually distinct subregions: an inner, non-motile chlorophyll-bearing structure; and a surrounding colourless 'jacket of material', which constantly varies in shape. They describe further how 'long protuberances extend from the jackets into the surrounding cytoplasm'. Wildman (1967) later reported that isolated chloroplasts lacking their envelope lose their motility, whereas those with an intact envelope retain it: some images clearly show chloroplasts with stromules. As part of the general study into cellular cytoplasmic streaming, Wildman and colleagues (Wildman *et al.*, 1962; Wildman, 1967) report that the protuberances segment into smaller, free-flowing structures visually indistinguishable from mitochondria – prompting them to suggest that these two organelle types are interconvertible. Whilst such a proposition might be dismissed outright in the post-genomic era, the observation that stromules might fragment is something that should not (Pyke and Howells, 2002).

Electron microscopy has also been extensively employed to investigate changes in plastid ultrastructure but this method obviously prevents any assessment of plastid motility, and is prone to artefacts. Nonetheless, several examples exist in which fixed samples exhibit what appear to be tubular protrusions from the plastid surface (Gray *et al.*, 2001). In order to preserve the cells in a condition as close as possible to their natural state, Bourett *et al.* (1999) used high-pressure freezing to describe the ultrastructure of stromules in rice leaf tissue. They describe protuberances of up to 4.1 μm in length that are contiguous with the stroma, and confirm

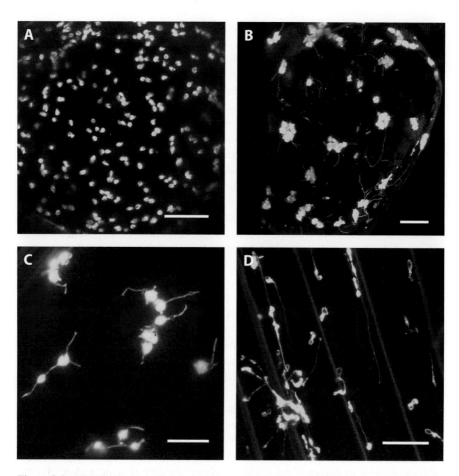

Figure 2.4   Variation in stromule morphology between different plastid types revealed by green fluorescent protein. (A) Chloroplasts in a pericarp cell of mature green tomato fruit. Chloroplasts in this tissue are regular in shape and produce few stromules. (B) Chromoplast morphology in a ripe tomato pericarp cell. Note the extensive stromules pervading the cytoplasm and apparently connecting individual plastid bodies. Chromoplast bodies are irregular in shape and are frequently distended by lycopene crystal deposits. (C) Non-pigmented plastids in a dark-grown immature (unripe) tomato fruit pericarp cell. This epifluorescence image reveals that plastids that do not differentiate into chloroplasts retain great plasticity in their shape; compare with A. (D) Highly complex and variable plastid morphology in the epidermal cells of the hypocotyl of an etiolated tobacco seedling. Stromules take on forms ranging from long tubules to loops, along with occasional bifurcations. Plastids in this etiolated tissue are very irregular in shape, suggesting that stromules may be an extreme of the continuum of plastid forms. The cell walls are also faintly autofluorescent, visible as parallel lines running diagonally down the image. Scale bars: A, B, 40 μm; C, 10 μm; D, 20 μm. A, B and D are maximum projections along a confocal Z-series.

this through immunolocalisation of the large subunit of Rubisco to both the main body of the chloroplast and the protuberances. They also provide one instance of two chloroplasts apparently interconnected by such a protuberance.

However, it was not until the advent of GFP and confocal microscopy that stromules could be investigated more systematically and in living tissue. The first use of this approach was reported by Köhler et al. (1997), who expressed plastid-targeted GFP in tobacco and petunia, and provided the first evidence that stromules were more extensive, both in length and in abundance, in some tissue types than others. They described tubules of between 350 and 850 nm in diameter and up to 15 μm in length. Furthermore, they demonstrated the transfer of GFP from one plastid to another along an interconnecting stromule by the use of selective photobleaching followed by monitoring the subsequent return of fluorescence (Köhler et al., 2000). This work led to the final acceptance of plastid protuberances in living tissue, and a variety of similar investigations have shown that stromules are a feature in all species so far examined using GFP, but that they are highly variable in form and abundance.

2.4.1 Stromules and plastid differentiation

Stromules have been observed in a number of higher plant taxa, but patterns of stromule distribution amongst different plastid types are becoming clear. In general, stromules are rarer (that is, fewer plastids produce stromules) and less extensive (stromules are shorter or less developed in shape) on chloroplasts than on other plastid types. The most extensive overview of stromule abundance in different tissues to date is that of Köhler and Hanson (2000), using transgenic tobacco carrying a constitutive plastid-targeted GFP construct. Chloroplasts in mesophyll and stomatal guard cells, which are amongst the largest (5–7 μm in length) and most regularly shaped plastids in the plant, showed very few stromules, with most plastids in a cell exhibiting none. In contrast, the achlorophyllous plastids in petal epidermal cells and roots appeared much less regular in shape and were generally smaller and highly variable in size (1.8–3 μm in diameter). Almost all plastids in these cells exhibited stromules, and root plastids of the meristematic zone frequently formed a circle around a non-fluorescent area reminiscent of the nucleus, with stromules pointing towards the cell periphery (Köhler and Hanson, 2000). A similar pattern of stromule distribution is visible in tomato, although basal cells of trichomes show stromules on around 30% of plastids (M. Waters and K. Pyke, unpublished observations, 2002), which is high when compared to M cells.

Stromule frequency and scope is, broadly speaking, negatively correlated with chloroplastic features. Nearly every leucoplast in onion bulb epidermal cells exhibits stromules (Gray et al., 2001). Köhler and Hanson (2000) presented images of plastid morphology from leaf explants that had been placed under callus-inductive conditions. Plastids from green leaf tissue adjacent to the dividing callus tissue showed few stromules, whereas plastids from pale green and colourless tissue showed increasing degrees of elongation and irregularity in shape. Stromules were correspondingly much more abundant and extensive. However, it might be argued that these changes in plastid morphology are merely a reflection of cellular dedifferentiation: in other words, are stromules related to the cell type in question, or are they a specific feature of the plastid state of development? This question has been partially addressed by growing tomato fruit, expressing plastid-targeted GFP, in the dark, thus preventing chloroplast development from proplastids. Normal tomato fruit reach a mature green stage before ripening commences. When the truss is covered, just following anthesis, the fruit enlarges as normal, but remains white. Plastids in the pericarp cells of such fruit exhibit much more frequent and extensive stromules than in the same cell type in light-grown, green fruit (Figure 2.4; M. Waters and K. Pyke, unpublished observations, 2004). The fruits then proceed to ripen as normal, and turn red even in the absence of light. Thus, plastids in cells that have begun, and are competent to complete, their normal path of development can show variable morphology that is dependent on the differentiation state of the plastid itself, and not on the cell type *per se*.

Stromules have been seen to form highly intricate networks, with plastid bodies apparently interconnected by stromules. When incubated under liquid suspension culture, tobacco cell plastids exhibited 'octopus or millipede' like morphologies, with plastid bodies frequently clustered around the nucleus (Köhler and Hanson, 2000); however, photobleaching experiments concluded that the majority of these plastids were not interconnected. Partial plastid networks have also been described in the ripe fruit of tomato (Pyke and Howells, 2002) that are not present in the unripe green fruit. Particularly extensive stromule formations that spread throughout the cell and that appear to link most plastids have been observed in tobacco epidermis (Arimura *et al.*, 2001), which contrasts with reports of plastid morphology in epidermal cells of tomato where stromules are relatively rare (Pyke and Howells, 2002). Occasional 'nodules' or vesicle-like entities with no obvious attachment to a plastid or stromule have also been reported (Arimura *et al.*, 2001; Pyke and Howells, 2002), reminiscent of Wildman's supposition that stromules may sever and form mitochondrion-like structures. A point to note, however, regarding the epidermal plastid 'networks' reported by Arimura *et al.* (2001) is that plastids were visualised using transient expression of GFP, delivered via particle bombardment of detached and dissected leaf tissue. It is quite possible that plastid morphology could change dramatically over the time course of a particle bombardment procedure, thus not accurately representing a genuine *in planta* characteristic. However, together with the general tendency for stromules to be rare in green tissue, results such as these do demonstrate that plastid morphology is highly variable and may be under the control of a large number of contributing environmental and genetic factors.

The precise role of stromules in plastid functioning is unknown, although various hypotheses have been put forward. The observation that GFP can be transported along stromules (Köhler *et al.*, 1997, 2000) suggests that they might be important in the transfer of macromolecules, such as imported proteins, either between plastid bodies or from peripheral locations in the cell towards the plastid body. The transfer of molecules between individual plastid bodies is probably not a primary role of stromules however, since the majority of stromules do not appear to join plastids together (Köhler *et al.*, 2000; Gray *et al.*, 2001). The existence of a size exclusion limit for stromules has yet to be shown, but it seems more than likely that they allow passage of a range of macromolecules and metabolites. It may be that nucleic acids can travel along stromules: the microinjection of individual plastid bodies with a plasmid encoding GFP led to the spread of fluorescence throughout the plastids within a cell (Knoblauch *et al.*, 1999), perhaps as a result of the movement of any combination of DNA, RNA or protein. It has also been proposed that stromules stretching towards the cell periphery may aid in the transduction of photoelectric signals perceived at the cell surface to the organelle membrane system itself (Tirlapur *et al.*, 1999). Furthermore, metabolic interactions with mitochondria and peroxisomes could be maximised through physical contact with stromules, especially if these organelles and stromules co-exist on the same actin microfilaments (Gray *et al.*, 2001).

Nevertheless, the most likely role for stromules is to provide further surface area for processes such as protein import and metabolite exchange, whilst minimising the plastid volume required to produce them (Gray *et al.*, 2001). It appears that plastids at a high density, such as in M cells, produce relatively fewer stromules than those more widely distributed throughout the cell, such as root or trichome cells. In the latter types of cells, the increased surface area of the plastid compartment in contact with the cytoplasm could help compensate for the lower plastid density, presumably maintained as such for reasons of economy. However, any apparent negative correlation of stromule frequency with plastid density is difficult to discern from the negative correlation with chlorophyll content, as the two factors are often related. More precise analysis of plastid density in tissues where chloroplast development has been disrupted will allow these two factors to be separated.

Understanding the true structure and functions of stromules will come about from further studies on a number of aspects of their development. Their proteomic profile, if different from that of the rest of the plastid envelope, will be most informative, as will a closer examination of what molecules can be transported along them. It is important that we understand how stromules move, including whether or not some form of internal motility system or 'plastoskeleton' is involved (Reski, 2002). Finally, a central issue is to what degree stromules are regulated: are they an indirect result of increased plastid membrane flexibility, or are they actively induced by signals and changes in gene expression? Such questions represent substantial challenges but ones which will need addressing before progress in this exciting field can be made.

2.5 Amyloplast differentiation

During the course of evolution, the plastid has acquired the role of storing various compounds and sequestering them away from the metabolic activities of the cytosol. In contrast to the chloroplast, storage plastids are heterotrophic organelles that convert photosynthate derived from source tissues into a long-term storage molecule of high specific energy, which can be drawn upon as required during plant development. Elaioplasts contain large quantities of oil, but are found in limited number of plant taxa such as in oilseeds and in the epidermal cells of the monocot families Liliaceae and Orchidaceae (Kirk and Tilney-Bassett, 1978). Chromoplasts accumulate carotenoids and thus also act as a specialised storage system, but they generally develop as a terminal plastid form that does not establish long-term storage of excess energy. The major storage form for excess photosynthate is starch, an insoluble, complex, semi-crystalline polymer of glucose. All starch is synthesised in the plastid compartment, and is produced in two ways: either in leaf chloroplasts as a transient store of excess photosynthate, or in heterotrophic tissues synthesised from photosynthate unloaded from the phloem, providing a more long-term storage location. This latter class of starch is stored in a specialised colourless plastid, the amyloplast (Figure 2.5A), which is of great economic and agricultural importance, since some 75% of human energy intake is attributable to starch produced by plants (Duffus, 1984). Amyloplasts are present in the endosperm of many seeds, most notably those of the cereal crops, as well as in tubers of potato and fruits such as bananas. In addition, amyloplasts are present in the columella cells in the root cap of most, if not all, plant species, where they are central to the perception of gravity (Kiss, 2000). However, these amyloplasts are highly specialised for a particular role, and probably represent only a superficial similarity to amyloplasts in storage tissues; indeed, it could be argued that their formation is regulated differently to that of other amyloplasts (see below).

Starch synthesis in amyloplasts occurs through the polymerisation of ADP-glucose, yielding highly branched amylopectin and relatively unbranched amylose, the latter composing 20–30% of the total (Smith *et al.*, 1997). The starch grain itself consists of a series of concentric rings of alternating semi-crystalline and amorphous zones, a structure resulting from regions of highly organised and poorly organised individual chains of amylopectin, respectively (Smith *et al.*, 1997). In wheat and

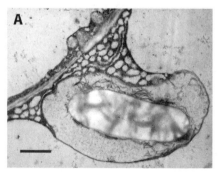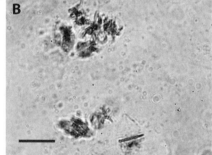

Figure 2.5 Amyloplast and chromoplast. (A) Electron micrograph of an amyloplast from the embryo of a pea seed (*Pisum sativum*). Courtesy of Alison Smith and Kay Denyer, John Innes Centre, Norwich, UK. Scale bar: 2 μm. (B) Chromoplasts from a ripe tomato pericarp cell. The dark bodies contain the red pigments lycopene and beta-carotene. Scale bar: 10 μm.

barley endosperm there are two major classes of starch grains: the A-type, of up to 45 μm in diameter; and the B-type, reaching up to 10 μm in diameter and forming later in the developing endosperm, the ratio of which can significantly influence the quality and suitable post-harvest application of the starch (Langeveld *et al.*, 2000). In potato tubers, amyloplasts are usually dominated by a single large starch grain (Kirk and Tilney-Bassett, 1978) whereas in those in the columella cells of the root cap contain several starch grains (MacCleery and Kiss, 1999). The basis of this variability in grain number and morphology is poorly understood, but can be influenced by a number of developmental as well as environmental factors.

Amyloplasts generally form from proplastids, but may also form from the dedifferentiation of chloroplasts (Thomson and Whatley, 1980). In red winter wheat, for example, plastids present in the coencytic endosperm remain as proplastids with occasional tubular cristae, but only start to deposit starch once cellularisation is complete (Bechtel and Wilson, 2003). Amyloplasts are also capable of redifferentiating into other plastid types, most famously in the re-greening of potato tubers where cell layers deep within the tuber undergo substantial chloroplast formation, albeit relatively slowly compared to meristematic proplastids (Ljubičić *et al.*, 1998). In terms of plastid division, Bechtel and Wilson (2003) speculate that the plastids divide in a novel manner in developing red winter wheat endosperm. In concordance with Langeveld *et al.* (2000), they observed that starch grains initiate within protusions (i.e. stromules) from the proplastid surface. Since very few amyloplasts with multiple grains were present, they inferred that the protrusions containing incipient starch grains break up into individual amyloplasts. They suggest that this may be the only possible mechanism for plastid division, given that binary fission would be difficult to complete with a large starch grain present in the plastid stroma (Bechtel and Wilson, 2003). However, this contrasts strongly with observations from potato stolons induced to undergo tuberisation by the addition of kinetin. In this tissue, amyloplasts clearly undergo binary fission, exhibiting dumb-bell shaped plastids with a well-defined central constriction, indicating that even plastids with bulky starch grains are capable of division (Mingo Castel *et al.*, 1991). Therefore, it may be that plastids destined to become amyloplasts undergo division at different stages, depending on species and tissue type, presumably in relation to the timing of cell division.

In terms of the regulation of amyloplast differentiation, the majority of work has been performed on tobacco BY-2 cell suspension cultures, which can be induced to undergo changes in cell division and plastid metabolism by the addition of phytohormones to the culture medium. BY-2 cells grown in the dark possess proplastids, in line with the undifferentiated status of the cell. The presence of auxin (2,4-dichlorophenoxyacetic acid, 2,4-D) promotes the rapid proliferation of BY-2 cells, and plastids remain in the proplastid state. Upon replacement of 2,4-D with the cytokinin benzyladenine (BA), BY-2 cells no longer proliferate rapidly but shift behaviour and begin to accumulate starch and form amyloplasts (Sakai *et al.*, 1992, 1999). The number of plastids per cell remains approximately constant, but all plastids have been converted to large amyloplasts within 48 h of transfer to BA-containing medium (Sakai *et al.*, 1992). This rapid change in plastid state is accompanied by changes in gene expression in both the plastid and nuclear genomes. For example, the plastomic Rubisco large subunit gene, *rbcL*, is dramatically down-regulated upon conversion of proplastids to amyloplasts, as are a number of other photosynthesis-related transcripts (Sakai *et al.*, 1992). Likewise, nuclear-localised starch synthesis genes such as ADP-glucose pyrophosphorylase small subunit (*AgpS*) are up-regulated by cytokinin and down-regulated by auxin in this system (Miyazawa *et al.*, 1999). Similarly, the addition of 2,4-D to 2,4-D-depleted medium induces BY-2 cells to undergo the reverse process: cells begin to proliferate and amyloplasts apparently revert to proplastids within 12–18 h of 2,4-D application, together with a concomitant decrease in *AgpS* mRNA levels (Miyazawa *et al.*, 2002). It would seem then that at least in the artificial environment of suspension cultures, the phytohormones auxin and cytokinin act antagonistically in directing plastid differentiation. This is, however, a situation far removed from that *in planta*, and provides only a simplistic understanding of the cellular changes that occur during amyloplast biogenesis. Indeed, it appears quite opposite to what one might predict with regards to the amyloplasts in root cap columella cells. It has been shown that the auxin indole-3-acetic acid is preferentially transported to these cells and is physiologically active there (Swarup *et al.*, 2001; Ottenschläger *et al.*, 2003). The studies on BY-2 cells imply that the plastids in root cap columella cells should not accumulate starch, and should remain as proplastids, thus suggesting that the genetic pathways that determine plastid status are more complex. In fact, it is of interest to understand how these plastids differentiate into amyloplasts whilst those in the adjacent quiescent centre of the root apex remain in the proplastid form. Moreover, it raises the general question of whether an amyloplast is anything more than a proplastid that contains the proteins required for substrate import and starch synthesis, a process which conceivably could be triggered in a number of different ways.

2.6   Root plastids

In terms of cell biology, plastids in root cells are very much the forgotten cousins of the plastid family. Along with plastids in the storage compartments of seeds and tubers, root plastids have come to typify the 'non-green plastids' in terms of biochemical analysis. Many aspects of their biochemistry and metabolism have been described in detail (Emes and Neuhaus, 1997; Debnam and Emes, 1999; Fox *et al.*, 2001), revealing major roles for root plastids as sites for the oxidative pentose phosphate pathway and nitrogen assimilation. However, since the subject was extensively reviewed in 1983 (Whatley, 1983a), little progress has been made in understanding details of the cell biology of root plastids and the extent to which they represent a distinct type of plastid. A major problem with root plastids has been their lack of pigmentation and hence the difficulty in visualisation. With GFP technology, root plastids can at least be seen in root cells with moderate ease and there is now potential to study their cell biology. Such studies reveal that root plastids are highly variable in morphology (Köhler and Hanson, 2000; C. Howells and K. Pyke, un-published data, 2004) and readily exhibit extensive stromules. The variation in root plastid morphology and the presence of stromules represents more of a continuum of form within this plastid type than in other types, where the plastid body and the stromule are distinct. There is some evidence that a gradient in root plastid mor-phology exists with distance from the root tip. Plastids derived from the root apical meristem pass through four stages as classified by Whatley (1983b) using electron microscopy. Proplastids from the root meristem contain some inner thylakoid-like structures, but soon lose these and become amyloplasts within 1 mm behind the meristem. Further beyond the root apex the plastids become highly pleomorphic (amoeboid) in shape, and then attain a discoid form with substantial pregranal thy-lakoid structures (Whatley, 1983b). Finally, in the most developmentally old cells the pregranal plastids appear to dedifferentiate into a form similar to the plastids found in the root meristem.

These observations imply that the ancestral, default state of plastid differentiation is the chloroplast, and that this developmental programme is restrained in root tissues, at least in higher plants. It is somewhat unclear as to whether root plastids truly differentiate into a distinct type or whether they represent a chloroplast-like differentiation pathway that is prevented from progressing to completion. Systems involving the COP9 signalosome appear to hold back plastid development in the root and prevent greening of roots in the dark (Kim *et al.*, 2002). However in light-grown roots, several species appear capable of some degree of chloroplast differentiation and produce roots that are visibly green, albeit at a rate much slower than chloroplast maturation in aerial tissues (Whatley, 1983a). In *Arabidopsis*, this greening is often confined to specific cell types within the root, most noticeably the cells immediately around the vasculature, in which chloroplast differentiation can occur readily in the light (C. Howells and K. Pyke, unpublished observation, 2004).

Of particular interest in root plastid biology is the potential for interaction be-tween the plastid compartment and symbiotic micro-organisms within the growing media. A noteworthy recent observation is that of interaction between root cell plas-tids, visualised by GFP, and the development of mycorhizzal arbscules within the root, which together form a symbiotic interface (Fester *et al.*, 2001). In such cases there is extensive proliferation of plastid networks with stromules that appear to in-teract with the fungal surface within the arbuscle. Such an observation suggests an important role for the root plastids in mediating symbiotic interactions, and modern microscopical and molecular techniques should enable this aspect of plastid cell biology to be investigated fully.

2.7 Chromoplasts in fruit and flowers

During the process of higher plant evolution, plants developed symbiotic interac-tions with other life forms. Such interactions proved highly beneficial to plants in two particular areas of plant function, namely pollination of floral structures by insects and seed consumption and subsequent dispersal by animals. In both of these situations the display of coloured plant structures as a form of attractant was required and a specialised form of plastid, the chromoplast (Figure 2.5B), evolved to carry out this function. The coloured pigments that accumulate in chromoplasts are mostly members of the carotenoid family, starting with the C_{40} molecule phytoene and undergoing a variety of complex reactions to give rise to other carotenoids including carotenes, lycopene, lutein, violaxanthin and neoxanthin (Camara et al., 1995; Cunningham and Gantt, 1998; Bramley, 2002). Although many coloured plant structures rely entirely on chromoplasts for their pigmentation, a significant number of petals and fruits contain pigmented chromoplasts often in addition to other pigments, either in cytosolic vesicles or in the vacuole (Kay et al., 1981; Weston and Pyke, 1999). Early work on chromoplast biology was largely centred around light microscopy and documentation of different types of chromoplasts in different tissues and species. More modern studies using electron microscopy have led to the clarification of five classes of chromoplasts based upon the frequency of different substructures related to pigment storage within the chromoplast (Thomson and Whatley, 1980; Camara et al., 1995).

1. Globular chromoplasts are relatively simple in structure and characterised by the accumulation of plastoglobules containing pigments in the stroma.
2. Crystalline chromoplasts accumulate crystals of lycopene or beta-carotene and are typified by the chromoplasts of tomato fruit.
3. Chromoplasts in the fibrillar and tubular class contain extensive bundled microfibrillar structures.
4. Membranous chromoplasts contained extended concentric membranes.
5. Reticulo-tubular chromoplasts contain a complex network of twisted fibrils filling the stroma.

Although such classification of chromoplast types may be convenient for documentation, the different classes are widely spread across different plant species and are present in a wide variety of organs, including fruits, petals, anthers, sepals arils, roots and leaves (Camara et al., 1995). The highly heterogeneous nature of chromoplasts within different organs and different species may simply reflect the extent to which differing profiles of carotenoids, flavonoids and other attractant pigments are stockpiled (Whatley and Whatley, 1987).

The major agricultural world market in ripening fruits, which involves chromoplast differentiation, has resulted in extensive research to elucidate the biochemical and molecular mechanisms of carotenoid biosynthesis and control during fruit chromoplasts differentiation. The majority of such studies have detailed chromoplast differentiation and carotenoid biosynthesis in ripening tomato fruit (*Lycopersicon esculentum*) (Fraser et al., 1994; Camara et al., 1995; Cunningham and Gantt, 1998) as well as a significant body of research on ripening in pepper (*Capsicum* spp.) (Hugueney et al., 1995a). This research has shown that the transition from green chloroplast to pigmented chromoplast is a clearly defined developmental process and not an uncontrolled breakdown of the chloroplast. The major observable changes that occur are the degradation of chlorophyll, the breakdown of thylakoid membrane complexes and the extensive synthesis and accumulation of carotenoid pigments, in particular lycopene and beta-carotene (Marano *et al.*, 1993; Dereure *et al.*, 1994; Grierson and Kader, 1996). Analysis of the *green flesh* mutant of tomato indicated that the breakdown of chlorophyll-containing thylakoid membrane and the formation of new chromoplast membranes are separate independent processes (Cheung *et al.*, 1993; Akhtar *et al.*, 1999). These processes are accompanied by expression of distinct nuclear genes, which are required for chromoplast differentiation and hence labelled as chromoplast specific (Lawrence *et al.*, 1993, 1997; Summer and Cline, 1999). Although chromoplasts retain plastid DNA, there is little evidence that plastid-encoded genes are important in chromoplast function. Indeed, plastid gene expression is reduced to a low level during chromoplast differentiation, apparently as a result of methylation of plastid DNA (Kobayashi *et al.*, 1990). Thus the nuclear genome dominates this particular path of plastid differentiation.

Although the signalling systems that initiate fruit ripening through ethylene signal transduction pathways have been described, the precise nature of the signals that cause chloroplasts to embark upon the chromoplast differentiation process remains unclear. The chromoplast's major function is as a specialised storage site to accumulate high levels of carotenoids and central to this induced biosynthesis is transcriptional control of gene expression (Pecker *et al.*, 1996). Several enzymes involved in carotenoid synthesis increase in level and activity dramatically, including phytoene synthase (Fraser *et al.*, 1994), 1-deoxy-D-xylulose 5-phosphate synthase (Lois *et al.*, 2000), phytoene desaturase (Fraser *et al.*, 1994) and a plastid terminal oxidase associated with phytoene desaturation (Josse *et al.*, 2000). A variety of tomato mutants with altered carotenoid metabolism have been very valuable in dissecting the process of carotenoid accumulation in tomato and elucidating the detailed metabolism involved (Ronen *et al.*, 1999, 2000), and a significant increased metabolic flux through the isoprenoid pathway leads to a 10–14-fold increase in lycopene content (Fraser *et al.*, 1994). In addition to increased expression of nuclear genes involved in carotenoid metabolism, a distinct subset of newly expressed proteins have been identified in developing chromoplasts that appear to be required for the differentiation process to progress. Amongst these proteins are enzymes involved in response to oxidative stress (Livne and Gepstein, 1988; Romer *et al.*, 1992) and a group of proteins, which are involved in carotenoid sequestration (Vishnevetsky *et al.*, 1999). One of these, fibrillin, appears to function as a structural protein in the biogenesis of fibril structures present in the fibrillar class of chromoplasts, in which carotenoids are sequestered internally and surrounded by a layer of polar lipids and coated with a layer of fibrillin (Dereure *et al.*, 1994).

As a framework to this molecular and biochemical development, the cell biology of the chloroplast to chromoplast transition is poorly documented. Microscopy-based studies of chromoplasts in fruits (Harris and Spurr, 1969a, b; Thomson and Whatley, 1980; Whatley and Whatley, 1987; Cheung *et al.*, 1993) have revealed elegant details of internal chromoplast morphology and details of pigment deposition in often irregular-shaped chromoplast bodies (Bathgate *et al.*, 1985; Thelander *et al.*, 1986; Whatley and Whatley, 1987). Details of the population dynamics of chloroplasts and chromoplasts have recently been revealed in tomato (Cookson *et al.*, 2003), where counts of plastid bodies indicate that the majority of plastid replication during the ripening process occurs during the chloroplast stage and in particularly just prior to the breaker stage when chlorophyll breakdown and carotenoid biosynthesis are initiated. As a result the large pericarp cells of the tomato fruit may contain up to 2000 chromoplast bodies observed as red pigmented bodies occasionally with needle-like lycopene crystals enveloped within the chromoplast membranes (Cookson *et al.*, 2003). An interesting component of the *high pigment 1* phenotype in tomato is an increase in the total amount of pigmented chromoplast bodies within the pericarp cell, resulting in more intensely reddened fruit (Cookson *et al.*, 2003). Use of plastid-targeted GFP in ripening fruit has revealed extensive stromule networking between chromoplasts, which raises the strong possibility that there may be molecular trafficking between them (Pyke and Howells, 2002). Such networking blurs, to some extent, the distinction between individual chromplasts and it may be better to consider such a population of chromoplasts in a ripe tomato pericarp cell to be interconnected to some extent rather than a disperse population of individual bodies.

Two other fruit chromoplast differentiating systems have been characterised to a limited extent, namely the ripening of orange citrus fruit (Thomson *et al.*, 1967; Mayfield and Huff, 1986; Iglesias *et al.*, 2001) and the ripening of pumpkins (Boyer, 1989). Whereas in tomato and pepper ripening, the chromoplast differentiation is regarded as terminal, in both orange and pumpkin fully differentiated orange chromoplasts can redifferentiate into green chloroplasts. Such a process appears to be under hormonal control, since the application of gibberelins can hasten and intensify the development of thylakoid membranes (Thomson *et al.*, 1967).

Chromoplasts in flower petals represent an interesting developmental system, similar to that of fruit ripening. In many species, young unexpanded petals contain chlorophyll within chloroplasts, which differentiate into pigmented chromoplasts (Weston and Pyke, 1999) or white unpigmented leucoplasts (Pyke and Page, 1998) as the petals expand and the flower opens. However in some species, chloroplasts are maintained in specific cell types within mature petals that are capable of photosynthesis (Weiss *et al.*, 1990; Vainstein and Sharon, 1993). The chlorophyll-containing chloroplasts in mature petunia corollas appear to be confined to the interior mesophyll tissue whereas pigment is confined to the epidermal cells (K. Pyke and T. Robbins, unpublished observation, 2004). As with fruit biology, an understanding of the biochemistry of carotenoid biosynthesis in petals, such as marigold (Moehs *et al.*, 2001), is far in advance of knowledge about the developmental cell biology of the plastids. Surveys of petal pigmentation in a wide variety of species reveal much variation in the nature of pigmentation by either chromoplasts in different petal cell types or vacuolar soluble pigments or both (Kay *et al.*, 1981). Although only a relatively few plant species have had their petal chromoplasts studied in detail, the majority appear to contain chromoplasts of the simple globular type (Kirk and

Tilney-Bassett, 1978). There is much potential with modern molecular and cytolog-ical techniques for a much greater exploration of the developmental cell biology of chloroplast to chromoplast differentiation in petals and fruit.

2.8   Future prospects

Since the last major work on plastids (Kirk and Tilney-Bassett, 1978), our under-standing of the molecular and biochemical processes by which plastids function has improved dramatically. Much of this new knowledge is considered in other chapters of this book. However, although the basic molecular framework within plastids and within nucleo-cytoplasmic systems that relate to the plastid are moderately well un-derstood, the dynamic nature of plastid development and the factors that influence development, replication and differentiation of plastids in different cell types remain unclear. It must be the challenge of the next decade of plastid research to uncover the precise control systems which promote plastid differentiation pathways in certain tissues but prevent similar pathways occurring in others. It is likely that such mech-anisms are subtle and complex since mutations in genes that appear superficially to be only involved with storage molecule synthesis can have dramatic effects on the plastid differentiation pathway. For instance, loss of *phytoene synthase 1* gene ac-tivity, which is required for the committing step in carotenoid biosynthesis, prevents chromoplast differentiation in tomato fruit (Fray and Grierson, 1993). The result-ing absence of a correct carotenoid complement may feedback on other aspects of plastid differentiation, which further complicates the issue.

   In the era of 'omics', it seems probable that both the proteome and the metabolome of different plastid types during differentiation pathways will be discussed in depth, and such studies will reveal much more detail about the delicate interplay between the plastid and the rest of the cell. Furthermore, the development of elegant micro-scopical techniques and reporter systems in the last decade has revealed a wealth of new information about plastid morphology and dynamics, and it would seem crucial that such techniques are extended to examine plastids in as wide a range of species as possible, rather than in the narrow range which is conventionally used. In this way we may establish a more suitable and less rigid framework for classifying the diversity of plastid form and function: one that is based upon a symphony of molec-ular, metabolic and morphological aspects and that treats the plastid as an integral and plastic part of the cell. Subsequently a greater understanding of the global role of plastids in the evolution and success of higher plants is likely to be achieved.

References

Akhtar, M.S., Golschmidt, E.E., John, I., Rodoni, S., Matile, P. and Grierson, D. (1999) Altered patterns of senescence and ripening in *gf*, a stay green mutant of tomato (*Lycopersicon esculentum* Mill.). *J. Exp. Bot.*, 50, 1115–1122.

Aluru, M.R., Bae, H., Wu, D. and Rodermel, S.R. (2001) The *Arabidopsis immutans* mutation affects plastid differentiation and the morphogenesis of white and green sectors in variegated plants. *Plant Physiol.*, 127, 67–77.

Arimura, S.-I., Hirai, A. and Tsutsumi, N. (2001) Numerous and highly developed tubular projections from plastids observed in tobacco epidermal cells. *Plant Sci.*, 169, 449–454.

Bathgate, B., Purton, M.E., Grierson, D. and Goodenough, P.W. (1985) Plastid changes during the conversion of chloroplasts to chromoplasts in ripening tomatoes. *Planta*, 165, 197–204.

Bauer, J., Hiltbrunner, A. and Kessler, F. (2001) Molecular biology of chloroplast biogenesis: gene expression, protein import and intraorganellar sorting. *Cell. Mol. Life Sci.*, 58, 420–433.

Baumgartner, B.J., Rapp, J.C. and Mullet, J.E. (1989) Plastid transcription activity and DNA copy number increase early in barley chloroplast development. *Plant Physiol.*, 89, 1011–1018.

Bechtel, D.B. and Wilson, J.D. (2003) Amyloplast formation and starch granule development in hard red winter wheat. *Cereal Chem.*, 80, 175–183.

Bourett, T.M., Czymmek, K.J. and Howard, R.J. (1999) Ultrastructure of chloroplast protuberances in rice laves preserved by high pressure freezing. *Planta*, 208, 472–479.

Boyer, C.D. (1989) Genetic control of chromoplast formation during fruit development of *Cucurbita pepo*. L. In *Current Topics in Plant Physiology, Vol. 2: Physiology, Biochemistry and Genetics of Non-green Plastids* (eds C.D. Boyer, J.C. Shannon and R.C. Hardison), American Society of Plant Physiologists, Rockville, MD, pp. 241–252.

Bramley, P.M. (2002) Regulation of carotenoid formation during tomato fruit ripening and development. *J. Exp. Bot.*, 53, 2107–2113.

Brutnell, T.P. and Langdale, J.A. (1998) Signals in leaf development. *Adv. Bot. Res.*, 28, 161–195.

Brutnell, T.P., Sawers, R.J.H., Mant, A. and Langdale, J.A. (1999) BUNDLE SHEATH DEFECTIVE2, a novel protein required for post-translational regulation of the *rbcL* gene of maize. *Plant Cell*, 11, 849–864.

Buchanan, B.B., Gruissem, W. and Jones, R.L. (2000) *Biochemistry and Molecular Biology of Plants*, American Society of Plant Physiologists, Rockville, MD.

Burgess, D. and Taylor, W. (1987) Chloroplast photooxidation affects accumulation of cytosolic mRNAs encoding chloroplast proteins in maize. *Planta*, 170, 520–527.

Cahoon, A.B. and Timko, M.P. (2000) Yellow-in-the-dark mutants of *Chlamydomonas* lack the CHLL subunit of light-independent protochlorophyllide reductase. *Plant Cell*, 12, 559–568.

Camara, B., Hugueney, P., Bouvier, F., Kuntz, M. and Moneger, R. (1995). Biochemistry and molecular biology of chromoplast development. *Int. Rev. Cytol.*, 163, 175–247.

Chaley, N. and Possingham, J.V. (1981) Structure of constricted proplastids in meristematic plant tissues. *Biol. Cell.*, 41, 203–210.

Chatterjee, M., Sparvoli, S., Edmunds, C., Garosi, P., Findlay, K. and Martin, C. (1996) *DAG*, a gene required for chloroplast differentiation and palisade development in *Antirrhinum majus*. *EMBO J.*, 15, 4194–4207.

Cheung, A.Y., McNellis, T. and Piekos, B. (1993) Maintenance of chloroplast components during chromoplast differentiation in the tomato mutant *green flesh*. *Plant Physiol.*, 101, 1223–1229.

Chory, J. and Peto, C.A. (1990) Mutations in the *DET1* gene affect cell-type-specific expression of light- regulated genes and chloroplast development in *Arabidopsis*. *Proc. Natl. Acad. Sci. U.S.A.*, 87, 8776–8780.

Cookson, P.J., Kiano, J., Fraser, P.D. *et al.* (2003) Increases in cell elongation, plastid compartment size and translational control of carotenoid gene expression underlie the phenotype of the *High Pigment-1* mutant of tomato. *Planta*, 217, 896–903.

Corriveau, J.L. and Coleman, A.W. (1988) Rapid screening method to detect potential biparental inheritance of plastid DNA and results for over 200 Angiosperm species. *Am. J. Bot.*, 75, 1443– 1458.

Cunningham, F.X. and Gantt, E. (1998) Genes and enzymes of carotenoid biosynthesis in plants. *Ann. Rev. Plant Physiol. Plant Mol. Biol.*, 49, 557–583.

Debnam, P.M. and Emes, M.J. (1999) Subcellular distribution of enzymes of the oxidative pentose phosphate pathway in root and leaf tissues. *J. Exp. Bot.*, 340, 1653–1661.

Dereure, J., Romer, S., D'Harlingue, A., Backhaus, R.A., Kuntz, M. and Camara, B. (1994) Fibril assembly and carotenoid overaccumulation in chromoplasts: a model for supramolecular lipoprotein structures. *Plant Cell*, 6, 119–133.

Duffus, C.M. (1984) Metabolism of reserve starch. In *Storage Carbohydrates in Vascular Plants* (ed. D.H. Lewis), Cambridge University Press, Cambridge, UK, pp. 231–252.

Emes, M.J. and Neuhaus, H.E. (1997) Metabolism and transport in non-photosynthetic plastids. *J. Exp. Bot.*, 48, 1995–2005.

Estévez, J.M., Cantero, A., Romero, C. *et al.* (2000) Analysis of the expression of *CLA1*, a gene that encodes the 1-deoxyxylulose 5-phosphate synthase of the 2-*C*-methyl-D-erythritol-4-phosphate pathway in *Arabidopsis. Plant Physiol.*, 124, 95–103.

Fester, T., Strack, D. and Hause, B. (2001) Reorganization of tobacco root plastids during arbuscle development. *Planta*, 213, 864–868.

Fox, S.R., Rawsthorne, S. and Hills, M.J. (2001) Fatty acid synthesis in pea root plastids is inhibited by the action of long-chain acyl coenzyme as on metabolite transporters. *Plant Physiol.*, 126, 1259–1265.

Fraser, P.D., Truesdale, M.R., Bird, C.R., Schuch, W. and Bramley, P.M. (1994) Carotenoid biosynthesis during tomato fruit development. *Plant Physiol.*, 105, 405–413.

Fray, R.G. and Grierson, D. (1993) Identification and genetic analysis of normal and mutant phytoene synthase genes of tomato by sequencing, complementation and co-suppression. *Plant Mol. Biol.*, 22, 589–602.

Fujie, M., Kuroiwa, H., Kawano, S. and Kuroiwa, T. (1994) Behaviour of organelles and their nucleoids in the shoot apical meristem during leaf development in *Arabidopsis thaliana* L. *Planta*, 194, 395–405.

Gray, J., Sullivan, J., Hibberd, J. and Hansen, M. (2001) Stromules: mobile protrusions and interconnections between plastids. *Plant Biol.*, 3, 223–233.

Grierson, D. and Kader, A.A. (1996) Fruit ripening and quality. In *The Tomato Crop: A Scientific Basis for Improvement* (eds J.G. Atherton and J. Rudich), Chapman and Hall, London, pp. 242–280.

Harak, H., Lagrange, T., Bisanz-Seyer, C., Lerbs-Mache, S. and Mache, R. (1995) The expression of nuclear genes encoding plastid ribosomal proteins precedes the expression of chloroplast genes during early phases of chloroplast development. *Plant Physiol.*, 108, 685–692.

Harris, W.M. and Spurr, A.R. (1969a) Chromoplasts of tomato fruits, I: ultrastructure of low-pigment and high beta mutants. Carotene analyses. *Am. J. Bot.*, 56, 369–379.

Harris, W.M. and Spurr, A.R. (1969b) Chromoplasts of tomato fruits, II: the red tomato. *Am. J. Bot.*, 56, 380–389.

Hugueney, P., Badillo, A., Chen, H.C. *et al.* (1995) Metabolism of cyclic carotenoids: a model for the alteration of this biosynthetic pathway in *Capsicum annuum* chromoplasts. *Plant J.*, 8, 417–424.

Hugueney, P., Bouvier, F., Badillo, A., D'Harlingue, A., Kuntz, M. and Camara, B. (1995) Identification of a plastid protein involved in vesicle fusion and/or membrane protein translocation. *Proc. Natl. Acad. Sci. U.S.A.*, 92, 5630–5634.

Iglesias, D.J., Tadeo, F.R., Legaz, F., Primo-Millo, E. and Talon, M. (2001) *In vivo* sucrose stimulation of colour change in citrus fruit epicarps: interactions between nutritional and hormonal signals. *Physiol. Plant*, 112, 244–250.

Jarvis, P., Chen, L.J., Li, H., Peto, C.A., Fankhauser, C. and Chory, J. (1998) An *Arabidopsis* mutant defective in the plastid general protein import apparatus. *Science*, 282, 100–103.

Jeong, W.J., Park, Y.-I., Suh, K., Raven, J.A., Yoo, O.J. and Liu, J.R. (2002) A large population of small chloroplasts in tobacco leaf cells allows more effective chloroplast movement than a few enlarged chloroplasts. *Plant Physiol.*, 129, 112–121.

Josse, E.-M., Simkin, A.J., Gaffe, J., Laboure, A.-M., Kuntz, M. and Carol, P. (2000) A plastid terminal oxidase associated with carotenoid desaturation during chromoplast differentiation. *Plant Physiol.*, 123, 1427–1436.

Juniper, B.E. and Clowes, F.A.L. (1965) Cytoplasmic organelles and cell growth in root caps. *Nature*, 208, 864–865.

Kay, Q.O.N., Daoud, H.S. and Stirton, C.H. (1981) Pigment distribution, light reflection and cell structure in petals. *Bot. J. Linn. Soc.*, 83, 57–84.

Keddie, J.S., Carroll, B., Jones, J.D. and Gruissem, W. (1996) The *DCL* gene of tomato is required for chloroplast development and palisade cell morphogenesis in leaves. *EMBO J.*, 15, 4208–4217.

Kim, T.-H., Kim, B.-Y. and von Arnim, A.G. (2002) Repressors of photomorphogenesis. *Int. Rev. Cytol.*, 220, 185–223.

Kirk, J.T.O. and Tilney-Bassett, R.AE. (1978) *The Plastids: Their Chemistry, Structure, Growth and Inheritance*, 2nd edn, Elsevier/North-Holland Biomedical Press, Amsterdam.

Kiss, J.Z. (2000) Mechanism of the early phases of plant gravitropism. *Crit. Rev. Plant Sci.*, 19, 551–573.

Kobayashi, H., Ngernprasirtsiri, J. and Akazawa, T. (1990) Transcriptional regulation and DNA methylation in plastids during transitional conversion of chloroplasts to chromoplasts. *EMBO J.*, 9, 307–313.

Knoblauch, M., Hibberd, J., Gray, J. and van Bel, A. (1999) A galinstan expansion femtosyringe for microinjection of eukaryotic organelles and prokaryotes. *Nat. Biotech.*, 17, 906–909.

Knoll, A.H. (1992). The early evolution of eukaryotes: a geological perspective. *Science*, 256, 622–627.

Köhler, R., Cao, J., Zipfel, W., Webb, W. and Hanson, M. (1997) Exchange of protein molecules through connections of higher plant plastids. *Science*, 276, 2039–2042.

Köhler, R. and Hanson, M. (2000) Plastid tubules of higher plants are tissue-specific and developmentally regulated. *J. Cell Sci.*, 113, 81–89.

Köhler, R., Schwille, P., Webb, W. and Hanson, M. (2000) Active protein transport through plastid tubules: velocity quantified by fluorescence correlation spectroscopy. *J. Cell Sci.*, 113, 3921–3930.

Kubis, S., Baldwin, A., Patel, R. *et al.* (2003) The *Arabidopsis ppi1* mutant is specifically defective in the expression, chloroplast import, and accumulation of photosynthetic proteins. *Plant Cell*, 15, 1859–1871.

Langeveld, S.M.J., Van Wijk, R., Stuurman, N., Kijne, J.W. and de Pater, S. (2000) B-type granule containing protrusions and interconnections between amyloplasts in developing wheat endosperm revealed by transmission electron microscopy and GFP expression. *J. Exp. Bot.*, 51, 1357–1361.

La Rocca, N., Rascio, N., Oster, U. and Rüdiger, W. (2001) Amitrole treatment of etiolated barley seedlings leads to deregulation of tetrapyrrole synthesis and to reduced expression of *Lhc* and *RbcS* genes. *Planta*, 213, 101–108.

Lawrence, S.D., Cline, K. and Moore, G.A. (1993) Chromoplast targeted proteins in tomato (*Lycopersicon esculentum* Mill.) fruit. *Plant Physiol.*, 102, 789–794.

Lawrence, S.D., Cline, K. and Moore, G.A. (1997) Chromoplast development in ripening tomato fruit: identification of cDNAs for chromoplast-targeted proteins and characterization of a cDNA encoding a plastid-localized low molecular weight heat shock protein. *Plant Mol. Biol.*, 33, 483–492.

Leister, D. (2003) Chloroplast research in the genomic era. *Trends Genet.*, 19, 47–56.

León, P. and Arroyo, A. (1998) Nuclear control of plastid and mitochondrial development in higher plants. *Ann. Rev. Plant Physiol. Plant Mol. Biol.*, 49, 453–480.

Livne, A. and Gepstein, S. (1988) Abundance of the major chloroplast polypeptides during development and ripening of tomato fruits. *Plant Physiol.*, 87, 239–243.

Ljubičić, J.M., Wrischer, M. and Ljubičić, N. (1998) Formation of the photosynthetic apparatus in plastids during greening of potato microtubers. *Plant Physiol. Biochem.*, 36, 747–752.

Lois, L.M., Rodriguez-Concepcion, M., Gallego, F., Campos, N. and Boronat, A. (2000) Carotenoid biosynthesis during tomato fruit development: regulatory role of 1-deoxy-D-xylulose 5-phosphate synthase. *Plant J.*, 22, 503–513.

MacCleery, S.A. and Kiss, J.Z. (1999) Plastid sedimentation kinetics in roots of wild-type and starch-deficient mutants of *Arabidopsis*. *Plant Physiol.*, 120, 183–192.

Mache, R., Zhou, D.-X., Lerbs-Mache, S., Harrak, H., Villain, P. and Gauvin, S. (1997). Nuclear control of early plastid differentiation. *Plant Physiol. Biochem.*, 35, 199–203.

Mandel, M.A., Feldmann, K.A., Herrera-Estrella, L., Rocha-Sosa, M. and León, P. (1996) *CLA1*, a novel gene required for chloroplast development, is highly conserved in evolution. *Plant J.*, 9, 649–658.

Marano, M.R., Serra, E.C., Orellano, E.G. and Carrillo, N. (1993) The path of chromoplast development in fruits and flowers. *Plant Sci.*, 94, 1–17.

Martin, W., Rujan, T., Richly, E. *et al.* (2002) Evolutionary analysis of *Arabidopsis*, cyanobacterial and chloroplast genomes reveals plastid phylogeny and thousands of cyanobacterial genes in the nucleus. *Proc. Natl. Acad. Sci. U.S.A.*, 99, 12246–12251.

Martínez-García, J.F., Huq, E. and Quail, P.H. (2000) Direct targeting of light signals to a promoter element-bound transcription factor. *Science*, 288, 859–863.

Mayfield, S.P. and Huff, A. (1986) Accumulation of chlorophyll, chloroplastic ptroteins and thylakoid membranes during reversion of chromoplasts to chloroplasts in *Citrus sinensis* epicarp. *Plant Physiol.*, 81, 30–35.

McFadden, G.I. (2001) Primary and secondary endosymbiosis and the origin of plastids. *J. Phycol.*, 37, 951–959.

Mingo Castel, A.M., Pelacho, A.M. and de Felipe, M.R. (1991) Amyloplast division in kinetin induced potato tubers. *Plant Sci.*, 73, 211–217.

Miyamura, S., Kuroiwa, T. and Nagata, T. (1990) Multiplication and differentiation of plastid nucleoids during development of chloroplasts and etioplasts from proplastids in *Triticum aestivum*. *Plant Cell Physiol.*, 31, 597–602.

Miyazawa, Y., Kutsuna, N., Inada, N., Kuroiwa, H., Kuroiwa, T. and Yoshida, S. (2002) Dedifferentiation of starch-storing tobacco cells: effects of 2,4-dichlorophenoxy acetic acid on multiplication, starch content, organellar DNA content, and starch synthesis gene expression. *Plant Cell Reprod.*, 21, 289–295.

Miyazawa, Y., Sakai, A, Miyagishima, S.-Y., Takano, H., Kawano, S. and Kuroiwa, T. (1999) Auxin and cytokinin have opposite effects on amyloplast development and the expression of starch synthesis genes in cultured Bright Yellow-2 tobacco cells. *Plant Physiol.*, 121, 461–469.

Moehs, C.P., Tian, L., Osteryoung, K.W. and DellaPenna, D. (2001) Analysis of carotenoid biosynthetic gene expression during marigold petal development. *Plant Mol. Biol.*, 45, 281–293.

Ni, M., Halliday, K.J., Tepperman, J.M. and Quail, P.H. (1998) PIF3, a phytochrome-interacting factor necessary for normal photoinduced signal transduction, is a novel basic helix-loop-helix protein. *Cell*, 95, 657–667.

Ni, M., Tepperman, J.M. and Quail, P.H. (1999) Binding of phytochrome B to its nuclear signalling partner PIF3 is reversibly induced by light. *Nature*, 400, 781–784.

Niklas, K.J. (1997) *The Evolutionary Biology of Plants*, The University of Chicago Press, Chicago.

Osteryoung, K.W. and McAndrew, R.S. (2001) The plastid division machine. *Ann. Rev. Plant Physiol. Plant Mol. Biol.*, 52, 315–333.

Ottenschläger, I., Wolff, P., Wolverton, C. *et al.* (2003) Gravity-regulated differential auxin transport from columella to lateral root cap cells. *Proc. Natl. Acad. Sci. U.S.A.*, 100, 2987–2991.

Pecker, I., Gabbay, R., Cunningham, F.X. and Hirschberg J. (1996) Cloning and characterisation of the cDNA for lycopene β-cyclase from tomato reveals decrease in its expression during fruit ripening. *Plant Mol. Biol.*, 30, 807–819.

Possingham, J.V. and Rose, R.J. (1976) Chloroplast replication and chloroplast DNA synthesis in spinach leaves. *Proc. R. Soc. Lond. B*, 193, 295–305.

Pyke, K. (1999) Plastid division and development. *Plant Cell*, 11, 549–556.

Pyke, K. and Howells, C. (2002) Plastid and stromule morphogenesis in tomato. *Ann. Bot.*, 90, 559–566.

Pyke, K. and Leech, R.M. (1992) Chloroplast division and expansion is radically altered by nuclear mutations in *Arabidopsis thaliana*. *Plant Physiol.*, 99, 1005–1008.

Pyke, K.A. and Page, A. (1998) Plastid ontogeny during petal development in *Arabidopsis*. *Plant Physiol.*, 116, 797–803.

Pyke, K., Zubko, M.K. and Day, A. (2000) Marking cell layers with spectinomycin provides a new tool for monitoring cell fate during leaf development. *J. Exp. Bot.*, 51, 1713–1720.

Reski, R. (2002) Rings and networks: the amazing complexity of FtsZ in chloroplasts. *Trends Plant Sci.*, 7, 103–105.

Robertson, E.J., Pyke K.A. and Leech R.M. (1995) *arc6*, a radical chloroplast division mutant of *Arabidopsis* also alters proplastid proliferation and morphology in shoot and root apices. *J. Cell Sci.*, 108, 2937–2944.

Rodermel, S. (2001) Pathways of plastid-to-nucleus signaling. *Trends Plant Sci.*, 6, 471–478.

Romer, S., D'Harlingue, A., Camara, B., Schantz, R. and Kuntz, M. (1992). Cysteine synthase from *Capsicum annum* chromoplasts. Characterization and cDNA cloning of an up-regulated enzyme during fruit development. *J. Biol. Chem.*, 267, 17466–17470.

Ronen, G., Carmel-Goran, L., Zamir, D. and Hirschberg, J. (2000) An alternative pathway to β-carotene formation in plant chromoplasts discovered by map-based cloning of *Beta* and *old-gold* colour mutations in tomato *Proc. Natl. Acad. Sci. U.S.A.*, 97, 11102–11107.

Ronen, G., Cohen, M., Zamir, D. and Hirschberg, J. (1999) Regulation of carotenoid biosynthesis during tomato fruit development: expression of the gene for lycopene epsilon-cyclase is down-regulated during ripening and is elevated in the mutant *Delta*. *Plant J.*, 17, 341–351.

Sakai, A., Kawano, S. and Kuroiwa, T. (1992) Conversion of proplastids to amyloplasts in tobacco cultured cells is accompanied by changes in the transcriptional activities of plastid genes. *Plant Physiol.*, 100, 1062–1066.

Sakai, A., Susuki, T., Miyazawa, Y., Kawano, S., Nagata, T. and Kuroiwa, T. (1998) Comparative analysis of plastid gene expression in tobacco chloroplasts and proplastids: relationship between transcription and transcript accumulation. *Plant Cell Physiol.*, 39, 581–589.

Sakai, A., Suzuki, T., Sasaki, N. and Kuroiwa, T. (1999) Plastid gene expression during amyloplast formation in cultured tobacco cells. *J. Plant Physiol.*, 154, 71–78.

Schimper, A.F.W. (1885) Untersuchungen über die Chlorophyllkörper und die ihnen homologen Gebilde. *Pringsheim Jahrbücher Wiss. Botanik*, 16, 1–247.

Sheen, J. (1999) C$_4$ gene expression. *Ann. Rev. Plant Physiol. Plant Mol. Biol.*, 50, 187–217.

Smith, A.M., Denyer, K. and Martin, C. (1997) The synthesis of the starch granule. *Ann. Rev. Plant Physiol. Plant Mol. Biol.*, 48, 67–87.

Smith, H. (2000) Phytochromes and light signal perception by plants – an emerging synthesis. *Nature*, 407, 585–591.

Strand, A., Asami, T., Alonso, J., Ecker, J.R. and Chory, J. (2003) Chloroplast to nucleus communication triggered by accumulation of Mg-protoporphyrin IX. *Nature*, 421, 79–83.

Summer, E.J. and Cline, K. (1999) Red bell pepper chromoplasts exhibit in vitro import competency and membrane targeting of passenger proteins from the thylakoid Sec and ΔpH pathways but not the chloroplast signal recognition particle pathway. *Plant Physiol.*, 119, 575–584.

Susek, R.E., Ausubel, F.M. and Chory J. (1993) Signal transduction mutants of *Arabidopsis* uncouple nuclear CAB and RBCS gene expression from chloroplast development. *Cell*, 74, 787–799.

Swarup, R., Friml, J., Marchant, A. *et al.* (2001) Localisation of the auxin permease AUX1 suggests two functionally distinct hormone transport pathways operate in the *Arabidopsis* root apex. *Genes Dev.*, 15, 2648–2653.

Thelander, M., Narita, J.O. and Gruissem, W. (1986) Plastid differentiation and pigment biosynthesis during tomato fruit ripening. *Curr. Top. Plant Biochem. Plant Physiol.*, 5, 128–141.

Thomson, W.W., Lewis, L.N. and Coggins, C.W. (1967) The reversion of chromoplasts to chloroplasts in Valencia oranges. *Cytologia*, 32, 117–124.

Thomson, W.W. and Whatley, J.M. (1980). Development of non-green plastids. *Ann. Rev. Plant Physiol.*, 31, 375–394.

Tirlapur, U., Dahse, I., Reiss, B., Meurer, J. and Oelmüller, R. (1999) Characterization of the activity of a plastid-targeted green fluorescent protein in *Arabidopsis*. *Eur. J. Cell Biol.*, 78, 233–240.

Vainstein, A. and Sharon, R. (1993) Biogenesis of petunia and carnation corolla chloroplasts: changes in the abundance of nuclear and plastid-encoded photosynthesis-specific gene products during flower development. *Physiol. Plant*, 89, 192–198.

Vishnevetsky, M., Ovadis, M. and Vainstein, A. (1999) Carotenoid sequestration in plants: the role of carotenoid associated proteins. *Trends Plant Sci.*, 4, 232–235.

Vothknecht, U.C. and Westhoff, P. (2001) Biogenesis and origin of thylakoid membranes. *Biochim. Biophys. Acta*, 1541, 91–101.

Voznesenskaya, E.V., Franceschi, V.R., Kiirats, O., Artyusheva, E.G., Freitag, H. and Edwards, G.E. (2002) Proof of C_4 photosynthesis without Kranz anatomy in *Bienertia cycloptera* (Chenopodiaceae). *Plant J.*, 31, 649–662.

Weiss, D., Shomer-Ilan, A., Vainstein, A. and Halvey, A.H. (1990) Photosynthetic carbon fixation in the corollas of *Petunia hybrida*. *Physiol. Plant*, 78, 345–350.

Weston, E.A. and Pyke, K.A. (1999) Developmental ultrastructure of cells and plastids in the petals of Wallflower (*Erysimum cheiri*). *Ann. Bot.*, 84, 763–769.

Whatley, J.M. (1983a) The ultrastructure of plastids in roots. *Int. Rev. Cytol.*, 85, 175–220.

Whatley, J.M. (1983b) Plastids in the roots of *Phaseouls vulgaris*. *New Phytol.*, 94, 381–391.

Whatley, J.M. and Whatley, F.R. (1987) When is a chromoplast? *New Phytologist*, 106, 667–678.

Wildman, S.G. (1967) The organization of grana-containing chloroplasts in relation to location of some enzymatic systems concerned with photosynthesis, protein synthesis, and ribonucleic acid synthesis. In *Biochemistry of Chloroplasts*, Vol. 2 (ed. T.W. Goodwin), Academic Press, London, pp. 295–319.

Wildman, S., Hongladarom, T. and Honda, S. (1962) Chloroplasts and mitochondria in living plant cells: cinephotomicrographic studies. *Science*, 138, 434–435.

Williamson, R.E. (1993) Organelle movements. *Ann. Rev. Plant Physiol. Plant Mol. Biol.*, 44, 181–202.

3   Plastid metabolic pathways

Ian J. Tetlow, Stephen Rawsthorne, Christine Raines
and Michael J. Emes

3.1   Introduction

Plastids are subcellular, self-replicating organelles present in all living plant cells,
and the exclusive site of many important biological processes, the most fundamental
being the photosynthetic fixation of CO_2 within chloroplasts. In addition, plastid
metabolism is responsible for generating economically important raw materials and
commodities such as starches and oils, as well as improving the nutritional status of
many crop-derived products. All plastids are enclosed by two membranes, the outer
and the inner envelope membrane. The outer membrane represents a barrier to the
movement of proteins, whilst the inner membrane is the actual permeability barrier
between the cytosol and the plastid stroma and the site of specific transport systems
connecting both compartments.

The classification of different plastid types is usually based on their internal structure and origin (for a review, see Kirk and Tilney-Bassett, 1978). *Proplastids*, or
eoplasts, are the progenitors of other plastids; these colourless plastids occur in the
meristematic cells of shoots, roots, embryos and endosperm and have no distinctive
morphology, varying in shape and sometimes contain lamellae and starch granules.
Chloroplasts are the site of the photochemical apparatus and possess a distinctive
internal membrane organization of thylakoid discs. The chlorophyll pigments and
light reactions of photosynthesis are associated with the thylakoid membrane system. These green, lens-shaped organelles are present in all photosynthetic tissues
and organs such as leaves, storage cotyledons, seed coats, embryos and the outer
layers of unripe fruits. *Chromoplasts* are red-, orange- and yellow-coloured plastids
containing relatively high levels of carotenoid pigments and are commonly found in
flowers, fruits, senescing leaves (also termed *gerontoplasts*) and certain roots. Chromoplasts often develop from chloroplasts, but may also be formed from proplastids
and amyloplasts (see below). Carotenoid synthesis and/or storage in chromoplasts
occur within osmiophillic droplets or plastoglobuli, filamentous pigmented bodies and crystals (Frey-Wissling and Kreutzer, 1958). Starch is often present early
in development and lost as the chromoplasts mature (Weier, 1942; Bouvier *et al.*,
2003). *Etioplasts* are found in leaf cells that are grown in continuous darkness, appearing yellow because of the presence of protochlorophyll, and are therefore not
a normal stage of development of chloroplasts. Etioplasts are structurally simple
possessing distinctive crystalline centres known as *prolamellar bodies*. Upon exposure to light, etioplasts rapidly differentiate into chloroplasts, during which the

protochlorophyll becomes converted into chlorophyll and the prolamellar body reorganizes into grana and stromal lamellae. *Leucoplasts* are colourless plastids that are distinct from proplastids in that they have lost their progenitor function. Within this group are amyloplasts, elaioplasts/oleoplasts and proteinoplasts, which are the sites of synthesis of starch, lipids and proteins respectively. Amyloplasts are characterized by the presence of one or more starch granules and are found in roots (where they may be involved in the detection of gravity) and storage tissues such as cotyledons, endosperm and tubers. Many of the primary metabolic pathways are shared within different types of plastids, but perform different functions within them. For example, starch made inside amyloplasts acts as a long-term store for the next generation, whereas starches produced in chloroplasts and leucoplasts act as temporary carbon stores. The specialized functions associated with some plastids are usually associated with the localization of the plastid within a specialized tissue/organ, for example chromoplasts found in petals or fruit pericarp.

Many of the commercially important products derived from plants are the direct result of metabolism within plastids, and the vast research effort expended in plant biology is a reflection of this: in particular in understanding CO_2 fixation in chloroplasts, and storage starch biosynthesis in heterotrophic plastids. An improved understanding of the key metabolic pathways in plastids that underpin the yield of many important crops is critical to the formulation of rational strategies for crop improvements in the future. This chapter focuses on recent developments in our understanding of the primary metabolic pathways in higher plants and the relationships of the plastidial pathways to metabolic activities outside the plastid via a suite of specialized metabolite transport proteins.

3.2 Carbon assimilation

3.2.1 The reductive pentose-phosphate pathway (Calvin cycle)

All photosynthetic organisms, including cyanobacteria, algae and higher plants, use the reductive pentose-phosphate pathway (RPPP) to convert CO_2 into organic molecules. In eukaryotic autotrophs these reactions take place in the chloroplast stroma and the RPPP utilizes the products of the light reactions of photosynthesis, ATP and NADPH, to fix atmospheric CO_2 into carbon skeletons that are used directly for starch and sucrose biosynthesis (Figure 3.1) (Woodrow and Berry, 1988; Geiger and Servaites, 1994; Quick and Neuhaus, 1997; Lawlor, 2002). The RPPP comprises 11 different enzymes, catalysing 13 reactions, and can be divided into three phases: carboxylation, reduction and regeneration. The first phase, carboxylation, is catalysed by the enzyme ribulose 1,5-bisphosphate carboxylase oxygenase (Rubisco), which brings about the fixation of CO_2 into the acceptor molecule, ribulose 1,5-bisphosphate (RuBP). The 3-phosphoglycerate (3-PGA) formed by this reaction is then utilized in the reductive stage to form triose-phosphates, glyceraldehyde 3-phosphate (G-3-P) and dihydroxyacetone phosphate (DHAP), via two

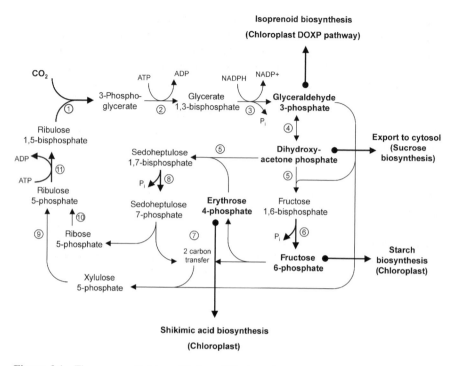

Figure 3.1 The enzyme Rubisco (1) fixes CO_2 into the acceptor molecule, ribulose 1,5-bisphosphate (RuBP), resulting in the formation of two molecules of 3-phosphoglycerate. In the reduction phase, 3-phosphoglycerate is phosphorylated by the enzyme phosphoglycerate kinase (2), forming 1,3-bisphosphoglycerate, which is then reduced by glyceraldehyde 3-phosphate dehydrogenase (3) to glyceraldehyde 3-phosphate consuming ATP and NADPH. Triose-phosphate isomerase (4) catalyses the reversible isomerization of glyceraldehyde 3-phosphate to dihydroxyacetone phosphate. In the regeneration phase of the cycle the CO_2 acceptor molecule ribulose 1,5-bisphosphate (RuBP) is produced from triose-phosphates through a series of sugar condensation and carbon rearrangement reactions. Condensation of the triose-phosphates (glyceraldehyde 3-phosphate and dihydroxyacetone phosphate) by aldolase (5) yields fructose 1,6-bisphosphate. This C6 sugar is then irreversibly hydrolyzed to the monophosphate form, fructose 6-phosphate by fructose 1,6-bisphosphatase (6). The enzyme transketolase then performs a C2 transfer from fructose 6-phosphate to glyceraldehyde 3-phosphate, forming xylulose 5-phosphate and erythrose 4-phosphate (7). Transketolase uses thiamine pyrophosphate as a prosthetic group to mediate the C2 transfer. The resulting erythrose 4-phosphate is combined with dihydroxyacetone phosphate, in a reaction again catalysed by aldolase (5), to form sedoheptulose-1,7-bisphosphate. This C7 product is hydrolyzed by sedoheptulose 1,7-bisphosphatase (8), yielding sedoheptulose-7-phosphate. Transfer of two carbons from sedoheptulose 7-phosphate to glyceraldehyde 3-phosphate by transketolase (7) produces ribose 5-phosphate and xylulose 5-phosphate. Ribose 5-phosphate and xylulose 5-phosphate are converted to ribulose 5-phosphate by ribose 5-phosphate isomerase (10) and ribulose phosphate epimerase (9) respectively. The final step converts ribulose 5-phosphate to the CO_2 acceptor molecule RuBP by the action of phosphoribulokinase (11) in an irreversible reaction utilizing ATP. Compounds exported from the RPPP are indicated in boldface.

reactions that consume ATP and NADPH. The final and most complex phase of the cycle involves a series of reactions that regenerate the CO_2 acceptor molecule, RuBP from triose-phosphates. The RPPP is autocatalytic, and for this reason five of every six molecules of triose-phosphate produced remain within the cycle to regenerate RuBP, the CO_2 acceptor molecule, otherwise the cycle would come to a halt. This means that one in every six molecules represents net product and can leave the cycle to be used to synthesize an array of compounds essential for plant growth and development. Two major pathways that utilize this output are those for the biosynthesis of sucrose and starch. The RPPP also supplies carbon compounds to an array of other metabolic pathways in the chloroplast, including erythrose 4-phosphate to the shikimate pathway for the biosynthesis of amino acids and lignin, and G-3-P for isoprenoid biosynthesis (Lichtenthaler, 1999). In addition, the RPPP shares enzymes and intermediates with the oxidative pentose-phosphate pathway (OPPP; see section below) and through this provides precursors for nucleotide metabolism and cell wall biosynthesis (Figure 3.1). In order to maintain a balance between the demands of the RPPP and outputs to other metabolic pathways a range of regulatory processes have evolved to ensure that a balance is maintained and that the pathway can respond both flexibly and rapidly to changing developmental and environmental conditions.

3.2.2   Regulation of the RPPP

The enzymes of the RPPP are subjected to a number of different regulatory processes that operate over different timescales: from rapid changes occurring over seconds to minutes through to mechanisms that bring changes over a period of days. Regulation occurring over seconds to hours involves changes in the catalytic properties of individual enzymes in response to changes in substrates, products and effector molecules. Over a longer time period, minutes to hours, activity of enzymes in the cycle is altered in response to environmental conditions (e.g. light, CO_2 levels) and can be regulated through modification of the activation state of the enzymes. These regulatory mechanisms provide flexibility in the operation of the RPPP so as to enable the enzymes to respond rapidly to changes in the environment encountered over the daily cycle. Over the longer term, changes in gene expression during development determine the level of the enzymes in the cycle and therefore determine the maximum potential catalytic activity of the cycle. In mature leaves, changes in gene expression will be brought about by prevailing environmental and metabolic conditions occurring over days, such as changes in nutrient status, light conditions and CO_2.

   Light is the major factor influencing the activity of enzymes in the RPPP. When plants are exposed to light after a period of darkness there is a lag phase before the activity of this cycle reaches a steady state level. This is because the catalytic activity of a number of enzymes in the RPPP is modulated by light via the thioredoxin/ferredoxin system and also by light-induced changes in pH and Mg^{2+} (Buchanan, 1980; Scheibe, 1991; Jacquot et al., 1997). The levels of dark activity and the kinetics of the light activation process vary for different enzymes in the pathway (Sassenrath-Cole and Piercy, 1992, 1994).

3.2.3   Regulation of enzymes – Rubisco

The enzyme Rubisco catalyses the first step in the RPPP and is responsible for fixing atmospheric CO_2 into the acceptor molecule RuBP. In addition to this carboxylation reaction, the Rubisco enzyme also catalyses a reaction utilizing O_2 as the substrate and producing glycollate, and this process, photorespiration, results in the release of CO_2 and ammonia (see section below). The carboxylase and oxygenase reactions are competitive and this, in addition to the very slow catalytic rate, makes Rubisco a rate-limiting enzyme in the RPPP. In C3 plants, photorespiration can reduce yields by around 30%. For this reason the molecular biology, structure and enzyme regulation of Rubisco have been studied extensively and many reviews covering these topics have been published previously (Spreitzer, 1993; Hartman and Harpel, 1994; Spreitzer and Salvucci, 2002; Parry et al., 2003). In brief, the Rubsico enzyme complex is composed of eight large subunits (LSU) encoded in the chloroplast genome, and eight small subunits (SSU) encoded by a nuclear multi-gene family. The assembly of this $LSU_8 SSU_8$ holoenzyme complex is mediated by chloroplast chaperonins. Furthermore, the activity of Rubisco is highly regulated involving a number of specific mechanisms including an additional nuclear-encoded chloroplast protein, Rubisco activase. In excess of 20 Rubisco X-ray crystal structures have been determined and more than 2000 LSU gene sequences and 300 SSU sequences are in the sequence database (GenBank).

Rubisco is inactive in the dark and in the light is converted to the active form by a process called *carbamoylation*, which involves the binding of CO_2 and Mg^{2+} to a lysine residue adjacent to the catalytic site. However, Rubsico can be inhibited by RuBP and related compounds as they can bind to the uncarbamoylated enzyme, thereby preventing activation. An additional protein, Rubisco activase, mediates light activation of the Rubisco enzyme, and during a dark/light transition, Rubisco activase removes inhibitors, RuBP and 2-carboxyarabinitol phosphate, bound to the Rubisco active site in the dark. The activity of Rubsico activase is sensitive to the stromal ADP/ATP ratio, and is activated directly by light via the ferredoxin/thioredoxin system (Portis, 2002). These activase regulatory mechanisms may enable the activity of Rubisco to be linked to both the activity of the electron transport chain and to the availability of sinks for triose-phosphate output from the RPPP. A major focus of research on the Rubsico enzyme has been aimed at elucidating information on the structure/function in order that its activity might be improved, particularly in relation to the carboxylation/oxygenation ratio. To date, this has not yielded any improvements; however, recent progress in manipulating Rubisco activity in plants via chloroplast transformation and the discovery of the regulatory importance of activase may together enable Rubisco activity to be improved in the future (Portis, 2002; Andrews and Whitney, 2003; Parry et al., 2003).

3.2.4   Thioredoxin regulation

The activity of a number of Calvin cycle enzymes has been shown to be regulated by light, mediated by the reducing power produced by the photosynthetic light reactions, which is then transferred from ferredoxin to thioredoxin catalysed by the enzyme ferredoxin/thioredoxin reductase (Buchanan, 1980; Scheibe, 1991; Jacquot et al., 1997; Schurman and Jacquot, 2000). Thioredoxin then binds to the inactive target enzyme and reduces the regulatory disulphide bond. The enzyme is activated by the associated change in conformation, and oxidized thioredoxin is released. In the leaf, light regulation of RPPP enzyme activity acts as an important on/off switch to prevent futile cycling of carbon in the dark. In addition, it is now thought that thiol regulation of the Calvin cycle also acts to modulate enzyme activity in response to transient alterations in the light environment, such as shading and sunflecks (Scheibe, 1991; Ruelland and Miginiac-Maslow, 1999).

A large number of thiroedoxins have been identified, but only thioredoxin m and thioredoxin f function in the chloroplast, whilst the thioredoxin h family are cytosol located (Meyer et al., 2002). Thioredoxin f is so called because the first enzyme identified as being activated by this protein was chloroplastic fructose 1,6-bisphosphatase (FBPase, E.C. 3.1.3.11). Additional targets of thioredoxin f have been identified, and those in the RPPP where the interaction with thioredoxin f has been demonstrated biochemically are sedoheptulose 1,7-bisphosphatase (SBPase, E.C. 3.1.3.37) and ribulose 5-phosphate kinase. The information now available from mutagenesis studies has revealed that the regulatory cysteine residues in the FBPase and SBPase protein sequences are located in different positions in the protein, and that for FBPase three cysteines appear to be involved whilst for SBPase only two regulatory cysteines have been identified (Figure 3.2). The feature that they have in common is that the redox active cysteines are distant from the catalytic site. This is in contrast with phosphoribulokinase (PRK, E.C. 2.7.1.19) where the cysteines involved in thiol regulation are some 39 amino acids apart and are located within the active site region of this protein. This work suggests that in each case thiol regulation has evolved independently in response to the appearance of oxygenic photosynthesis (Buchanan, 1991; reviewed in Jacquot et al., 1997; Schurman and Jacquot, 2000). The enzyme glyceraldehyde 3-phosphate dehydrogenase (GAPDH, E.C. 1.2.1.13) has often been considered to be thioredoxin-regulated, but, although the activity of this enzyme is increased in the light, no biochemical evidence demonstrating a direct role for thioredoxin has yet been provided. It is now known that GAPDH forms part of a multi-protein stromal complex, and this may be involved in mediating light activation of GAPDH (see Section 3.2.5).

Very recently, using a proteomic approach, a number of new potential thioredoxin targets in the chloroplast have been identified, including enzymes in the 1-deoxy-D-xylulose 5-phosphate (DOXP) isoprenoid biosynthetic pathway, starch degradation and tetrapyrrole biosynthesis. Additional enzymes in the RPPP have also been identified including transketolase (E.C. 2.2.1.1) and ribulose 3-phosphate epimerase

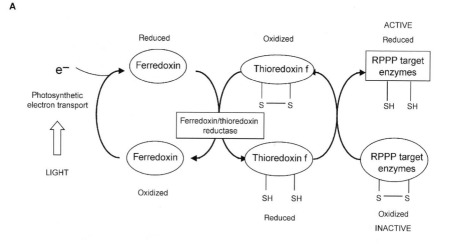

Figure 3.2 (A) The chloroplast ferredoxin/thioredoxin reduction system. Light-driven electron transport produces electrons that are passed via ferredoxin to ferredoxin/thioredoxin reductase, and it is this enzyme that reduces thioredoxin f, converting the disulphide bridge to two thiol groups. Thioredoxin f then activates the target proteins, FBPase, SBPase, PRK and activase by reducing a disulphide bridge, formed between two cysteine groups in the protein, and converting it to two thiol groups, thereby changing the conformation of the protein. (B) The regulatory cysteine residues on the target RPPP enzymes identified using mutagenesis.

(E.C. 5.1.3.1). These findings are interesting and raise the possibility that thioredoxin f may be involved in regulating flux of carbon out of the RPPP. However, biochemical analysis is needed to confirm a functional role for thioredoxin in regulating the activity of these enzymes (Balmer *et al.*, 2003).

3.2.5 Multi-protein complexes

Several lines of evidence have suggested that enzymes involved in the RPPP associate to form functional multi-protein complexes and that this may provide an additional level of regulation of enzyme activity (Harris and Koniger, 1997). A stromal complex, involving ribulose 5-phosphate kinase and GAPDH, has been identified in a green alga and spinach, and incubation of this complex with dithiothreitol and NADPH was shown to induce dissociation of the complex, accompanied by activation of the two component enzymes (Lazaro *et al.*, 1986; Clasper *et al.*, 1991). More recently, PRK and GAPDH have been shown to form a complex with the novel chloroplast protein, CP12, in higher plants, algae and cyanobacteria (Wedel *et al.* 1997; Wedel and Soll, 1998). This complex is approximately 600 kDa and is composed of a dimer of CP12 proteins, each of which binds one PRK dimer and one GAPDH heterotetramer. The catalytic properties of PRK and GAPDH were altered significantly when these enzymes formed the PRK/CP12/GAPDH complex, and in higher plants formation of this complex is modulated by dark/light transitions (for review, see Gontero *et al.*, 2002; Scheibe *et al.*, 2002). Taken together, these data suggest that the role of this complex may be to link activity of light-driven electron transport to carbon metabolism; however, no *in planta* data is yet available on the function of this protein complex *in vivo*.

The predicted protein sequence of CP12 contains two conserved motifs, one closer to the N-terminus and other at the C-terminal end of this protein, each with the potential to form an intramolecular loop via disulphide bonds between cysteine (Cys) residues. Mutagenesis studies have indicated that the N-terminal cysteine pair is involved in PRK binding and that binding of NADP and GAPDH is dependent on the C-terminal cysteines (Wedel and Soll, 1998). It is interesting to speculate that these cysteine pairs provide a mechanism by which the CP12 protein, and its interaction with PRK and GAPDH, could be modulated by redox through the thioredoxin system. However, the relationship, if any, between thioredoxin-mediated redox regulation and CP12 is not clear and the presence of CP12 in cyanobacteria, which lack the thioredoxin system, indicates that CP12 has the potential to operate independently in higher plants. The physiological role of the CP12 complex *in planta* is not yet known but the available *in vitro* data has provided the first evidence that stromal multi-enzyme complexes can have both novel and important regulatory roles. In addition to the PRK/CP12/GAPDH complex there is also good evidence that additional multi-enzyme complexes in the stroma are present. However, we know very little about the nature of these in terms of stoichiometry, stability or the influence of stromal factors (Harris and Koniger, 1997; Jebanathirajah and Coleman, 1998).

3.2.6 Regulation of RPPP gene expression

The genes encoding the enzymes of the RPPP are encoded in the nuclear genome, with the exception of the large subunit of Rubisco. As a result, the proteins are synthesized in the cytosol, with N-terminal extensions termed transit sequences/peptides, which direct their import into the chloroplast stroma (see Chapter 5). The expression of the RPPP enzymes appears to be coordinated during leaf development and chloroplast maturation (Raines *et al.*, 1991). Light plays a key role in triggering chloroplast development and in regulating the expression of RPPP genes, and much of this control has so far been found to occur at the level of gene transcription. Plants grown in darkness have very low levels of RPPP mRNAs, but on exposure to light there is a rapid induction of mRNA synthesis within 1 h, with maximum levels finally attained after 1–2 days of a normal light/dark growth regime. The light response of the RPPP genes is mediated via a number of photoreceptors, including phytochrome and the blue-light receptor cryptochrome, but this has only been demonstrated for

Rubisco SSU and GAPDH. When mature light-grown plants are placed in the dark for 1–2 days, RPPP mRNA levels decrease, and re-illumination causes mRNA levels to increase to pre-dark levels within hours. These data suggest that the leaf has been primed to respond by previous completion of light-induced chloroplast development. It has now been shown that signals that come from the chloroplast (see Chapter 9) are important in maintaining expression of nuclear-encoded chloroplast proteins, including RPPP enzymes (Strand *et al.*, 2002; Jarvis, 2003).

In mature leaves RPPP gene expression is sensitive to environmental and metabolic signals, providing long-term mechanisms for the plant to regulate primary carbon fixation (Stitt and Krapp, 1999; Stitt and Hurry, 2002). High levels of glucose and sucrose have been shown to be associated with reduced levels of a number of Calvin cycle mRNAs, including those encoding the SSU of Rubisco, SBPase and FBPase. This feedback mechanism, which again appears to act at the level of transcription, might be important for source/sink regulation in the plant (Krapp and Stitt, 1994; Rogers *et al.*, 1998). The signalling pathway involved in glucose repression of gene expression is not known, although it has been suggested that the enzyme hexokinase (E.C. 2.7.1.1) might be involved. It has also been shown that photosynthesis genes can respond at the transcript level to nutrient status, namely inorganic nitrogen (N) and phosphorus (P) levels and that this is modulated by carbohydrate status (Neilsen *et al.*, 1998; Stitt and Krapp, 1999). These results suggest that this interaction between carbohydrate and nutrient status may function as a long-term strategy in the control of primary carbon metabolism. However, there is at present no evidence linking changes in metabolic flux within the RPPP with changes in the expression of RPPP genes or proteins. Further support for this has come from the studies of the RPPP antisense plants (Kossmann *et al.*, 1994; Haake *et al.*, 1998; Harrison *et al.*, 1998; Olcer *et al.*, 2001).

3.2.7   Limitations to carbon flux through the RPPP

The maximum rate of CO_2 uptake from the atmosphere into the RPPP, and the subsequent flow of carbon through the pathway, is determined by the slowest step (enzyme reaction) in this pathway. The catalytic and regulatory properties of each individual enzyme in the RPPP have been determined *in vitro*, but, although this is important information, it does not allow the limitation that any individual enzyme exerts on the whole system to be predicted *in vivo*. Metabolic control analysis is an alternative approach that asks questions about the whole system: how much does the flow of carbon in the Calvin cycle vary as the activity of an individual enzyme is changed (Fell, 1997)?

This can be expressed quantitatively:

$$C = \frac{dJ/J}{dE/E}$$

where C is the flux control coefficient, J is the original flux through the pathway, dJ is the change in flux, E is the original enzyme activity and dE is the change in enzyme activity.

The flux control coefficient (C) can have values from 0 to 1, where 0 means no control and 1 means complete control. This approach has been taken to address the question of which enzymes control CO_2 fixation through the RPPP.

Antisense technology has been used to produce transgenic plants in which the levels of specific individual enzymes have been changed. The results obtained from the analysis of these plants have revealed new and interesting information on limitations to carbon flow in the RPPP. Flux control (C) values for Rubisco were between 0.2 and 1.0, depending on the environmental conditions in which the plants were grown or analysed in (for a review, see Stitt and Schulze, 1994). These data provide evidence suggesting that Rubisco is not the only enzyme-limiting carbon fixation in all environmental conditions. Interestingly, it has been shown that both SBPase and transketolase can have C values in excess of 0.5, indicating that these enzymes can limit the rate of carbon fixation (Raines *et al.*, 2000; Henkes *et al.*, 2001; Olcer *et al.*, 2001). In contrast, FBPase, PRK and GAPDH have C values that are never greater than 0.3, and are usually much less than this, indicating that the activity of any one of these enzymes will have little control over the rate of carbon fixation through the RPPP (Kossmann *et al.*, 1994; Paul *et al.*, 1995; Price *et al.*, 1995; Banks *et al.*, 1999). Under many conditions it is likely that control of flux through the RPPP is poised such that Rubisco limitation and regenerative capacity are balanced and that the extent of control exerted by any one of these enzymes will vary depending on developmental stage and environmental conditions (reviewed in Raines, 2003). These findings are in keeping with results obtained from two very different modelling approaches (Poolman *et al.*, 2000, 2003; Von Caemmerer, 2000). The practical use of these data is that they predict that it might be possible to increase photosynthetic carbon fixation by increasing the activity of Rubisco, SBPase or transketolase. Recently, support for this hypothesis has come from the analysis of transgenic plants expressing a bifunctional cyanobacterial FBPase/SBPase enzyme where increased photosynthetic capacity was observed together with increased growth (Miyawaga *et al.*, 2001).

Although considerable information is available on the regulatory properties of enzymes within the pathways that receive carbon from the RPPP, very little is known about how the regulatory mechanisms within the RPPP control the allocation of this carbon to the different biosynthetic pathways. Recently, data from the analysis of antisense plants with reduced levels of individual RPPP enzymes have clearly shown that changes in carbon flux through the cycle can alter allocation. The most striking example of this is the shift in allocation of carbon towards starch, and the dramatic reductions in the levels of aromatic amino acids and intermediates in the shikimic acid pathway, resulting from reductions in transketolase activity (Henkes *et al.*, 2001). In contrast, plants with reductions in another Calvin cycle enzyme, SBPase, maintained sucrose production at the expense of starch, changing the sink/source balance of the plant (Olcer *et al.*, 2001). More recently, it has been shown that reduced levels of Rubisco activity result in a decrease in the levels of secondary metabolites and in changes of the amino acid/sugar ratio, and the nicotine/chlorogenic acid ratio (Matt *et al.*, 2002). These antisense studies have revealed the importance of the levels of individual enzymes not only in the control of primary carbon flux but also in the allocation of carbon from the RPPP.

3.2.8   Integration and regulation of allocation of carbon from the RPPP

Triose-phosphates produced by the linear part of the RPPP are utilized by a number of pathways to synthesize carbon compounds essential for plant growth and development. It is well documented that only one out of every six molecules of triose-phosphate produced by the RPPP is net product, and available for export. A proportion of this excess triose-phosphate in the form of DHAP is transported from the chloroplast to the cytosol through the triose-phosphate/inorganic phosphate translocator (TPT; see section on metabolite transporters below), and is used to synthesize sucrose. There is a strict stoichiometry in the transport through the TPT, and for every one molecule of DHAP transported out of the chloroplast; one molecule of inorganic phosphate (Pi) is moved from the cytosol into the chloroplast. A number of enzymes in the cytosol are involved in regulating sucrose biosynthesis, cytosolic FBPase, sucrose phosphate synthase (SPS, E.C. 2.4.1.14) and sucrose phosphatase (E.C. 3.1.3.24). The synthesis of starch takes place in the chloroplast utilizing fructose 6-phosphate (Fru6P) produced in the RPPP from the triose-phosphate DHAP and G-3-P. The main regulator of starch biosynthesis is the enzyme adenosine 5′ diphosphate glucose pyrophosphorylase (AGPase, E.C. 2.7.7.27), the activity of which is regulated by products of the RPPP, 3-PGA and Pi. AGPase is also regulated at the level of gene expression, and high levels of sucrose increase transcription of the large subunit of the enzyme (see section on starch biosynthesis below). It is the combination of the mechanisms regulating the enzymes in both the cytosol and the chloroplast that regulates the amount of DHAP that goes to sucrose versus starch synthesis. Transgenic plants expressing altered activities of SPS, cytosolic FBPase and AGPase have provided further evidence of the importance of enzyme regulation in balancing the relative amounts of starch and sucrose that are synthesized.

3.2.9   Isoprenoid biosynthesis

The triose-phosphate G-3-P can also exit the RPPP to enter the chloroplastic isoprenoid biosynthetic pathway. Until recently isoprenoids were believed to be synthesized solely through the mevalonic pathway located in the cytosol. However, a chloroplastic pathway for the synthesis of isoprenoids is also present, where G-3-P is combined with phosphoenolpyruvate (PEP) to form DOXP, catalysed by DOXP synthase (E.C. 2.2.1.7); rearrangement of this molecule results in the formation of the isoprenoid units, isopentenyl diphosphate and dimethylallyl phosphate, which are then used to produce isoprene, plastoquinone, carotenoids, chlorophyll side chains, hormones (brassinosteroids, cytokinins, gibberellins, abscisic acid) (Lichtenthaler, 1999). The regulatory mechanisms operating to control the rate of carbon flux from the RPPP into the DOXP pathway are unknown. Interestingly, application of transcriptomic and proteomic approaches suggest that post-transcriptional mechanisms such as thioredoxin-mediated enzyme regulation may be involved (Balmer *et al.*, 2003; Laule *et al.*, 2003).

3.2.10   Shikimic acid biosynthesis

Erythrose 4-phosphate produced in the RPPP can be used for the synthesis of aromatic amino acids in the shikimate biosynthetic pathway in the chloroplast (Herrmann and Weaver, 1999). The first step in this pathway is the condensation of erythrose 4-phosphate and PEP to form 3-deoxy-D-arabino-heptulosonate 7-phosphate (DAHP) through the action of the enzyme DAHP synthase (E.C. 2.5.1.54). Uncontrolled exit of erythrose 4-phosphate to the shikimate pathway has been shown to result in a cessation of activity of the RPPP, and plants become chlorotic and die (reviewed in Geiger and Servaites, 1994). Entry of carbon into the pathway is regulated by inhibition of activity of DAHP synthase by binding of arogenate, a precursor of tyrosine and phenylalanine. Additionally, biochemical evidence has been produced demonstrating that reduced thioredoxin is essential for activation of DAHP synthase (Entus *et al.*, 2002). This is an interesting finding and raises the possibility that light activation of this enzyme is part of a mechanism regulating flux out of the RPPP.

3.2.11   OPPP and RPPP

The OPPP in plants provides NADPH, ribose for nucleic acid synthesis and erythrose 4-phosphate for the biosynthesis of aromatic amino acids and their derivatives, polyphenols and lignins. The plastidic OPPP and RPPP share some common enzymes and intermediates; given that these two pathways function in opposite directions, it is of interest to consider how this flux is regulated in order to prevent futile cycling of intermediates (Figure 3.3). Details on the control and regulation of the plastidial OPPP are dealt with subsequently (see below).

3.3   Photorespiration

Photorespiration is the light-dependent evolution of CO_2 that occurs in C3 leaves and leads to inhibition of photosynthesis by oxygen. Unlike true respiration, photorespiration is energy-wasting, and results in a decrease in the efficiency of net carbon assimilation by the plant. The CO_2 released in photorespiration is a product of the bifunctional nature of Rubisco and the generation of glycollate 2-phosphate in the oxygenase reaction, which is quickly converted to glycollate within the chloroplast. The regeneration of glycerate 3-phosphate from 2 moles of glycollate

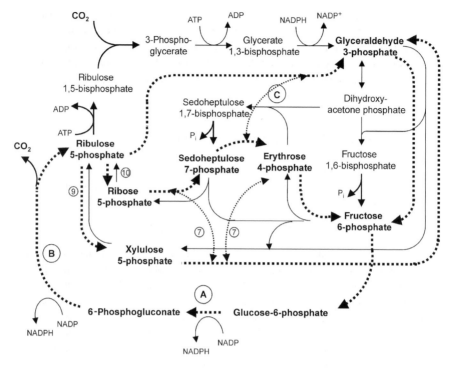

Oxidative steps of the OPPP

Figure 3.3 The relationship between the plastid OPPP and RPPP. The OPPP can be considered to have two phases: the oxidative steps that convert Glc6P to 6-phosphogluconate, which in turn is converted to ribulose 5-phosphate catalysed by glucose 6-phosphate dehydrogenase (**A**) and 6-phosphogluconate dehydrogenase (**B**), producing two molecules of NADPH. The second, non-oxidative, phase utilizes enzymes and intermediates common to both the OPPP and the RPPP; ribulose 5-phosphate is used to produce xylulose 5-phosphate and ribose 5-phosphate by the action of ribulose 5-phosphate epimerase (9) and ribose 5-phosphate isomerase (10). Transketolase (7) then catalyses the C2 transfer reaction forming G-3-P and sedoheptulose 7-phosphate. The action of the enzyme transaldolase, unique to the OPPP, Fru6P and, the major OPPP product, erythrose 4-phosphate are produced. The flow of carbon in the RPPP is shown by the continuous line; OPPP flow is denoted by the thick broken lines and the transaldolase (**C**) and transketolase (7) reactions are shown by thin broken lines.

via peroxisomal and mitochondrial metabolism leads to the formation of ammonia and the loss of CO_2. Ammonia and CO_2 arise directly from the activity of glycine decarboxylase (E.C. 1.4.4.2), a multi-complex enzyme comprising four different proteins, and a major component of the mitochondria in C3 plants, capable of oxidizing glycine at very high rates. Assimilation of the ammonia resulting from the decarboxylation of glycine during photorespiration plays a major role in N metabolism within the chloroplast (the photorespiratory nitrogen cycle). Despite the importance of photorespiration in plant carbon and N metabolism, the process is not strictly confined to the plastids, and is therefore not described in detail in this chapter. However, details of the assimilation of photorespiratory ammonia by the glutamine synthetase/glutamate synthase cycle within the chloroplast are discussed in Section 3.4. The control and regulation of the photorespiratory pathway has been discussed in detail by Leegood *et al.* (1995).

3.4   Nitrogen assimilation and amino acid biosynthesis

The assimilation of inorganic nitrogen (N, in the form of nitrate, NO_3^-) from the soil occurs in the cytosol of plant cells and is catalysed by the inducible enzyme nitrate reductase (E.C. 1.6.6.1):

$$NO_3^- + NAD(P)H + H^+ \rightarrow NO_2^- + NAD^+ + H_2O$$

Recent work by Huber and co-workers has shown that nitrate reductase activity is highly regulated within the cytosol, by protein phosphorylation and binding with 14-3-3 proteins (Bachmann *et al.*, 1996a, b). The nitrite (NO_2^-) produced in the cytosol is transported into the plastid, presumably by a homologue of the NAR gene family, which in *Chlamydomonas reinhardtii* encodes a high-affinity chloroplast nitrite transporter (Galván *et al.*, 2002). All subsequent reactions of primary assimilation and amino acid synthesis take place within the plastids of both photosynthetic and non-photosynthetic tissues. Plastidial nitrite reductase (NiR, E.C. 1.6.6.4) is an inducible enzyme which catalyses the reduction of NO_2^- to ammonia, which is toxic and must be assimilated into non-toxic metabolites. The catalytic activity of NiR is severalfold greater than nitrate reductase activity, and suggests that nitrate reductase is a rate-limiting step in NO_3^- assimilation. NiR is regulated by light, but not to the same extent as nitrate reductase. Transcription of NiR mRNA is rapidly induced by nitrate, in some cases within 5 h, is unaffected by carbohydrates, but is inhibited by the amino acids glutamate, glutamine and asparagine (Vincentz *et al.*, 1993). Analysis of transgenic tobacco plants over-expressing the NiR gene suggests that synthesis of the NiR protein may be down-regulated by ammonium ions at the post-transcriptional level (Crete *et al.*, 1997). The initial steps in ammonia assimilation inside the plastid are catalysed by glutamine synthetase (GS, E.C. 6.3.1.2) and glutamate synthase (glutamine:oxoglutarate aminotransferase, GOGAT), with the net production of one molecule of glutamate for each molecule of ammonia assimilated (Mifflin and Lea, 1980). GS catalyses the ATP-dependent conversion of glutamate to glutamine and has a very high affinity for ammonia (typical K_m of around 3–5 μM). In leaves two major forms of GS exist: a minor isoform, not present in all plant tissues, is located in the cytosol (GS_1); and a chloroplast isoform (GS_2), which is the predominant activity. The appearance of GS_2 in leaves is regulated by light, possibly via phytochrome and a blue-light receptor. For example, pea GS_2 mRNA increases some 20-fold following illumination of etiolated shoots (Lam *et al.*, 1996). The activity of GS_2 in wheat leaves increases with leaf age and

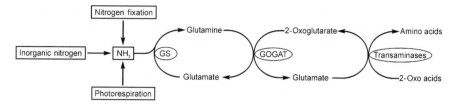

Figure 3.4 Assimilation of ammonia in the plastids of higher plants via the glutamine synthetase/glutamate synthase (GS/GOGAT) cycle.

correlates with photosynthetic and photorespiratory capacity. GOGAT is responsible for the transfer of the amide group of glutamine to 2-oxoglutarate, yielding two molecules of glutamate, one of which may be used to regenerate glutamine via the assimilation of ammonia (the GS/GOGAT cycle; see Figure 3.4). In leaves the major isoform of GOGAT utilizes reduced ferredoxin (which is a soluble component of photosystem I associated with the chloroplast thylakoid membrane) as a source of reductant; this ferredoxin-dependent GOGAT (E.C. 1.4.7.1) is located solely in the chloroplast. A minor plastidial form of GOGAT (usually comprising between 3 and 5% of total GOGAT activity) uses NADH in place of ferredoxin as the electron donor, and has been cloned and studied extensively in *Arabidopsis*, alfalfa and rice (Gregerson *et al.*, 1993; Ishiyama *et al.*, 1998; Trepp *et al.*, 1999; Schoenbeck *et al.*, 2000). NADH-dependent GOGAT (E.C.1.4.1.14) genes encode functional domains that are highly homologous to the large and small subunits of bacterial NADH-dependent GOGAT (Gregerson *et al.*, 1993). mRNA levels of *Arabidopsis* NADH-dependent GOGAT are enhanced in roots as opposed to leaves (Coschigano *et al.*, 1998), and in alfalfa roots NADH-dependent GOGAT was immunolocalized to amyloplasts (Trepp *et al.*, 1999). In *Arabidopsis* expression of NADH-dependent GOGAT mRNA appears to be coordinated with the expression of GS_1, suggesting a role in the primary assimilation of nitrate-derived ammonia in root cells. Recent isolation of a NADH-dependent GOGAT null mutant in *Arabidopsis* (*glt1*-T) indicates a role for NADH-dependent GOGAT in non-photorespiratory ammonium assimilation (Lancien *et al.*, 2002). Transgenic alfalfa with decreased expression of NADH-dependent GOGAT showed impaired nitrogen fixation in the root nodules and altered metabolic fluxes of carbon and nitrogen compounds between photosynthetic and nodule tissues (Cordoba *et al.*, 2003). Ferredoxin-dependent GOGAT genes have been isolated from a wide range of plant species (e.g. maize, tobacco and spinach), where a single gene was identified in every species except *Arabidopsis*, in which two copies of the gene were detected, and shown to be expressed (*GLU1* and *GLU2*) (see Lam *et al.*, 1996). All ferredoxin-dependent GOGAT genes show a high degree of similarity. In all plants studied, ferredoxin-dependent GOGAT mRNAs increased dramatically following illumination of etiolated leaves (for review, see Lam *et al.*, 1996; Temple *et al.*, 1998), and the involvement of phytochrome in the light induction process has been demonstrated in tomato (Becker *et al.*, 1993). Maize leaf ferredoxin-dependent GOGAT showed diurnal changes under light/dark cycles, unlike the NADH-GOGAT, which showed no light response in enzyme activity or gene expression (Suzuki *et al.*, 2001). The GS/GOGAT cycle is the major route of ammonia assimilation in the leaves of C3 plants, rapidly removing potentially toxic ammonium ions and generating amino acids. Much of the ammonia entering the GS/GOGAT pathway is derived from either NO_3^- reduction, symbiotic N_2 fixation, or in leaves, from photorespiration. During photorespiration the decarboxylation of glycine produces ammonia in stoichiometric amounts with the photorespiratory CO_2 evolved. In C3 plants the rate of this ammonia production can be as high as 20 times the rate of primary nitrate assimilation, potentially making photorespiration a major source of ammonia for assimilation via the GS/GOGAT cycle. Studies of mutants of *Arabidopsis* and barley indicate that the ferredoxin-dependent GOGAT in chloroplasts is essential for the reassimilation of photorespiratory ammonia, since mutants lacking the enzyme accumulate large quantities of ammonia under photorespiratory conditions and eventually die. However, under non-photorespiratory conditions the activity of the NADH-dependent GOGAT appears to be sufficient for the assimilation of ammonia derived from nitrate reduction. A similar situation occurs in barley mutants lacking the chloroplast GS, where under photorespiratory conditions ammonia accumulates and photosynthesis is severely inhibited (see Blackwell *et al.*, 1988). All other amino acids are derived from glutamate or glutamine from the GS/GOGAT reactions, as well as other N-containing compounds in the cell such as nucleic acids, cofactors, chlorophyll and secondary metabolites. The GS/GOGAT cycle, therefore, is positioned at the interface of N and carbon metabolism. N assimilation places a high demand for energy and reducing power on the tissue. The ATP requirement could be met by photophosphorylation in chloroplasts, or by glycolytic activity within the organelle (Qi *et al.*, 1994), or import from the cytosol in non-green plastids (Schünemann *et al.*, 1993).

In heterotrophic plastids there is substantial evidence that the reductant required for the assimilation of nitrite to glutamate is generated by the OPPP (see section below). Experiments with isolated pea root leucoplasts have shown that the oxidation of glucose 6-phosphate (Glc6P) (transported into the plastids by a Glc6P-specific hexose-phosphate translocator, see metabolite transport section below) by the OPPP is tightly coupled to the reactions catalysed by NiR and GOGAT (Bowsher *et al.*, 1989, 1992). The plastidial OPPP operates in a cyclic manner, oxidizing each mole of Glc6P (which is imported in heterotrophic plastids, and generated internally in chloroplasts) to 3 moles of CO_2 and 1 mole of triose-phosphate, which, in heterotrophic tissues, is exported in exchange for the Glc6P imported by the hexose-phosphate transporter, so maintaining the stoichiometric balance (Borchert *et al.*, 1993; Hartwell *et al.*, 1996). The ratio of NADPH/NADP strongly inhibits the first enzyme of the OPPP, glucose 6-phosphate dehydrogenase (G6PDH; E.C. 1.1.1.49); since NADPH is required for N assimilation and other associated biosynthetic processes, a balance needs to be struck as too high a concentration will inhibit its own production (Wright *et al.*, 1997). However, it appears that in green tissue, there are at least two isozymes of plastidial G6PDH, P1 and P2 (Wendt *et al.*, 2000;

see Section 3.7). Interestingly, the P2 isoform is far less sensitive to inhibition by NADPH, and consequently may be able to function during illumination. The same isoform is expressed in roots, where there is little or no expression of P1-G6PDH, suggesting that the P2 form may have an important role in sustaining reductive biosynthesis in heterotrophic cells.

Ferredoxin is the immediate source of reducing power for NiR and GOGAT, and electrons are transferred from NADPH via a ferredoxin-NADP reductase (FNR, E.C. 1.18.1.2). The properties and primary sequences of leaf and root FNRs are substantially different, reflecting their different roles within the different plastid types (Aoki and Ida, 1994).

Studies in pea and maize root plastids showed that the activities of both the OPPP enzymes G6PDH and 6PGDH (6-phosphogluconate dehydrogenase, E.C. 1.1.1.43) increased during nitrate assimilation, indicating a close coupling between the pathways generating and utilizing reductant (Emes and Fowler, 1983; Redinbaugh and Campbell, 1998). In addition to the changes in activities of the OPPP enzymes, transcript levels of 6PGDH accumulated rapidly and transiently in response to low concentrations of external nitrate (Redinbaugh and Campbell, 1998). In pea roots, both ferredoxin and FNR are induced by nitrate assimilation (Bowsher *et al.*, 1993). Maize roots contain two forms of ferredoxin, one being constitutive, whilst the other is rapidly transcribed following the application of nitrate (Matsumara *et al.*, 1997). Furthermore, in rice roots nitrate assimilation induces ferredoxin and the appearance of mRNAs for NiR and FNR (Aoki and Ida, 1994). Interestingly, analysis of the promoter sequences for NiR (Tanaka *et al.*, 1994), and the inducible forms of FNR (Aoki *et al.*, 1995), ferredoxin (Matsumara *et al.*, 1997) and G6PDH (Knight *et al.*, 2001), reveals a common regulatory element known as a NIT-2 motif, which points to the coordinated expression of root plastid enzymes involved in N assimilation, and is so far consistent with experimental observations.

N and carbon metabolism in plants needs to be tightly coordinated, since the assimilation of N demands carbon skeletons (in the form of 2-oxoglutarate), and both ATP and reductant are required for the reduction of nitrate, through ammonium, into glutamine and glutamate. In bacterial carbon and N metabolism the PII signal transduction protein plays an important role in coordinating these pathways by performing both sensing and signalling functions (Magasanik, 2000). Under conditions of high N, cellular levels of glutamine (which acts as a signal for cellular N status) promote the deuridylylation of PII, ultimately causing the inactivation of GS. PII simultaneously causes the inhibition of the transcription of N-regulated genes (via an N-sensitive regulon) including GS. When glutamine levels drop and the carbon and energy (ATP) status is perceived as adequate, PII is uridylylated, GS is activated and transcription of the N-sensitive regulon is promoted. This signal transduction mechanism also operates in photosynthetic bacteria, and the PII protein was first identified in higher plants in *Arabidopsis* (Hsieh *et al.*, 1998). Analysis of PII homologues in higher plants revealed that this sensor/signal transduction protein is present in a wide range of plants, including castor bean, tomato and rice (Moorhead and Smith, 2003). Molecular and biochemical analyses of the plant PII indicates that the protein may be phosphorylated rather than uridylylated, and has a predicted transit peptide (Smith *et al.*, 2003b), indicating a plastidial location. Biochemical analysis has located PII inside the chloroplast in *Arabidopsis* (Hsieh *et al.*, 1998), suggesting that this protein plays an important role in coordinating carbon and N metabolism in the plastid.

3.5 Synthesis of fatty acids

In plants the *de novo* synthesis of fatty acids is carried out in the plastids. The enzymology of fatty acid synthesis has been well characterized and the process is carried out by a multi-subunit fatty acid synthetase (FAS) complex (Slabas and Fawcett, 1992). The precursor for fatty acid synthesis is acetyl-coenzyme A (acetyl-CoA), and this is carboxylated by a plastidial acetyl-CoA carboxylase (ACCase, EC 6.4.1.2) to form the malonyl-CoA that is then used by the FAS complex. Most higher plants possess a prokaryotic-type ACCase (type II) in their plastids (Sasaki *et al.*, 1995). This is a multi-subunit enzyme with each subunit possessing a separate function. The Poaceae (grasses) represent an exception to this in that they have a eukaryotic, type I ACCase, which is a large multifunctional protein that possesses all of the separate subunit functions (Konishi *et al.*, 1996). Interestingly, the type II enzyme is strongly resistant to the herbicides of the aryloxyphenoxypropionate and cyclohexanedione type, which are inhibitors of the plastidial type I enzyme and this forms the basis for the use of these chemicals as Graminaceous weedkillers (Sasaki *et al.*, 1995). In addition to the plastidial ACCases all plants have a cytosolic, type I isoform (Sasaki *et al.*, 1995). Whether plants outside of the Poaceae also possess a plastidial type I enzyme is uncertain. There is no gene encoding a plastid-targeted isoform of ACCase I in *Arabidopsis* although the closely related species *Brassica napus* (L.) (oilseed rape or canola) does possess such a gene (Schulte *et al.*, 1997). A type I ACCase has been localized to the plastid in *B. napus* embryos (Roesler *et al.*, 1997) and about 10% of the propionyl-CoA carboxylase activity that is associated with the type I enzyme is localized in the plastids of developing embryos of this species (Sellwood *et al.*, 2000).

It is known that acetyl-CoA cannot be transported across the plastid envelope (Weaire and Kekwick, 1975; Roughan *et al.*, 1979) and it must therefore be generated *de novo* inside the plastid to enable fatty acid synthesis to occur. As described in the above section, plastids from heterotrophic tissues can contain a complete glycolytic pathway that, in combination with a plastidial isoform of the pyruvate dehydrogenase complex (E.C. 1.2.4.1), enables conversion of imported Glc6P into acetyl-CoA. However, some chloroplasts are reported to have restricted glycolytic activity due to low activities of phosphoglyceromutase and/or enolase (see above). This restriction in glycolysis would require import of metabolites such as PEP and/or pyruvate in order to support synthesis of acetyl-CoA via the plastidial pyruvate dehydrogenase complex. Acetate uptake by the plastid, followed by activation to acetyl-CoA via the action of acetyl-CoA synthetase (EC 6.2.1.1), also represents a possible metabolic route, but three lines of evidence suggest that this is unlikely. Firstly, detailed $^{14}CO_2$ feeding experiments with whole leaves suggest that the acetate pool in intact leaf tissues is small and the turnover of label through this pool is not consistent with its involvement with *de novo* fatty acid synthesis (Bao *et al.*, 2000). Secondly, the expression pattern of acetyl-CoA synthetase in developing siliques of *Arabidopsis* is wholly inconsistent with a role in fatty acid synthesis in the developing embryos (Ke *et al.*, 2000). Thirdly, reduced expression of acetyl-CoA synthetase through antisense RNA down-regulation did not affect lipid content in leaves (Behal *et al.*, 2002). These data are supported by analysis of a knockout mutation of a gene encoding a subunit of the plastidial pyruvate dehydrogenase complex of *Arabidopsis* (Lin *et al.*, 2003). This knockout is lethal in homozygotes and therefore implies that plastidial synthesis of acetyl-CoA via the pyruvate dehydrogenase complex is essential and cannot be complemented by plastidial acetyl-CoA synthetase activity.

Investigations of the sources of carbon to support plastid fatty acid synthesis are largely based upon an *in vitro* approach in which the enzymology within isolated plastids and their ability to import and then utilize a given substrate *in vitro* are determined. These studies have concentrated upon plastids isolated from storage and other non-photosynthetic tissues, largely because previous dogma that acetate was the carbon source for fatty acid synthesis in chloroplasts was so widely accepted, although the use of pyruvate as a carbon source for fatty acid synthesis by isolated chloroplasts has been reported (Springer and Heise, 1989; Eastmond and Rawsthorne, 1998). This strategy provides a wealth of information on what may happen *in vivo* but ultimately requires a genetic approach to test hypotheses derived from such *in vitro* data. Given this caveat, these previous studies have shown that a range of metabolites is used for fatty acid synthesis by plastids from heterotrophic tissues of different plant species (Smith *et al.*, 1992; Kang and Rawsthorne, 1994; Qi *et al.*, 1995; Eastmond and Rawsthorne, 2000), including Glc6P, triose-phosphate, malate, PEP (Kubis *et al.*, in press) and pyruvate (Figure 3.5). While many studies have focused on the uptake and utilization of carbon for fatty acid synthesis from a single metabolite, it is clear that multiple metabolites can be used simultaneously as carbon sources (Kang and Rawsthorne, 1996; Eastmond and Rawsthorne, 2000) and this may be more representative of what actually occurs *in vivo*. Collectively from the above experiments it is known that the relative rate of utilization of each of these metabolites by isolated plastids depends upon the organ and species from which the plastids have been isolated and is also affected by developmental stage within organs. These differences are determined in part by variation in the activities of specific translocator proteins on the plastid envelope. For example, pyruvate uptake by plastids from castor endosperm is weak compared to the transporter-mediated import of malate for fatty acid synthesis (Eastmond *et al.*, 1997). Also, the rate of incorporation of exogenous pyruvate into fatty acids by isolated plastids increases during development in oilseed rape embryos, reflecting the measured increase in pyruvate uptake activity (Eastmond and Rawsthorne, 2000). Not only can changing expression of translocators determine the extent to which

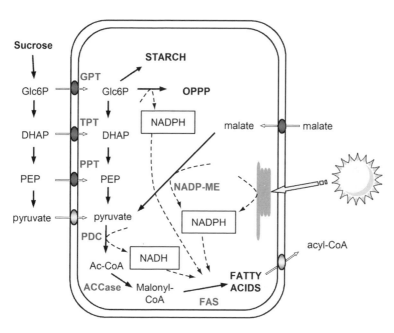

Figure 3.5   Provision of carbon substrates and reducing power for fatty acid synthesis in plastids. The potential routes of glycolytic metabolism in either the cytosol or the plastid and the interaction between them via the Glc6P (GPT), triose-phosphate (TPT), PEP (PPT) and the as-yet uncharacterized pyruvate transporter are illustrated. Pyruvate is metabolized to fatty acids by the actions of the pyruvate dehydrogenase complex (PDC), acetyl-CoA carboxylase (ACCase) and the fatty acid synthetase complex (FAS). The export of fatty acids from the plastid and their activation to acyl-CoAs in the cytosol is largely uncharacterized. Reducing power in the form of NADH or NADPH can be provided to fatty acid biosynthesis by the activities of the oxidative pentose-phosphate pathway (OPPP), NADP-malic enzyme (NADP-ME), the PDC, and by light energy. Plastids such as those from castor endosperm are likely to be dependent on NADP-ME and PDC (Smith *et al.*, 1992; Eastmond *et al.*, 1997), those from oilseed embryos on the OPPP and light energy (Eastmond and Rawsthorne, 2000; Schwender *et al.*, 2003), whilst chloroplasts are essentially light-dependent. ⇨: transporters; ⟶: carbon flux; ---▶: reductant flux. Glc6P: glucose 6-phosphate; DHAP: dihydroxyacetone phosphate; PEP: phosphoenolpyruvate; Ac-CoA: acetyl-coenzyme A.

metabolites are used to support plastidial fatty acid synthesis but the translocators themselves can be subject to regulation during it. An example of this is the Glc6P translocator, which is inhibited by acyl-coenzyme As, the end products of plastidial fatty acid synthesis and acyl chain export (Fox *et al.*, 2000; Johnson *et al.*, 2000; 2002).

The evidence for involvement of specific cytosolic metabolites as carbon sources for fatty acid synthesis *in vivo* is poor at present but advances are being made in the approaches to address this issue. The use of knockout mutants in *Arabidopsis* is one, although recent work on the *cue1* mutant has suggested that such an approach may not be sophisticated enough. *In vitro* work suggests that cytosolic PEP could be a substrate for fatty acid synthesis and knocking out PEP import would be a route to test this. It transpired that the *cue1* mutants contain a mutated gene for the phosphoenolpyruvate/phosphate translocator (PPT1) that causes interveinal chlorosis in the leaves (Streatfield *et al.*, 1999). This was originally believed to be due to a block in PEP import that prevented aromatic amino acid synthesis through the plastid-localized shikimate pathway (Streatfield *et al.*, 1999). The supply of aromatic amino acids to mutant plants in culture complemented the phenotype of the mutant, and the lipid content of the leaves and seeds of these plants was normal, leading to the conclusion that a loss of PEP import did not affect fatty acid synthesis and therefore PPT1 was not involved in this process (Streatfield *et al.*, 1999). However, the presence of a second PPT gene (Knappe *et al.*, 2003b) and the fact that many other routes for carbon import into the plastid exist in *B. napus* (Figure 3.5), and therefore almost certainly *Arabidopsis*, suggest that it is hard to draw clear conclusions from such studies. Nevertheless, seeds of *Arabidopsis* store significant amounts of oil, and this species therefore represents a good model in which to study fatty acid, and hence storage oil synthesis. The use of RNAi and over-expression methods to alter the activity of plastidial transporter proteins specifically in the developing seed would be an approach that removes the potentially pleiotropic effects that may occur in whole plant development as seen in *cue1*. However, manipulation of the activities of multiple transporters and/or plastidial enzymes may be required and then these should be combined with measurements of metabolism *in vivo* when possible. Measurement of carbon metabolism under close-to-*in vivo* conditions can be achieved using ^{13}C-metabolite feeding in combination with non-magnetic resonance (NMR) and mass spectrometry (MS) based techniques. Schwender and colleagues (Schwender and Ohlrogge, 2002; Schwender *et al.*, 2003) have isolated developing embryos from *B. napus* and have grown them in culture using ^{13}C-labelled sugars and amino acids in order to study their metabolism. This has enabled a preliminary map of metabolic fluxes for intact tissue in which fatty oil synthesis predominates. Direct glycolytic flux to fatty acids from hexose-phosphates predominates over indirect flux involving the OPPP by an estimated 9:1 ratio (Schwender and Ohlrogge, 2002). Moreover, measurements of metabolic markers for the pyruvate or PEP pools in the cytosol and/or plastid in feeding experiments with ^{13}C-alanine and ^{13}C-glucose have revealed that plastidial PEP is a major source of carbon for fatty acid synthesis (J. Schwender, personal communication, 2003). However, it is not yet possible to determine the extent to which the plastidial PEP pool is derived from PEP import by the PPT, implicating cytosolic glycolytic flux in carbon flux to fatty acid synthesis, or whether import of hexose- or triose-phosphate followed by plastidial glycolysis also contributes to this process. The use of ^{13}C-metabolite feeding studies has also enabled the question of the source of reducing power for fatty acid synthesis in the developing oilseed to be addressed.

Fatty acid synthesis requires 2 moles of reducing equivalents for each cyclical addition of a mole of malonyl-CoA onto a growing fatty acyl chain (Slabas and

Fawcett, 1992). In a true heterotrophic tissue this reducing power must come from oxidative metabolism inside the plastid. This could be provided by the plastidial OPPP (as above, in relation to reductant supply for N assimilation) or through metabolism of imported substrates such as malate into acetyl-CoA (Figure 3.5). In the latter case sequential enzyme steps inside the plastid involving NADP-dependent malic enzyme (EC 1.1.1.40) and then pyruvate dehydrogenase would yield precisely 2 moles of reducing equivalents and a single mole of acetyl-CoA (Smith *et al.*, 1992). The extent to which the plastidial OPPP is involved in supporting fatty acid synthesis is a matter of debate. *In vitro* evidence using isolated plastids from *B. napus* embryos has revealed that the activity of OPPP can be increased by supplying a substrate such as pyruvate, and that incorporation of carbon from pyruvate into fatty acids is increased when exogenous Glc6P is provided to supply the OPPP (Kang and Rawsthorne, 1996; Eastmond and Rawsthorne, 2000). Metabolic flux measurements with whole isolated *B. napus* embryos under conditions close to those *in vivo* support these *in vitro* data and reveal that 38% (confidence range of 22–45%) of the reducing power for fatty acid synthesis may be derived from the OPPP (Schwender *et al.*, 2003). These authors conclude that the remainder of the reducing power may come from photosynthetic electron transport, supporting earlier studies of carbon metabolism and oxygen exchange that concluded that light energy was, or could be, utilized for fatty acid synthesis in chlorophyllous seeds (Browse and Slack, 1985, Aach *et al.*, 1997; King *et al.*, 1998; Willms *et al.*, 2000).

The light dependence of fatty acid synthesis in chloroplasts has been reported previously (Sauer and Heise, 1983; Eastmond and Rawsthorne, 1998) and this demonstrates a clear link between fatty acid synthesis and the provision of reducing power from the photosynthetic electron transport chain. This relationship is not straightforward. Measured changes in the amounts of intermediates in fatty acid synthesis during light/dark transitions provide evidence that plastidial ACCase may represent a control point in fatty acid synthesis during such a transition (Post-Beittenmiller *et al.*, 1991, 1992). This earlier observation has been followed by reports that ACCase type II in chloroplasts is controlled by redox status and is more active under light conditions (Sasaki *et al.*, 1997; Kozaki and Sasaki, 1999; Kozaki *et al.*, 2000). The latter observations imply that a tight control may exist to limit flux of carbon into fatty acid synthesis in the chloroplasts when reductant supply is also limited.

Two key questions remain in understanding plastidial fatty acid synthesis. Firstly, which (and there may be several) carbon precursors are imported by plastids, including chloroplasts, and used for fatty acid synthesis *in vivo*. Studies with *Arabidopsis* have the potential to answer this question but previous *in vitro* work would suggest that there are species- and organ-specific factors that cannot be addressed solely using this model species. Secondly, how is the demand for reducing power to support fatty acid synthesis satisfied by the balance of oxidative metabolism and light. Clear distinction between truly photosynthetic and heterotrophic tissues makes light dependence a straightforward issue, but the division between reducing power provided by metabolism of imported substrates and from the OPPP is more difficult to resolve. Furthermore, green seed tissues represent a mixture of photosynthetic and heterotrophic metabolism, making dissection of the latter question even more complex.

3.6   Starch metabolism

Starch is an insoluble polymer of glucose residues produced by the majority of higher plant species, and is a major storage product of many of the seeds and storage organs produced agriculturally and used for human consumption. All starches are synthesized inside plastids, but their function therein will depend upon the particular type of plastid, and the plant tissue from which they are derived. Transient starches synthesized in chloroplasts during the day are degraded at night to provide carbon for non-photosynthetic metabolism. Starch produced in tuberous tissues also acts as a carbon store, and may need to be accessed as environmental conditions dictate, whilst storage starches in developing seeds are a long-term carbon store for the next generation. The starch granule is a complex structure with a hierarchical order composed of two distinct types of glucose polymer: amylose, comprising largely of unbranched α-$(1\rightarrow4)$-linked glucan chains; and amylopectin, a larger, highly branched glucan polymer typically constituting about 75% of the granule mass, produced by the formation of α-$(1\rightarrow6)$-linkages between adjoining straight glucan chains. The polymodal distribution of glucan chain lengths within amylopectin allows the chains to form double helices that can pack together in organized arrays, which are the basis of the semi-crystalline nature of much of the matrix of the starch granule (for reviews of starch structure, see Buléon et al., 1998; Thompson, 2000). Granule formation may be largely a function of both the semi-crystalline properties of amylopectin, e.g. length of the linear chains, and the frequency of α-$(1\rightarrow6)$-linkages (French, 1984; Myers et al., 2000). The crystalline structure of starch granules is highly conserved in plants at the molecular level (Jenkins et al., 1993), as well as at the microscopic level, where alternating regions of semi-crystalline and amorphous material, commonly known as growth rings, are present in all the higher plant starches studied to date (Hall and Sayre, 1973; Pilling and Smith, 2003). The synthesis of this architecturally complex polymer is achieved through the coordinated interactions of a suite of starch biosynthetic enzymes, including some which had traditionally been associated with starch degradation. The complement of these starch metabolic enzymes, which is a reflection of the starch biosynthetic pathway, is well conserved between plastids/tissues that make different types of starches, for example, transitory starch (made in chloroplasts) and storage starch (made in amyloplasts). With few exceptions, the various isoforms of the many starch metabolic enzymes can be found in both chloroplasts and amyloplasts. In addition, the amino acid sequences of the various enzymes involved in starch metabolism are highly conserved (Jespersen et al., 1993; Smith et al., 1997).

We will now consider the pathway of starch biosynthesis in more detail, examining the key reactions taking place in the various plastids that make starch, and then examine recent progress in understanding the process of starch degradation in plastids. Finally, we will consider recent research that begins to address the question of how the pathway as a whole may be coordinated and regulated in plastids.

3.6.1 The formation of ADPglucose by ADP glucose pyrophosphorylase

In all plant and green algal tissues capable of starch biosynthesis, ADP glucose pyrophosphorylase (AGPase, E.C. 2.7.7.27) is the enzyme responsible for the production of ADPglucose, the soluble precursor and substrate for starch synthases. AGPase catalyses the following reversible reaction:

$$Glc1P + ATP \leftrightarrow ADPglucose + PPi$$

In order for the biosynthesis of starch to occur, it is thought that *in vivo* the reaction catalysed by AGPase is displaced from equilibrium by the action of alkaline inorganic pyrophosphatase (APPase, E.C. 3.6.1.1), an enzyme displaying high catalytic activities in different plastid types (Gross and ap Rees, 1986; Weiner *et al.*, 1987). It is thought that APPase removes the PPi generated by the AGPase reaction during the synthesis of ADPglucose. The AGPase reaction is the first committed step in the biosynthesis of both transient starch in chloroplasts/chromoplasts and storage starch in amyloplasts. AGPase from higher plants is heterotetrameric, consisting of two large (AGP-L) subunits and two small (AGP-S) catalytic subunits encoded by at least two different genes (Preiss and Sivak, 1996). Until recently it was generally accepted that AGPase is exclusively located in plastids. However, this is not the case in cereal endosperm. Biochemical evidence indicates the presence of at least two distinct AGPase enzymes in the endosperms of maize (Denyer *et al.*, 1996b), barley (Thorbjørnsen *et al.*, 1996), rice (Sikka *et al.*, 2001) and wheat (Tetlow *et al.*, 2003c), which have been shown to correspond to plastidial and cytosolic isoforms of AGPase. In the developing endosperms of maize, barley and rice, the cytosolic isoform accounts for 85–95% of the total AGPase activity, although recent studies in wheat endosperm indicate that the relative proportions of cytosolic and plastidial AGPase in a tissue may be influenced by environmental factors (Tetlow *et al.*, 2003c; K. Denyer and I.J. Tetlow, unpublished observations, 2003). There is, however, no evidence that a cytosolic isoform of AGPase exists in the storage tissues of non-graminaceous species (Beckles *et al.*, 2001). If AGPase is expressed in the cytosol then the ratio of measurable ADPglucose to UDPglucose (which is present exclusively in the cytosol) would be expected to be higher in such tissues, compared with species where UDPglucose pyrophosphorylase (UGPase) and AGPase are in discrete subcellular compartments. Beckles *et al.* (2001) made a thorough analysis of a large number of monocotyledonous and dicotyledonous species and found that the ADPglucose/UDPglucose ratio was significantly higher in cereal endosperms compared to other species and tissues. Since the majority of the AGPase activity in cereal endosperms appears to be cytosolic, we must assume that most of the storage starch biosynthesis in these tissues occurs through import of ADPglucose. In cereal endosperms, therefore, the control of carbon partitioning into storage starch

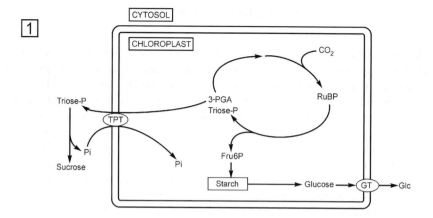

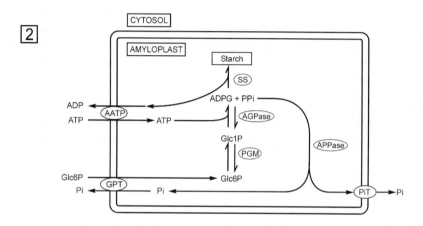

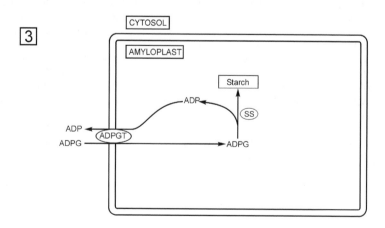

effectively lies outside the amyloplast, which, it has been argued, may be a more efficient way of channelling photosynthetic carbon into starch in the amyloplasts, rather than other competing metabolic pathways inside the plastid (Beckles *et al.*, 2001) and reducing the requirement for ATP generation and inorganic pyrophosphate (PPi) recycling in the plastid. The formation of ADPglucose in the cytoplasm of monocotyledonous storage tissues requires the coupling of PPi-consuming reactions, such as UGPase, with the AGPase reaction (Kleczkowski, 1994). In this respect, the pathways of storage starch biosynthesis in monocots and dicots are very different, and probably require different modes of regulation (see Figure 3.6).

AGPase is a major rate-controlling step in starch biosynthesis in different plant tissues, and under some conditions, its activity is the most significant factor determining the rate of starch accumulation. This has been shown convincingly in the leaves of *Arabidopsis* where a mutation in a gene encoding one of the subunits of the enzyme reduces starch accumulation to a quarter of normal values in leaves (Lin *et al.*, 1988) and experiments with transgenic potato plants, involving antisense-RNA inhibition (Müller-Rober *et al.*, 1992). AGPase activity is controlled by allosteric regulation, post-translational modification, and, on a longer timescale, transcriptional regulation. The chloroplast AGPase, which synthesizes ADPglucose from the glucose 1-phosphate (Glc1P) produced from photosynthesis, is tightly regulated by metabolite concentrations, being activated by micromolar amounts of 3-PGA and inhibited by Pi (Ghosh and Preiss, 1966). The ratio of these two allosteric effectors is believed to play a key role in the control of starch synthesis in photosynthetic tissues (Preiss, 1991). There is conflicting evidence concerning the relative responsiveness of AGPases from cereal endosperms to allosteric effectors. However, evidence from wheat (Gómez-Casati and Iglesias, 2002; Tetlow *et al.*,

---

Figure 3.6 Starch biosynthesis in plastids of higher plants. (**1**) Transient starch synthesis in chloroplasts. Excess carbon fixed during photosynthesis, which is not exported by the triose-phosphate/phosphate translocator (TPT) for sucrose production in the cytosol, is used for starch biosynthesis. The carbon for starch synthesis is derived from the Calvin cycle, whilst energy for the production of the precursor ADPglucose is probably derived from photophosphorylation. Starch degradation at night ultimately produces glucosyl monomers and maltose, which are exported by specific glucose transporters (GT) and a maltose transporter, respectively. (**2**) Starch biosynthesis in heterotrophic plastids requires the coordinate operation of at least three metabolite transporters. In most species carbon enters the plastid as Glc6P via the Glc6P/Pi antiporter (GPT) in exchange for some of the Pi generated from the formation of ADPglucose (ADPG) and is converted to Glc1P by a plastidial phosphoglucomutase (PGM). ATP is supplied to the plastid by the ATP/ADP transporter (AATP) in exchange for the ADP generated by the formation of starch by isoforms of SS. ADPglucose is synthesized from the Glc1P and ATP by AGPase, and this reaction is displaced from equilibrium by the removal of PPi by the enzyme APPase. To prevent Pi accumulation, which would inhibit AGPase and therefore starch synthesis, the extra Pi produced by APPase that is not exported by the GPT is probably removed by a Pi uniporter (PiT). (**3**) Storage starch synthesis in cereal endosperms. ADPglucose is generated in the cytosol by an isoform of AGPase and transported by a specific form of NST, the ADPglucose transporter (ADPGT) in exchange for ADP generated by the SSs. Starch biosynthesis in cereal endosperm amyloplasts may also take place via import of hexose-phosphates as shown in (**2**).

2003c) and barley (Kleczkowski *et al.*, 1993) suggests that measurable activity (the majority of which is cytosolic) is much less sensitive to 3-PGA activation and Pi inhibition than other forms of AGPase. Heterologously expressed AGP-L and AGP-S subunits of the barley cytosolic AGPase showed insensitivity to allosteric effectors (Doan *et al.*, 1999), as did the plastidial AGPase from wheat endosperm amyloplasts (Tetlow *et al.*, 2003c). However, the plastidial AGPase from the storage tissues of dicots appears to be as sensitive to the allosteric effectors as their counterparts in the chloroplast (Hylton and Smith, 1992; Ballicora *et al.*, 1995). The sensitivity of plastidial AGPase to allosteric regulation in other plastid types, such as leucoplasts and chromoplasts, is unknown. Thus, monocots may have evolved a cytosolic AGPase that is insensitive to allosteric activation by 3-PGA to suit the needs of endosperm metabolism, and distinct from the activator-sensitive AGPase required to coordinate starch synthesis with photosynthetic activity in the leaves. It is possible that the high yielding cereals selected for by plant breeding/agriculture over the centuries has also resulted in endosperm AGPases with reduced sensitivity to allosteric effectors. Interestingly, when plant tissues were transformed with an *Escherichia coli* gene encoding a version of AGPase insensitive to allosteric regulation (by fructose 1,6-bisphosphate), this led to a dramatic increase in starch biosynthesis in both cultured tobacco cells and potato tubers (Stark *et al.*, 1992).

Plants possess multiple genes encoding either the AGP-L or the AGP-S subunits, or both, and these are differentially expressed in different plant organs. This means that the AGPase subunit composition may vary in different parts of the same plant in tissues such as potato (La Cognata *et al.*, 1995), rice (Nakamura and Kawaguchi, 1992) and barley (Villand *et al.*, 1992a). The multiple genes encoding the AGP-L subunits show strong specificity in their expression, for example, being restricted to either leaf or root and endosperm in both barley and wheat (Olive *et al.*, 1989; Villand *et al.*, 1992a, b) or induced under specific conditions, such as increased sucrose levels in potato (Müller-Rober *et al.*, 1990). Multiple isoforms of the AGP-S subunit in bean show organ-specific expression patterns: one form is expressed only in leaves, the other in both leaves and cotyledons (Weber *et al.*, 1995). Different cDNAs encoding the AGP-S subunit in maize have distinct tissue expression patterns (Giroux and Hannah, 1994; Prioul *et al.*, 1994). The AGPase expressed early in wheat endosperm development may be a homotetramer of AGP-S (Ainsworth *et al.*, 1995). The differential expression of subunits in different tissues may produce AGPases with varying degrees of sensitivity to allosteric effectors, which are suited to the particular metabolic demands of a given plant tissue/organ.

Post-translational modification of AGPase involving thioredoxin has recently been proposed following the observation of partial inactivation of the recombinant potato enzyme by the formation of intramolecular disulphide bonds between the N-termini of the AGP-S subunits (Fu *et al.*, 1998). Redox control of AGPase through sucrose supply in potato tubers has recently been proposed whereby reductive activation causes channelling of carbon to starch and away from respiratory metabolism (Tiessen *et al.*, 2002). The proposed regulatory site is a cysteine at position 82 (Cys^{82}) on the AGP-S; it is thought that under oxidizing conditions the Cys^{82} residues from

AGP-Ss form intramolecular disulphide bonds, resulting in an inactive dimer. The Cys[82] is highly conserved amongst other forms of AGP-S, with the notable exception of the cytosolic isoform of AGP-S from monocots. Recent work has demonstrated that this phenomenon is relatively widespread, and includes photosynthetic as well as non-photosynthetic tissues from a number of species (Hendriks *et al.*, 2003).

3.6.2 Elongation of the glucan chain by starch synthases

The starch synthases (SS, E.C. 2.4.1.21) catalyse the transfer of the glucosyl moiety of the soluble precursor ADPglucose to the reducing end of a pre-existing α-(1→4)-linked glucan primer to synthesize the insoluble glucan polymers amylose and amylopectin. Plants possess multiple isoforms of SSs, containing up to five isoforms that are categorized according to conserved sequence relationships. The isoforms within each of the major classes of SS genes are highly conserved, from the green algae through the dicots and monocots (see Ball and Morell, 2003). The major classes of SS genes can be broadly split into two groups: the first group primarily involved in amylose synthesis, and the second group confined to amylopectin biosynthesis.

3.6.3 Amylose biosynthesis

The first group of SS genes contains the granule-bound starch synthases (GBSS), and includes GBSSI and GBSSII. GBSSI is encoded by the *Waxy* locus in cereals, functioning specifically to elongate amylose (De Fekete *et al.*, 1960; Nelson and Rines, 1962) and found as an abundant ~60-kDa polypeptide, essentially completely within the granule matrix (one of the granule-associated proteins). The *Waxy* mutants lack amylose and have starches comprised solely of amylopectin. Additional evidence that GBSSI synthesizes amylose within the granule matrix *in vivo* came from transgenic potatoes in which GBSSI was specifically reduced by expression of antisense RNA, leading to a dramatic reduction in the amylose content of the tubers (Visser *et al.*, 1991; Kuipers *et al.*, 1994; Tatge *et al.*, 1999). In addition to its role in amylose biosynthesis, GBSSI was also found to be responsible for extension of long glucans within the amylopectin fraction in both *in vitro* and *in vivo* experiments (Delrue *et al.*, 1992; Maddelein *et al.*, 1994; Van de Wal *et al.*, 1998). Expression of GBSSI appears to be mostly confined to storage tissues, and a second form of GBSS (GBSSII), encoded by a separate gene, is thought to be responsible for amylose synthesis in leaves and other non-storage tissues that accumulate transient starch (Nakamura *et al.*, 1998; Fujita and Taira, 1998; Vrinten and Nakamura, 2000). An interesting aspect to the control of polymer (amylose) elongation has been observed in the leaves of sweet potato (*Pomoea batatas*) where GBSSI transcript abundance and protein levels were shown to be under circadian clock control, as well as being modulated by sucrose levels (Wang *et al.*, 2001).

Kinetic analyses of GBSS and other SSs revealed that the K_m values for the *in vitro* substrates was up to 10-fold higher for GBSS than for the soluble SSs.

On the basis of these data, it was proposed that plastidial ADPglucose concentrations could therefore have a large impact on the amylose/amylopectin ratios of starches (Clarke *et al.*, 1999). One of the unique properties of GBSSI is its requirement for malto-oligosaccharides in order to synthesize amylose, which comes from *in vitro* experiments with isolated starch granules from pea embryos (Denyer *et al.*, 1996a). It is thought that malto-oligosaccharides must be able to diffuse into the granule matrix and GBSSI exclusively synthesizes amylose by elongating the malto-oligosaccharide primers (for a recent review of amylose synthesis, see Denyer *et al.*, 2001).

3.6.4   Amylopectin biosynthesis

The second group of SS genes contains the remaining SSs (designated SSI, SSII, SSIII and SSIV) that are exclusively involved in amylopectin synthesis, and whose distribution within the plastid between the stroma and starch granules varies between species, tissue and developmental stage. The individual SS isoforms from this group probably play unique roles in amylopectin biosynthesis. The study of SS mutants in a number of systems has been helpful in the assignment of *in vivo* functions/roles for the soluble and granule-associated SS isoforms in amylopectin synthesis. Although valuable information about the roles of the SS isoforms *in vivo* is being derived from mutants lacking specific isoforms, and analysis of plants appears to show that each isoform performs a specific role in amylopectin synthesis, such data should be treated with caution as in some cases there are pleiotropic effects on mutations on other enzymes of starch synthesis (see later). However, not all the SSs have characterized mutants, and for this reason, the role of SSI, for example, in starch biosynthesis remains unclear, as no mutations in this gene have been reported to date. All three of the major amylopectin-synthesizing SS isoforms (SSI, SSII and SSIII), divided on the basis of their amino acid sequences, have been identified in potato tuber (Edwards *et al.*, 1995; Abel *et al.*, 1996; Marshall *et al.*, 1996; Kossmann *et al.*, 1999) and maize endosperm (Gao *et al.*, 1998; Harn *et al.*, 1998; Knight *et al.*, 1998), but appear to be widely distributed in higher plants in both leaf and storage tissues. The proposed function of the individual SS proteins in amylopectin biosynthesis has also been determined, in part, by *in vitro* experiments with purified native or recombinant proteins. A proportion of each of the three soluble SS isoforms (SSI, SSII and SSIII) is partitioned between the starch granule and the stroma. The mechanism by which specific proteins become granule-associated still remains unclear.

SSI is primarily responsible for the synthesis of the shortest glucan chains, i.e. those with a degree of polymerization (DP) of 10 glucosyl units or less (Commuri and Keeling, 2001), and further extension of longer chains is achieved by the activities of SSII and SSIII isoforms, each of which act on progressively longer glucan chains (see Figure 3.7). Two classes of SSII genes are found in monocots: SSIIa and SSIIb; the role of the latter in starch biosynthesis is unknown as no mutants have been identified. SSIIa predominates in cereal endosperms, whilst SSIIb is mostly confined to photosynthetic tissues. Loss of SSIIa (in monocots) and SSII (in dicots)

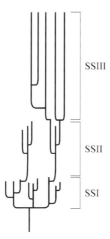

Figure 3.7 Diagrammatic representation of α-glucan chain elongation by isoforms of SS in plastids of higher plants and green algae. All starch synthesizing plastids contain the SS isoforms, SSI, SSII, SSIII and GBSS I/II (not shown), which respectively utilize progressively longer α-glucan chains as substrates for elongation in the synthesis of amylopectin.

results in reduced starch content and amylopectin chain-length distribution, altered granule morphology and reduced crystallinity, suggesting that the SSII forms have similar roles in starch biosynthesis across different species boundaries. In monocots SSIIa plays a specific role in the synthesis of the intermediate-size glucan chains of DP 12–24 by elongating short chains of DP ≤10, and its loss/down-regulation has a dramatic impact on both the amount and the composition of starch, despite the fact that SSIIa is a minor contributor to the total SS activities in cereal endosperms, as opposed to SSI and SSIII. Analysis of the effects of SSII in potato tubers suggests that it plays a similar role in storage starch biosynthesis in dicots (Edwards *et al.*, 1999). The extent of the participation of the different SS proteins in starch synthesis may vary from one species to another, and between different parts of the plant. For example, suppression of SSIII activity in potato has a major impact on the synthesis of amylopectin, resulting in modified chain-length distribution and decreased starch synthesis (Edwards *et al.*, 1999), whilst the maize (*dull1*) mutant, lacking SSIII activity, has a subtle phenotype that can only be observed in a background also containing the waxy mutation (Gao *et al.*, 1998). Sequences for the SSIV (also designated SSV) appear in a wide range of higher plants in EST databases, although to date, no role has been assigned for this class of SS in the process of starch biosynthesis.

3.6.5 *Branching of the glucan chain by starch branching enzymes*

Starch branching enzymes (SBEs, E.C. 2.4.1.18) generate α-(1→6)-linkages by cleaving internal α-(1→4) bonds and transferring the released reducing ends to C6 hydroxyls to form the branched structure of the amylopectin molecule. As with the elongation of glucan chains by SSs, SBE activity is also a function of multiple isoforms, some of which are tissue- and/or developmental-specific in their expression patterns. Analysis of the primary amino acid sequences of higher plant SBEs reveals two major classes: SBEI (also known as SBE B) and SBEII (also known as SBE A). The two classes of SBE differ in terms of the length of the glucan chain transferred *in vitro*: SBEII proteins transfer shorter chains than their SBEI counterparts (Guan and Preiss, 1993; Takeda *et al.*, 1993). In monocots the SBEII class is made up of two closely related but discrete gene products, SBEIIa and SBEIIb (Rahman *et al.*, 2001). The additional forms of SBEII are the result of important gene duplication events that occurred after the divergence of the monocots and dicots, giving rise to additional members of gene families with specialized roles and localization; within the starch pathway this also includes the SSIIa and SSIIb isoforms involved in amylopectin biosynthesis (see above). SBEIIb plays an indispensable role in amylopectin biosynthesis by forming short glucan chains with DP ≤ 13; manipulation of SBEIIb activity in rice endosperm led to dramatic alterations in amylopectin structure and altered physicochemical properties of starches (Nakamura *et al.*, 2003).

The different SBE isoforms show varied temporal and spatial patterns of expression and partitioning within plastids. SBEI and SBEIIa are expressed in both leaves and storage/endosperm tissues, but at different levels depending upon the species. For example, in pea embryo, both forms (SBEI and SBEII, also called SBE B and SBE A, respectively) are present at comparable levels in the stroma, whereas in potato tuber, SBEI is predominant and SBEII expressed at low levels (Jobling *et al.*, 1999). In monocots, SBEIIb is expressed only in the endosperm (throughout the development of this tissue) and reproductive tissues, whereas SBEI is strongly expressed during later stages of both maize and wheat endosperm development (Gao *et al.*, 1996; Morell *et al.*, 1997), and SBEIIa is more highly expressed in leaves than endosperm.

To date, only mutations in SBEII isoforms give clear phenotypes, and in monocots this is confined to SBEIIb mutants. Down-regulation or elimination of SBEI activity in both monocots and dicots appears to have minimal effects on starch synthesis and composition in photosynthetic and non-photosynthetic tissues (Flipse *et al.*, 1996; Blauth *et al.*, 2002). Interestingly, *Arabidopsis*, which primarily synthesizes transient leaf starch, does not appear to possess a SBEI gene, despite the fact that the gene is clearly expressed in other dicots (Khoshnoodi *et al.*, 1998). Recent analysis of a maize SBEIIa mutant showed a clear phenotype in the leaf starch, but showed no alterations in the storage starch of the endosperm (Blauth *et al.*, 2001). This observation suggests a primary role for SBEIIa in leaf (transient) starch synthesis, and either no critical role for SBEIIa in amylopectin biosynthesis in the endosperm, or else a role that can easily be compensated for by other SBEs in its absence. There is also evidence for proteins regulating the co-expression of starch metabolic enzymes. The *floury-2* (*flo2*) mutation in rice showed reductions in expression and protein content of SBEI (to 10% of wild-type levels); in addition, significant reductions in the expression of SBEIIb, SSI, GBSS, pullulanase-type debranching enzyme and starch phosphorylase (Pho) were also observed, suggesting that the wild-type *Flo2* gene encodes a regulatory protein responsible for simultaneously modulating the expression of a number of starch biosynthetic genes (Satoh *et al.*, 2003).

SBEII isoforms are partitioned between the plastid stroma and the starch granules. In maize endosperm, the granule-associated forms of SBEII comprise up to 45% of total measurable SBEII activity (Mu-Forster *et al.*, 1996). SBEI has not been detected within starch granules, and is presumably confined to the stroma. However, a form of SBEI, termed SBEIC, is located exclusively in the starch granules; this large 152-kDa protein contains two SBEI-like domains, and may be a result of a *trans*-splicing event between a *SBEI*-like mRNA and a *SBEI* transcript (Båga *et al.*, 2000). SBEIC may perform the same role in the starch granule as SBEI does in the stroma; however, this granule-associated form of SBEI appears to be confined only to monocots. As with the granule-associated SSs (above), the factors/mechanisms involved in partitioning the SBEII and SBEIC proteins to the starch granules remain undetermined.

In vitro analysis of heterologously expressed maize SBEs by Seo *et al.* (2002) has shed further light on the roles of the different SBE isoforms in the construction of the starch granule, which would not have been possible by analysing mutations in single SBE genes. Expression of the three maize SBE genes in a yeast strain lacking the endogenous yeast glucan branching enzyme showed that SBEI was unable to act in the absence of SBEIIa or SBEIIb, and that SBEII may act before SBEI on precursor polymers. Both of the maize SBEII isoforms heterologously expressed by Seo *et al.* (2002) could complement the lack of yeast glucan branching enzyme, and produce glucans with unique chain distributions and branch frequencies. These data suggest that SBEI does not play a central role in this *in vitro* system, leaving the role of SBEI in the starch biosynthetic pathway still an open question.

3.6.6 *The role of debranching enzymes in polymer synthesis*

Analysis of low-starch mutants that accumulate a water-soluble polysaccharide termed *phytoglycogen* has been described in a wide range of higher plants, including *Arabidopsis* and maize, as well as *Chlamydomonas* (James *et al.*, 1995; Mouille *et al.*, 1996; Zeeman *et al.*, 1998b), and indicates that starch synthesis involves debranching enzymes (DBEs, E.C. 3.2.1.41 and E.C. 3.2.1.68) in addition to SSs and SBEs. Two groups of DBEs exist in plants: isoamylase-type and pullulanase-type (also known as limit-dextrinases), which efficiently hydrolyze (debranch) α-(1→6)-linkages in amylopectin and pullulan (a fungal polymer of maltotriose residues), respectively. The *Arabidopsis* genome contains three isoamylase-type DBEs and one pullulanase-type DBE. Both groups of DBEs in higher plants share a common N-terminal domain whose function is yet to be elucidated. The decrease/loss of isoamylase-type DBE activity is thought to be responsible for the accumulation of phytoglycogen rather than starch in mutant plants and algae, and it is thought that, in rice endosperm, residual pullulanase-type DBE activity modulates these phenotypic effects (Nakamura *et al.*, 1997). In maize endosperm the pullulanase-type DBE activity is thought to have a bifunctional role, assisting in both starch synthesis and degradation (Dinges *et al.*, 2003), and has been shown to be controlled by redox regulation in spinach leaf chloroplasts and barley endosperm amyloplasts (Beatty *et al.*, 1999; Schindler *et al.*, 2001). In wheat the expression of a cDNA for the isoform of an isoamylase-type DBE (*iso1*) is maximal in developing endosperm and undetectable in mature grains, which suggests a biosynthetic role for isoamylase in this tissue.

The precise roles for the isoamylase-type and pullulanase-type DBEs in starch biosynthesis are not yet known. Two models have been proposed that could define a role for the DBEs in starch synthesis and phytoglycogen accumulation. The glucan-trimming (pre-amylopectin trimming) model proposes that glucan trimming is required for amylopectin aggregation into an insoluble granular structure. DBE activity would be responsible for the removal of inappropriately positioned branches (pre-amylopectin) generated at the surface of the growing starch granules, which would otherwise prevent crystallization. As such, the debranched structure would favour the formation of parallel double helices, leading to polysaccharide aggregation. Recent observations which show that the surface of the immature granules contains numerous short chains are consistent with this model (Nielsen *et al.*, 2002). An alternative to the glucan-trimming model proposes that the DBEs function in a 'clearing' role, removing soluble glucan from the stroma, thereby removing a pool of substrates for the amylopectin synthesizing enzymes (SSs and SBEs). This model could also explain the accumulation in phytoglycogen at the expense of amylopectin observed in DBE mutants (Zeeman *et al.*, 1998b).

3.6.7   Starch degradation in plastids

Starch degradation is part of the overall process of starch turnover that occurs in all starch containing plastids to varying degrees. Much of the research on starch degradation has focused on understanding the diurnal fluctuations of starch in leaves, whereby the starch synthesized in leaves during the day is degraded at night, and the carbon exported from the chloroplasts used to meet various metabolic demands of the plant. In common with the starch biosynthetic pathway (above), most, if not all, of the enzymes involved in the pathway of starch degradation are known, but the details of its operation and regulation are poorly understood. Little (Fondy *et al.*, 1989) or no (Zeeman *et al.*, 2002) starch turnover has been reported in leaves during the day, suggesting that the process of starch degradation is switched on, or strongly up-regulated, during the night, and switched off/down-regulated in the light by as yet unknown/ undetermined mechanisms.

The process of starch degradation requires an initial hydrolytic attack on the intact starch granule, followed by debranching (hydrolysis) of α-$(1\rightarrow6)$-linkages to produce linear glucan chains, and finally, the degradation of the linear chains to glucosyl monomers. There is a range of plastidial enzymes with starch degrading capabilities that may participate in the process of starch degradation and turnover. α-Amylase (E.C. 3.2.1.1) and other endoamylases hydrolytically cleave α-$(1\rightarrow4)$-glucosyl bonds, resulting in the production of a mixture of linear and branched malto-oligosaccharides and, ultimately, glucose, maltose, maltotriose and a range of branched α-limit dextrins. In addition, β-amylase (E.C. 3.2.1.2) catalyses the hydrolysis and removal of successive maltose units from the non-reducing end of the α-glucan chain. Alternatively, α-(1→4)-glucosyl bonds may be cleaved phosphorolytically by starch phosphorylase (E.C. 2.4.1.1) to produce Glc1P from successive glucosyl residues at the non-reducing end of a α-glucan chain. It is worthy of note that the majority of endoamylase and starch phosphorylase activity in leaves is located in the cytosol and vacuoles (Stitt and Steup, 1985; Ziegler and Beck, 1986). The function of these cytosolic/vacuolar starch degrading enzymes is unknown. Only the plastidial forms of these putative starch degrading enzymes (possessing a transit peptide) will be considered to be potentially part of the plastidial starch degradation pathway.

The initial hydrolytic attack on the intact, semi-crystalline starch granule is thought to be via endoamylases (Steup *et al.*, 1983; Kakefuda and Preiss, 1997). This idea was tested recently by Smith *et al.* (2003a) using *Arabidopsis*, whose genome contains a single α-amylase, which is predicted to be plastidial owing to the presence of a putative transit peptide. Analysis of a knockout mutant for the putative plastidial α-amylase showed that mutant plants had normal rates of starch degradation, indicating that the initial hydrolysis must be catalysed by another endoamylase or as yet unidentified protein(s). Mutations at the *sex1* locus in *Arabidopsis* result in leaf starch accumulation and an inability to degrade starch at night. The mutation has been mapped to a gene encoding a homologue of the potato R1 protein (Yu *et al.*, 2001), a starch–water dikinase that phosphorylates glucose residues on amylopectin (Ritte *et al.*, 2002). Interestingly, there are few or no phosphate groups in the amylopectin from the *sex1* mutants. It was recently hypothesized by Smith *et al.* (2003a) that either the R1 protein or the presence of the phosphate groups on amylopectin is necessary for the action of an enzyme(s) that catalyses the initial attack on the starch granule.

It is not yet clear which of the various DBEs present in plastids is responsible for the hydrolysis of α-(1→6)-linkages during starch degradation. A mutant of *Arabidopsis*, lacking one form of DBE (the *dbe1* mutant), shows complete degradation of starch and phytoglycogen during the night (Zeeman *et al.*, 1998b), indicating that this DBE alone is not necessary for the debranching step during starch degradation.

Both β-amylase and starch phosphorylase activities are present in plastids (Zeeman *et al.*, 1998a; Lao *et al.*, 1999), and each could be responsible for the degradation of linear glucan chains to glucosyl monomers *in vivo*. Analysis of the *Arabidopsis* genome sequence predicts four plastidial β-amylases and one plastidial starch phosphorylase. A plastidial starch phosphorylase knockout mutant in *Arabidopsis* showed normal rates of leaf starch synthesis and degradation during the diurnal cycle (Smith *et al.*, 2003a), suggesting that starch phosphorylase is not necessary for starch degradation, and that one or more of the isoforms of β-amylase probably plays an important role in the process. Evidence in support of this idea comes from studies of a down-regulated form of plastidial β-amylase in *Arabidopsis*, which showed reduced rates of starch degradation at night (Scheidig *et al.*,

2002). Maltose produced by the activity of the β-amylases may be converted to glucose by plastidial maltases, although to date, none have been identified. The chloroplast envelope is permeable to maltose (Rost *et al.*, 1996), and recent identification and analysis of a maltose transporter (MEX1) indicates that MEX1 is the major route by which the products of starch degradation are exported from the chloroplast at night in higher plants (Niittyä *et al.*, 2004). Plastids also contain a glucose transporter at the inner envelope membrane (Weber *et al.*, 2000), and chloroplasts have been shown to export glucose during starch degradation (Schäfer *et al.*, 1977; Schleucher *et al.*, 1998), suggesting this is one other route of carbon export following starch degradation. Maltotriose released from the glucan chain by the action of β-amylases, and which is unavailable for further degradation by these enzymes, could be utilized by disproportionating enzyme (D-enzyme, E.C. 2.4.1.25) that transfers two of the glucosyl units from maltotriose onto a longer glucan chain, making them available to the β-amylases, and the resulting glucosyl monomer available for export from the plastid. Knockout mutants of D-enzyme show reduced rates of nocturnal starch degradation (Critchley *et al.*, 2001), indicating that this reaction plays a part in the pathway of starch degradation.

3.6.8   Post-translational regulation of starch metabolic pathways

The above sections describe the key components of the likely pathway of starch synthesis and degradation in the plastids of higher plants, and where known, how individual proteins/reactions in each pathway may be regulated *in vivo*. However, description and appreciation of the main reactions in each pathway does not explain how the starch granule is synthesized or degraded, nor account for the varied patterns of starch turnover in different plant tissues with essentially the same complement of starch metabolic enzymes. The mechanisms underlying the distinct structure of amylopectin are still unknown, and attempts to synthesize the molecule *in vitro* or to reconstitute the system have not been successful. It is the coordination of these expressed proteins inside different types of plastids that allows the controlled synthesis and degradation of this architecturally complex polymer, and our rudimentary knowledge of these processes is discussed below.

A potentially important mechanism for coordinating the multiple actions of different proteins involved in starch polymer synthesis (and degradation) has recently been proposed involving protein–protein interactions. Initially, this idea arose from the analysis of the pleiotropic effects of various mutations in genes of starch biosynthesis, and more directly by *in vitro* experiments with plastids. For example, the *amylose extender* (*ae*) mutant of maize lacks SBEIIb, but also shows loss of activity of SBEI and altered properties of an isoamylase-type DBE (Colleoni *et al.*, 2003). Mutations in maize that affect both a pullulanase-type DBE (*zpu1-204*) and an isoamylase-type DBE (*su1-st*) both cause a loss in SBEIIa activity, although the amount of SBEIIa protein is apparently unchanged (James *et al.*, 1995; Dinges *et al.*, 2001; Dinges *et al.*, 2003). The *ae* mutation in rice endosperm (lacking SBEIIb) caused a dramatic reduction in the activity of soluble starch synthase I (SSI; Nishi *et al.*, 2001). In addition to the pleiotropic loss of SBEIIa activity in the *zpu1-204* mutation, a reduction in β-amylase activity and a shift in β-amylase migration on native gels has also been observed (Colleoni *et al.*, 2003). In both the *zpu1-204* and *su1-st* mutants, the inactive SBEIIa polypeptide accumulated to seemingly normal levels, suggesting the possibility of post-translational modifications and altered interactions with the DBEs. A recent study in barley endosperm suggests that starch granule associated proteins form protein complexes, as loss of SSIIa activity was shown to abolish binding of SSI, SBEIIa and SBEIIb within the granule matrix, with no apparent loss in the affinity of these enzymes for amylopectin/starch (Morell *et al.*, 2003). It has been speculated that the coordination of debranching, branching and SS activities required for starch synthesis might be accomplished by physical association of the enzymes in a complex(s) within the amyloplast (Ball and Morell, 2003). Thus, the various mutations in different components of a putative protein complex would disrupt or alter the complex and cause a loss or reduction in biosynthetic capacity, and at least partially explain some of the pleiotropic effects associated with a number of well-characterized mutants in cereal endosperms. Recent experiments with isolated amyloplasts from wheat endosperm have shown that some of the key enzymes of the starch biosynthetic pathway form protein complexes that are dependent upon their phosphorylation status (Tetlow *et al.*, 2004). Phosphorylation of SBEI, SBEIIb and starch phosphorylase by plastidial protein kinase(s) resulted in the formation of a protein complex between these enzymes, which was lost following dephosphorylation. The role of protein complex formation between these starch biosynthetic enzymes in the process of starch synthesis is not fully understood, but it is thought that protein complexes of this kind improve the efficiency of polymer construction as the product of one reaction becomes a substrate for another within the complex. Such schemes have recently been hypothesized following analysis of heterologously expressed starch biosynthetic proteins in yeast cells (Seo *et al.*, 2002). The findings of Tetlow *et al.* (2004) may point to a wider role for protein phosphorylation and protein complex formation in the regulation of starch synthesis and degradation in plastids.

Direct regulation of enzyme activity and/or regulation by protein complex formation in many cases involve the phosphorylation of a target protein, followed by the formation of a complex with 14-3-3 proteins, and this appears to be a general mechanism for regulating enzymes and pathways in eukaryotic systems (see Chung *et al.*, 1999, for a general review of plant 14-3-3 proteins). The 14-3-3 proteins are a structurally highly conserved group of proteins, and isoforms have been identified in chloroplasts from pea leaves (Sehnke *et al.*, 2000), although there is currently no evidence for the presence of 14-3-3 proteins in other plastid types. The involvement of 14-3-3 proteins in the regulation of transient starch metabolism has recently been proposed by Sehnke *et al.* (2001) who showed that a form of 14-3-3 protein (from the ε subgroup) was present in *Arabidopsis* leaf starch. Suppression of granule-associated 14-3-3 proteins resulted in leaf starch accumulation (Sehnke *et al.*, 2001). These workers speculated that a possible target for the 14-3-3 proteins in *Arabidopsis* was SSIII, implying phosphorylation of SSIII at a conserved putative 14-3-3 binding motif, causing inactivation of the enzyme. Few of the major starch metabolizing enzymes have well-conserved 14-3-3 binding motifs (one notable exception is β-amylase), which suggests that if 14-3-3 proteins are involved in forming protein–protein interactions between starch metabolizing enzymes in plastids, then binding to the target phosphoproteins must be at as yet, uncharacterized binding sites.

3.7 Glycolysis

Glycolysis and the OPPP are two interrelated metabolic pathways by which carbohydrate is converted to pyruvate and malate (major respiratory substrates of the mitochondrion), and both pathways share several common intermediates (Glc6P, Fru6P, and G-3-P). Glycolysis, like the OPPP described below, was regarded as being exclusively localized in the cytosol. It is now realized that many, if not all, reactions are duplicated in plastids, where distinct isoforms are found. Most of the major plastid types analysed possess full glycolytic sequences, including, for example, chloroplasts (Liedvogel and Bäuerle, 1986), amyloplasts (Entwistle and ap Rees, 1988), fruit and petal chromoplasts (Thom *et al.*, 1998; Tetlow *et al.*, 2003a) and cauliflower leucoplasts (Journet and Douce, 1985). However, some chloroplasts and root leucoplasts lack one or several enzymes of the lower half of glycolysis, for example, enolase (E.C. 4.2.1.11) and phosphoglyceromutase (E.C. 5.4.2.1) (Stitt and ap Rees, 1979; Trimming and Emes, 1993). It is likely that the lack of enzyme activities in some tissues represents tissue or developmental stage-specific differences in metabolism. Details of the organization and regulation of plastidial glycolysis are dealt with in a review by Plaxton (1996).

Recent research has indicated that the subcellular location of plant glycolysis extends beyond the cytosolic and plastidic compartments, and that the glycolytic pathway may also be localized to other subcellular regions, in particular, those with a high ATP demand. In *Arabidopsis* a proportion of the entire glycolytic pathway is associated with the mitochondria via attachment to the cytosolic face of the outer mitochondrial membrane (Giegé *et al.*, 2003). It was postulated that this arrangement allows the direct provision of cytosolic pyruvate to the mitochondrion at the site of consumption as a respiratory substrate. It has not yet been determined whether the whole, or any substantial part, of the glycolytic pathway is associated with membranes in plastids, as opposed to being in the soluble phase. However, individual glycolytic enzymes have been shown to associate with plastids. In spinach chloroplasts, hexokinase I was shown to adhere to the outer envelope membrane, where it was proposed that the glucose exiting the plastid during the dark period (arising from starch degradation) could be more efficiently phosphorylated (Wiese *et al.*, 1999). Since chloroplasts cannot readily import the resulting Glc6P (see Kammerer *et al.*, 1998), it could feed into glycolysis or sucrose biosynthesis.

3.8   The oxidative pentose–phosphate pathway

The OPPP is a major source of reducing power (NADPH) and metabolic intermediates for a wide range of biosynthetic processes occurring in higher plants, including fatty acid synthesis and N assimilation (see above sections). Heterotrophic plastids are not able to synthesize NADPH by photosynthesis, and so the OPPP is an important source of NADPH in these organelles. The OPPP consists of two sections: an oxidative and a non-oxidative section. The oxidative section produces ribulose 5-phosphate (a substrate for nucleotide biosynthesis) and reducing power (NADPH); the latter can be used for the synthesis of fatty acids or amino acids. The reversible, non-oxidative section of the pathway (catalysed by transketolase, transaldolase (E.C. 2.2.1.2), pentose-phosphate isomerase (E.C. 5.3.1.6) and pentose-phosphate epimerase (E.C. 5.1.3.1) is also the source of carbon skeletons for the synthesis of nucleotides, aromatic amino acids and phenylpropanoids and their derivatives (for a review of the shikimic acid pathway and its link with the OPPP, see Herrmann and Weaver, 1999). The ribulose 5-phosphate produced from the oxidative section may be converted, in the non-oxidative section, to ribose 5-phosphate by ribose 5-phosphate isomerase (E.C. 5.3.1.6) and to xylulose 5-phosphate by ribulose 5-phosphate epimerase. The interconversions of sugar-phosphates catalysed by the non-oxidative reactions of the OPPP eventually produce Fru6P and triose-phosphate, which could also enter glycolysis. Further reactions of the non-oxidative section of the pathway lead to interconversion of C3 through to C7 sugar phosphates, which are also intermediates of the RPPP (see above section). In contrast to its location in other eukaryotes, the OPPP is not confined to the cytosol (ap Rees, 1985). Some, if not all, the enzymes of the OPPP are found in both cytosol and plastids; the precise distribution of activities varies to different degrees, depending on species and stage of development.

Studies from a wide range of photosynthetic and non-photosynthetic tissues have shown that the first two reactions of the OPPP (G6PDH and 6PGDH), comprising the oxidative section, have a dual location in plastids and the cytosol, generating NADPH for biosynthetic and assimilatory pathways. Evidence in support of the subcellular duplication of the oxidative section of the pathway has arisen from the identification of separate genes encoding discrete cytosolic and plastidic isozymes of G6PDH (Graeve et al., 1994; von Schaewen et al., 1995; Wendt et al., 2000; Knight et al., 2001) and 6PGDH (Redinbaugh and Campbell, 1998; Krepinsky et al., 2001). However, the subcellular compartmentation of the non-oxidative section of the OPPP is less clear. Results of careful cell fractionation studies in spinach leaves suggested that the non-oxidative reactions are restricted to plastids (Schnarrenberger et al., 1995), and is consistent with earlier studies in leaves and roots of a variety of plant tissues. In cases where the non-oxidative reactions are confined to the plastids, those pathways that rely on intermediates from the OPPP (e.g. ribose 5-phosphate for nucleotide biosynthesis) may be dependent on export from the plastid. A subsequent study in the leaves and roots of tobacco found a significant proportion of the non-oxidative enzymes outside the plastids (Debnam and Emes, 1999), and was consistent with similar cell fractionation studies in castor bean endosperm (Nishimura and Beevers, 1979), cauliflower buds (Journet and Douce, 1985) and soybean root nodules (Hong and Copeland, 1990), which all indicated that this part of the pathway was largely cytosolic. On the other hand, cell fractionation studies with tobacco mesophyll chloroplasts indicated that the majority of transketolase was inside the chloroplasts (Henkes et al., 2001). The likely explanation for these contradictory observations in different tissues is that the compartmentation and distribution of the enzymes of the OPPP are not fixed, and are influenced by species, ontogeny and environmental factors. Analysis of the *Arabidopsis* genome by Eicks *et al.* (2002) indicated three genes encoding ribose 5-phosphate isomerase, one of which was plastidial (i.e. possessed a putative transit peptide), and four genes for ribulose 5-phosphate epimerase, only one of which was a putative plastidial isoenzyme. The isoforms of ribulose 5-phosphate epimerase and ribose 5-phosphate isomerase that lacked any obvious N-terminal plastid-targeting sequences were presumably cytosolic isoforms. On the other hand, two isoforms each of transketolase and transaldolase appeared to be plastidial enzymes, with no cytosolic forms detected. This work suggests that the cytosolic OPPP in *Arabidopsis* can only proceed to the stage of interconvertible pentose-phosphates. The assumed function of the *Arabidopsis* xylulose 5-phosphate/Pi translocator recently cloned by Eicks *et al.* (2002) is to provide the plastidial OPPP with cytosolic carbon in the form of xylulose 5-phosphate.

Interactions between the cytosolic and plastidial OPPPs is probably facilitated by a group of recently discovered transport proteins which are part of a family of phosphate-translocators located at the inner envelope membranes of plastids (Flügge, 1999; Eicks *et al.*, 2002). The major transporters within this group are capable of exchanging pentose-phosphates (see above example) and hexose-phosphates, as well as PEP between the OPPP of the cytosol and plastid, and are discussed in more detail in the section on plastid transport systems below.

An alternative approach to resolving the question of compartmentation of the OPPP is the analysis of complete genome sequences, and is discussed in more detail in a recent review of the OPPP by Kruger and von Schaewen (2003). The identification of transit peptides on proteins indicates a plastidial location, and this analysis may be further refined by analysis of expression patterns of putative cytosolic and plastidial proteins during development and under different environmental conditions. Such analysis in *Arabidopsis* suggests that both transketolase and transaldolase may be confined to plastids, although the identification of cDNAs encoding two cytosolic isozymes of transketolase in *Craterostigma plantagineum* certainly indicates that the organization of the pathway differs between species (Bernacchia *et al.*, 1995).

As discussed by Kruger and von Schaewen (2003), significant progress has been made by applying steady-state labelling techniques using [^{13}C]glucose, followed by detection of the metabolic products using NMR or MS. The benefit of such approaches is that they allow the resolution of intracellular fluxes *in situ*. Such studies may shed light on the compartmentation of the enzymes of the OPPP *in vivo*. For example, NMR studies examining the redistribution of [1-^{13}C]glucose in both maize root tips and tomato cell cultures suggest that the oxidative steps of the OPPP are active only in plastids, whereas transaldolase is active in both plastids and cytosol (Dieuaide-Noubhani *et al.*, 1995; Rontein *et al.*, 2002).

An additional level of complexity in the organization of the OPPP is realized from biochemical analyses, and more recently, the isolation of multiple copies of genes from cDNA libraries, which indicates that many enzymes of the OPPP are represented by multiple isoforms. For example, the *Arabidopsis* genome potentially contains six forms of G6PDH and three isozymes of 6PGDH (Kruger and von Schaewen, 2003), and two cytosolic forms of transketolase were identified in the resurrection plant *C. plantagineum* (Bernacchia *et al.*, 1995). It is thought that one aspect of the duplication of genes encoding enzymes of the OPPP is a requirement for the enzymes to function in different subcellular environments, i.e. the cytosol and plastid. However, multiple genes may be differentially expressed within a given cellular compartment, in different tissues, at different developmental stages, and in response to different environmental factors. Altered gene expression of the multiple isoforms of enzymes from the OPPP is often a response to an altered demand for NADPH or OPPP intermediates for various biosynthetic processes. Different isoforms of OPPP enzymes may also have altered regulatory properties or sensitivities to effectors/substrates. For example, the two forms of plastidial G6PDH in potato have markedly different sensitivities to $NADPH/NADP^+$. Plastidic G6PDH is inactivated by a reversible dithiol–disulfide interconversion of two conserved regulatory cysteine residues (Wenderoth *et al.*, 1997), and the two forms in potato have different sensitivities to this form of regulation. The P2 isozyme, which is expressed throughout the plant, is strikingly less sensitive to both forms of regulation than the P1 enzyme, which is not detectable in non-photosynthetic tissues (von Schaewen *et al.*, 1995; Wenderoth *et al.*, 1997; Wendt *et al.*, 2000; Knight *et al.*, 2001). However, it is more difficult to explain the differential expression of other isozymes of enzymes of the OPPP that have no obvious regulatory properties.

3.9   Plastid metabolite transport systems

All plastids possess two functionally and physically distinct membranes that act as a physical barrier between plastid metabolism and other cellular processes. Communication between events occurring within the organelle and the surrounding plant cell is achieved via a diverse array of proteins residing on and within the two envelope membranes, whose protein complement is unique and reflects the metabolic functions of the plastid type. Many plastids contain duplicate primary metabolic pathways such as glycolysis and the OPPP that are also present in the cytosol (Entwistle and ap Rees, 1988; Thom *et al.*, 1998; Tetlow *et al.*, 2003a). In order to surmount the deficiencies in activities of one or more enzymes in primary metabolic pathways, the plastid must interact with the cytosolic counterparts of the pathways and utilize a number of metabolite transporters. For example, import of glycolytic intermediates by plastids lacking a complete/functional glycolytic pathway is necessary for the synthesis of the acetyl-CoA required for fatty acid biosynthesis (Liedvogel and Bäuerle, 1986) and the DOXP (synthesized from pyruvate and G-3-P) required for plastidial isoprenoid biosynthesis (Lichtenthaler *et al.*, 1997). The outer envelope membrane is freely permeable to low-molecular-weight compounds up to a molecular weight of about 10,000 (Flügge and Benz, 1984) because of the presence of porins/aquaporins. However, it is the inner membrane which is the site at which specific transport of metabolites occurs, and the major metabolite transporters are described in detail below.

3.9.1 The triose-phosphate/Pi translocator

The triose-phosphate/Pi translocator (TPT) of chloroplasts has an essential role during photosynthesis by mediating the export of fixed carbon in the form of triose-phosphates and 3-PGA from the chloroplasts into the cytosol. The exported photosynthates may then be used for the synthesis of sucrose or amino acids and the phosphate released during these processes is returned into the chloroplasts via the TPT to allow further photosynthesis (see Figure 3.6). The chloroplast TPT was the first phosphate translocator to be characterized biochemically (Fliege *et al.*, 1978), and the spinach TPT was the first plant membrane system whose primary amino acid sequence was determined (Flügge *et al.*, 1989). All TPT amino acid sequences share a high similarity to each other (for a review of the phosphate translocators in plastids, see Flügge, 1999). Research on the TPT has helped formulate ideas about how the synthesis of end products is controlled during photosynthesis. The TPT is a dimer composed of two identical subunits each with a molecular weight of about 30,000, with one binding site and belonging to the group of translocators with a 6 + 6 helix folding pattern, similar to the mitochondrial transporter proteins (Flügge, 1985). All TPTs are nuclear-encoded and possess N-terminal transit peptides that direct the protein to the chloroplasts. The TPT is the major protein in the envelope of chloroplasts, comprising up to 15% of the total protein (Flügge and Heldt, 1989). Under physiological conditions the TPT catalyses the strict 1:1 counter-exchange of Pi with 3-PGA or triose-phosphates. C3 compounds such as PEP or 2-phosphoglycerate, which have the phosphate group at C2, are transported with low efficiency (Fliege *et al.*, 1978). A ping-pong reaction mechanism is thought to occur during the counter-exchange of substrates by the TPT, whereby one substrate is transported across the membrane and leaves the active site before the second substrate binds and is transported. Under certain conditions the chloroplast TPT may catalyse the unidirectional transport of Pi, but at rates 2–3 orders of magnitude lower than that of the antiport reaction (Fliege *et al.*, 1978; Neuhaus and Maass, 1996), and probably by a channel-like uniport mechanism. Since the TPT is involved with photosynthetic carbon metabolism, expression of the TPT gene is observed only in photosynthetically active tissues (Schulz *et al.*, 1993).

The role of the TPT in photosynthetic metabolism has been assessed using antisense approaches. Antisense potato plants were generated whereby the activity of the TPT was reduced (Riesmeier *et al.*, 1993). Transformants showed no alterations in photosynthetic rates, growth or tuber development. However, leaf metabolism was altered in the antisense TPT plants. Levels of stromal 3-PGA increased and levels of Pi decreased compared to wild-type plants, causing an expected increase in leaf starch in transformants, presumably because of allosteric activation of the chloroplast AGPase. Tobacco antisense TPT plants also accumulated leaf starch, and it was demonstrated that transformants unexpectedly mobilized leaf starch during photosynthesis, showed increased rates of amylolytic starch breakdown, and an increased capacity for glucose export across the chloroplast envelope (Häusler *et al.*, 1998). Mobilization of leaf starch allows transformants to compensate for the deficiency in TPT activity. Evidence in support of this idea comes from potato plants with antisense repression of both the TPT and AGPase (reducing the ability of the plant to make leaf starch). The double transformants showed severe phenotypic effects as they were unable to export sufficient carbon during photosynthesis, and did not have an adequate carbon store (starch) in the leaf to support metabolic activities during the dark period (Hattenbach *et al.*, 1997).

3.9.2 Transport of phosphoenolpyruvate

The recently discovered PEP/Pi translocator (PPT; Fischer *et al.*, 1997) serves a number of functions in plastid metabolism. Mesophyll chloroplasts of C4 plants possess a PPT that mediates the export of PEP from the chloroplasts as substrate for the PEP carboxylase (E.C. 4.1.1.31) in the cytosol and the resulting Pi is returned to the chloroplasts via the PPT. The PPT exchange activity has also been detected in a wide range of heterotrophic tissues (for references, see Flügge, 1999).

PEP serves different functions in different types of plastids, acting as a precursor for fatty acid biosynthesis (see above section on fatty acid biosynthesis) or amino acid synthesis. The efficient coordination of primary and secondary biosynthetic pathways often requires the input of intermediates, reductant, or ATP from other metabolic pathways, which have to be imported directly and/or generated within the organelle by oxidative processes such as glycolysis and the OPPP. The shikimate pathway is just such an example, synthesizing aromatic amino acids and providing precursors for the synthesis of defence and wound repair compounds such as phenolic acids, suberin and lignin as well as many pigments, UV protectants and membrane constituents. The first step of the shikimate pathway (the formation of DAHP) requires input from two primary metabolic pathways in the form of erythrose 4-phosphate (from the OPPP) and PEP (from glycolysis), suggesting operation of this pathway must be tightly linked to primary carbohydrate metabolism. With the exception of the oleoplasts of lipid-storing tissues, most chloroplasts and heterotrophic plastids are unable to convert 3-PGA into PEP via the plastidic glycolytic pathway, as the low activities of phosphoglucomutase and/or enolase means this pathway cannot proceed further than 3-PGA (Stitt and ap Rees, 1979; Miernyk and Dennis, 1992). These systems rely on a supply of PEP from the cytosol. Furthermore, work with the *cue1* mutant of *Arabidopsis* has demonstrated that the shikimate pathway operating in chloroplasts is supplied principally by PEP transported from the cytosol by a specific PEP translocator (Streatfield *et al.*, 1999). The *cue1* mutant is deficient in the PPT gene, shows a severe phenotype and is unable to produce anthocyanins as a product of secondary metabolism (Streatfield *et al.*, 1999; see Section 3.5).

Expression analysis of the PPT indicated that PPT transcripts were detected in both photosynthetic and heterotrophic tissues; however, PPT-specific transcripts were more abundant in non-photosynthetic tissues (Fischer *et al.*, 1997). Recently, two PPT genes were identified in *Arabidopsis* (Knappe *et al.*, 2003b).

3.9.3 Hexose-phosphate/Pi antiporters

Phosphorylated intermediates, particularly hexose-phosphates, are central to many of the primary metabolic pathways occurring inside plastids. In chloroplasts, hexose-phosphates are generated internally from the intermediates of the Calvin cycle. Non-green plastids of heterotrophic tissues are normally unable to generate hexose-phosphates from C3 compounds owing to the absence of FBPase activity (Entwistle and ap Rees, 1990). In heterotrophic plastids, therefore, hexose-phosphates need to be imported for use in a number of important pathways such as the OPPP, and starch and fatty acid biosynthesis – the importance of which depends on the plastid type (see Figure 3.8). For a recent review of carbon transporters in heterotrophic plastids, see Fischer and Weber (2002).

The results of transport measurements using reconstituted plastid membrane systems, and with isolated organelles from a wide range of plant tissues, have shown that hexose-phosphate transport is mediated by a phosphate translocator importing hexose-phosphate in exchange for Pi or C3 sugar phosphates. In most

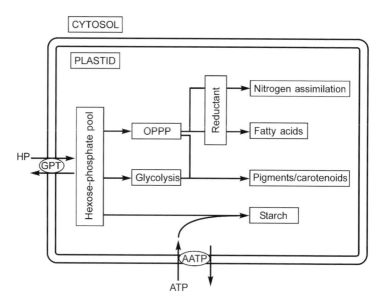

Figure 3.8 The central role of hexose-phosphates in anabolic and catabolic pathways in heterotrophic plastids. Hexose-phosphates (HP) generated in the cytosol are transported into the plastid by a Glc6P/Pi translocator (GPT). The imported hexose-phosphates are a source of precursors for a number of biosynthetic and assimilatory pathways, as well as for oxidative metabolism that generates necessary cofactors and intermediates for a wide range of metabolic activities.

non-photosynthetic plastids studied to date, including amyloplasts from cauliflower and potato (Neuhaus *et al.*, 1993; Schott *et al.*, 1995; Naeem *et al.*, 1997), pea root leucoplasts (Emes and Fowler, 1983; Emes and Traska, 1987; Borchert *et al.*, 1989) and chromoplasts from fruits and flower petals (Thom *et al.*, 1998; Tetlow *et al.*, 2003a), Glc6P is the preferred hexose-phosphate taken up in exchange for Pi. Chloroplasts from guard cells are also able to transport Glc6P; these particular chloroplasts are like non-green plastids in that they lack FBPase activity (Overlach *et al.*, 1993) and therefore any starch that is formed within these organelles must arise from hexose-phosphates. The presence of a Glc6P transporter in guard cell chloroplasts is probably a reflection of the fact that starch turnover occurs during opening and closing of stomata. The ability to transport Glc6P appears to be a feature of heterotrophic plastids. However, Glc6P transport capacity can also be induced in chloroplasts following feeding detached spinach leaves with glucose (Quick *et al.*, 1995). The glucose feeding experiment of Quick *et al.* (1995) induced a switch in the function of chloroplasts from carbon-exporting (source) to carbon-importing (sink) organelles that synthesized unusually large quantities of starch with an accompanying capacity for Glc6P transport. The rapid conversion from autotrophy to heterotrophy by glucose feeding may indicate a role for sugars in signalling this switch. In amyloplasts from wheat endosperm, however, Glc1P rather than Glc6P is the preferred hexose-phosphate precursor for starch synthesis, although the highest rates of starch biosynthesis were obtained with exogenous ADPglucose (Tetlow *et al.*, 1994). When envelope membranes from wheat endosperm amyloplasts were reconstituted into proteoliposomes, the reconstituted transport system was able to catalyse the transport of Glc1P in a 1:1 stoichiometric exchange with Pi (Tetlow *et al.*, 1996), indicating that some tissues may possess another type of hexose-phosphate transporter. Dicotyledonous storage tissues, such as potato tuber, which do not possess a cytosolic AGPase (see above), must synthesize ADPglucose for starch synthesis within the amyloplast by importing hexose-phosphate and ATP. The importance of Glc6P import in amyloplasts of dicotyledonous storage tissues is highlighted by results from studies of potato tubers lacking a plastidial phospho-glucomutase (E.C. 2.7.5.1, which converts imported Glc6P to Glc1P for use by the plastidial AGPase; see Figure 3.6), showing reduced starch accumulation (Fernie *et al.*, 2001).

cDNAs coding for Glc6P/Pi translocators from heterotrophic tissues (maize endosperm, pea roots and potato tubers) have been isolated and characterized *in vitro* (Kammerer *et al.*, 1998). The Glc6P/Pi transporters operate as antiporters exchanging (importing) Glc6P for Pi or C-3 sugar phosphates with a 1:1 stoichiometry. The plastid Glc6P/Pi transporter cloned from pea roots is unable to transport Glc1P. Molecular analysis of the Glc6P/Pi antiporter indicates that it shares only 36% homology to the TPT of leaves, and belongs to the large group of solute transporters exhibiting 2×6 transmembrane helices (Kammerer *et al.*, 1998), but no substantial similarity to the inducible Glc6P/Pi exchanger from *E. coli* (Island *et al.*, 1992).

The consequences of hexose-phosphate-driven starch biosynthesis in heterotrophic plastids is an imbalance in the stromal phosphate status. For every molecule of hexose-phosphate converted to ADPglucose by the plastidial AGPase, two molecules of Pi are released by the action of plastidial APPase on the PPi produced as a by-product of the AGPase reaction (see Figure 3.6). Since the Glc6P (Glc1P)/Pi transporters catalyse a strict 1:1 exchange of hexose-phosphate with Pi, then Pi could potentially build up within the stroma of the starch synthesizing plastid and inhibit starch synthesis by its inhibitory effect on the AGPase reaction (see above). Plastids probably possess a mechanism for removing excess Pi by its unidirectional release. Analysis of the unidirectional release of Pi from cauliflower bud amyloplasts revealed that the rate of Pi release was sufficient to account for the export of the entire Pi liberated during starch synthesis (Neuhaus and Maass, 1996). Furthermore, Pi did not accumulate in wheat endosperm amyloplasts synthesizing starch from exogenous Glc1P and ATP, indicating these organelles also possess an as-yet unidentified mechanism to remove the excess Pi produced within the stroma (Tetlow *et al.*, 1998). This problem does not occur in chloroplasts, or in amyloplasts of monocotyledonous species where ADPglucose is synthesized in the cytosol (see Figure 3.6).

3.9.4   Pentose-phosphate transport

Early reports suggested that pentose-phosphates can be transported into both chloroplasts (Bassham *et al.*, 1968) and heterotrophic plastids (Hartwell *et al.*, 1996), where in the latter case they are able to support NO_2^- reduction. The recent discovery of another member of the phosphate-translocator family of plastid inner envelope membrane proteins that has the capacity to transport pentose-phosphates indicates the increased potential for interactions between OPPP reactions in the cytosol and in the plastid (Eicks *et al.*, 2002). The reconstituted translocator preferentially catalyses the counter-exchange of xylulose 5-phosphate, triose-phosphate and Pi, and is termed the xylulose 5-phosphate/phosphate translocator (XPT). In *Arabidopsis* the XPT is encoded by a single gene that is distinct from other phosphate transporter genes, and has homologues in a number of other plants (Eicks *et al.*, 2002; Knappe *et al.*, 2003a). A functional XPT in the plastid membrane allows the exchange of pentose-phosphates between the plastid and the cytosol, thus facilitating the production of NADPH and biosynthetic precursors via the OPPP independently of one another in each compartment.

3.9.5   The plastidic ATP/ADP transporter

Heterotrophic plastids require uptake of ATP to energize various important anabolic reactions. For example, starch and fatty acid biosynthesis in isolated storage plastids are more or less totally dependent upon the presence of exogenous ATP (Neuhaus *et al.*, 1993; Kang and Rawsthorne, 1994; Tetlow *et al.*, 1994). All higher plant plastids analysed to date possess an inner envelope protein that mediates ATP/ADP exchange (Winkler, 1991). Möhlmann *et al.* (1994) showed that the plastidic ATP/ADP transporter (AATP) plays an important role in controlling the rate of end-product synthesis in storage tissues by demonstrating that both Glc6P-driven starch synthesis and acetate-dependent fatty acid synthesis in isolated cauliflower bud amyloplasts compete for the ATP imported into the organelle. However, in chloroplasts the rate of ATP import is not sufficient to support photosynthetic CO_2 fixation (Robinson and Wiskich, 1977). Biochemical characterization of the AATP in heterotrophic plastids indicated that the molecular nature of the protein must differ substantially from the functional equivalent in mitochondria, which imports ADP in strict counter-exchange with ATP. This is because the primary role of the AATP in heterotrophic plastid metabolism is to import ATP for its consumption in the type of anabolic pathways described above.

The AATP was first cloned from an *Arabidopsis thaliana* cDNA library by Kampfenkel *et al.* (1995). The isolated cDNA encoded a highly hydrophobic membrane protein with 12 predicted transmembrane domains and showed 66% similarity to the ATP/ADP transporter from the pathogenic bacterium *Rickettsia prowazekii*. The AATP is a nuclear-encoded protein with an N-terminal transit peptide allowing it to be targeted and integrated into the inner plastid envelope membrane, and processed into a mature active protein (Neuhaus *et al.*, 1997). The similarities between the plastidial AATP and the bacterial proteins meant that this was the first plant solute transporter to be used in a functional heterologous bacterial expression system (Tjaden *et al.*, 1998b). The biochemical features of the recombinant AATP were analysed and found to be identical to those of the AATP in isolated plastids (Tjaden *et al.*, 1998a). Apparent affinities of the AATPs from different sources for ATP and ADP are all in the micromolar range, and the transporters are absolutely specific for ATP and ADP. Two isoforms of the AATP have been isolated from *Arabidopsis* (AATP1 and AATP2); both have similar biochemical properties, e.g. high affinity for their substrates (Tjaden *et al.*, 1998b), but the distinct physiological role that each isoform plays in this organism is unclear.

The significance of the AATP in the provision of ATP for starch biosynthesis in heterotrophic plastids (amyloplasts) has been investigated in transgenic plants. Over-expression of AATP1 in transgenic potatoes led to a substantial increase in starch yield and an increase in the ratio of amylose to amylopectin (Tjaden *et al.*, 1998a). This observation is consistent with previous findings in isolated heterotrophic plastids showing a limitation in ATP supply, impacting on starch synthesis (Möhlmann *et al.*, 1994). It was reasoned that elevation of the stromal ATP concentration leads to an increased ADPglucose content, which in turn leads to higher starch synthesis. The importance of increasing the plastidial ADPglucose pool has been demonstrated by the heterologous expression of an allosterically unregulated bacterial AGPase, which enhanced starch synthesis (Stark *et al.*, 1992). Conversley, analysis of the AATP antisense tubers showed a decrease in starch content and reduced amylose/amylopectin ratio (Tjaden *et al.*, 1998a). The alterations in the quality of tuber starch (amylose/amylopectin ratio) in these transgenic plants is explained by the altered stromal ADPglucose concentrations (high in sense plants and low in antisense plants) effected by the altered delivery of ATP to the amyloplast. The altered ADPglucose levels affect the activities of GBSS (low affinity for

ADPglucose, synthesizing amylose) and SSs (higher affinity for ADPglucose, synthesizing amylopectin), resulting in higher amylose contents where the ADPglucose levels are predicted to be higher, in the sense plants. This case illustrates that changes in AATP activity have a profound effect on both starch yield and composition in storage tissues. No information is yet available on the effects of altering the expression of the AATP on the yield/composition of storage products in other heterotrophic plastids, such as oleoplasts in oil-rich storage tissues.

3.9.6   2-Oxoglutarate/malate transport

The chloroplast 2-oxoglutarate/malate transporter is involved in the transport of carbon skeletons into the plastid for the synthesis of glutamate and plays an important role in the pathway of amino acid biosynthesis. The glutamate derived from the GS/GOGAT cycle is then released into the cytosol via the glutamate/malate translocator. The 2-oxoglutarate/malate translocator was cloned by Weber *et al.* (1995), and, like the AATP (above), shares some structural similarities (e.g. a 12-helix motif) with plasma membrane transporters from prokaryotes and eukaryotes, and functions as a monomer.

3.9.7   The transport of ADPglucose into plastids

Examination of the subcellular distribution of AGPases in cereal endosperms (see Section 3.6.1) suggests that activity is predominantly cytosolic during the most active period of grain-filling from around 10 days after pollination (DAP), although the plastidial form is still present. This implies that ADPglucose import is the major route of carbon entry into the amyloplast during the most active period of starch synthesis in the endosperms of cereals such as wheat. Studies with intact amyloplasts of wheat (Tetlow *et al.*, 1994) and maize endosperms (Shannon *et al.*, 1996; Möhlmann *et al.*, 1997) have shown that isolated plastids are able to synthesize starch from exogenously supplied ADPglucose, as well as from hexose-phosphates and ATP (substrates for the AGPase reaction within the plastid). In isolated wheat endosperm amyloplasts the highest rates of starch synthesis were observed when ADPglucose was supplied to intact organelles, some four- to five-fold in excess of the rate observed with G1P (Tetlow *et al.*, 2003b). Indeed, direct transport of ADPglucose to the SSs within the plastid may be a more efficient way of partitioning carbon into starch than the separate import of hexose-phosphates and ATP, each of which could be used in other metabolic pathways within the organelle (see Figure 3.6, and other sections in this chapter). The requirement to transport ADPglucose into the amyloplast cannot be met by the AATP because of the substrate specificity of this protein (see above). The Bt1 mutant of maize lacks a 44-kDa protein in the amyloplast envelope, and the developing kernels have an increased ADPglucose content (Sulivan and Kaneko, 1995; Shannon *et al.*, 1996). It has been proposed that the Bt1 locus of maize encodes an ADPglucose transporter (Shannon *et al.*, 1998), though direct evidence that this is its actual function is lacking. Recently, the ADPglucose transporter has been partially purified from wheat endosperm envelope membranes, and cross-linking experiments with radiolabelled azido-ADPglucose shows the transporter has a mass of 38 kDa (Tetlow *et al.*, 2003b). When the partially purified protein is solubilized in detergent and reconstituted into liposomes, it is able to catalyse the counter-exchange of ADPglucose with AMP, ADP or ATP. The transporter does not bind UDPglucose or other uridylates, which is consistent with previous findings using isolated plastids (Tetlow *et al.*, 2003b). The cross-linking of radiolabelled azido-ATP to the partially purified ADPglucose transporter could be reduced by pre-incubations with counter-exchange substrates, and pre-incubations with ADPglucose or ADP caused greatest inhibition of cross-linking, suggesting the transporter has the highest affinity for these substrates and predominantly utilizes them *in vivo* (Tetlow *et al.*, 2003b).

A detailed kinetic analysis of the ADPglucose transporter from wheat endosperm amyloplasts was undertaken using reconstituted amyloplast envelope membranes (Tetlow *et al.*, 2003b). This study showed that the time-dependent transport of ADP-[U-^{14}C]glucose into proteoliposomes was essentially dependent upon the presence of a preloaded counter-exchange substrate inside the proteoliposome; rates of ADP-[U-^{14}C]glucose transport were greatest with AMP as a counter-exchange substrate followed by ADP and ATP, respectively. The previously reconstituted maize amyloplast ADPglucose transporter also showed highest rates of ADPglucose transport when AMP was provided as the counter-exchange substrate (Möhlmann *et al.*, 1997). The ADPglucose transporter in plastid membranes may share similarities with other nucleotide sugar transporters (NSTs), which tend to utilize the corresponding nucleotide monophosphate as the counter-exchange substrate. The functions of the few NSTs that have been characterized are transport of nucleotide sugars into the ER and Golgi apparatus, largely for glycoconjugate synthesis (for a review, see Abeijon *et al.*, 1997). However, analysis of the substrate dependence of AMP and ADP import into proteoliposomes preloaded with ADPglucose by the ADPglucose transporter of wheat endosperm amyloplasts showed an almost eightfold greater affinity for ADP than AMP (unpublished results). This suggests that *in vivo*, ADP may be the preferred counter-exchange substrate for the ADPglucose transporter. This also seems likely when the pathway of starch biosynthesis in cereals is considered (Figure 3.6), whereby ADP is generated as a by-product of the SS reaction inside the amyloplast. The transport of ADPglucose by an NST-like transporter in the amyloplasts of monocots also circumvents the problem of Pi build-up in the plastid because of an imbalance in the exchange stoichiometry, which is a feature of ADPglucose synthesis inside the plastid, for example, in the dicots (see Figure 3.6, and above section).

The ability of plastids to import ADPglucose may not be restricted to monocotyledonous tissues that possess a cytosolic AGPase. For example, amyloplasts isolated from potato tubers (Naeem *et al.*, 1997) and sycamore (*Acer pseudoplatanus*) cell suspension cultures (Pozueta-Romero *et al.*, 1991) have been shown to import exogenously supplied ADPglucose. The distribution of the ADPglucose transporter within different plants and tissues is currently unknown, as is the function of the protein in tissues apparently lacking the capacity to synthesize ADPglucose outside the plastid. There is some evidence that the expression of the ADPglucose transporter coincides with the extra-plastidial production of ADPglucose in wheat endosperm. The ADPglucose transporter can be identified by cross-linking to radiolabelled azido-ADPglucose, and it is not detected until 10 DAP (Emes *et al.*, 2003). The radioactive cross-linker increases (and presumably the amount of ADPglucose transporter) from 10 DAP up to 40 DAP, coinciding with the major period of grain-filling and consistent with observed changes in cytosolic AGPase expression in wheat and barley endosperms (Ainsworth *et al.*, 1995; Doan *et al.*, 1999).

In vivo starch biosynthesis in cereal endosperms probably occurs as a result of ADPglucose synthesis in both amyloplasts and cytosol. Amyloplast AGPase in barley, maize, rice and wheat varies from 2 to 30% of the total activity, but it is unclear which pathway of ADPglucose synthesis predominates *in vivo*, though there is evidence of developmental regulation. If the ADPglucose transporter is the primary route for carbon entry into the amyloplast then it is likely to have a major impact on the ratios of amylose and amylopectin in storage starches of cereals in much the same way as the AATP in potato tubers (Tjaden *et al.*, 1998a).

3.10   Conclusion

This chapter has dealt with recent developments in a number of aspects of primary metabolism within plastids. The control and regulation of individual metabolic pathways, and the interactions/coordination between them is implemented at many levels, and these different modes of regulation are often the same, irrespective of the particular pathway. Transcriptional control of gene expression occurs over longer time frames, during plant growth and development, or in response to environmental changes. Modification of the activation state of enzymes through effector molecules (allosteric regulation), or through post-translational modifications (e.g. protein phosphorylation, redox modulation) offers short- to mid-term regulation, enabling flexibility in the operation of the various pathways in order to respond to immediate cellular and environmental changes. One such example is redox control of enzymes positioned at key points in major metabolic pathways, such as the OPPP (G6PDH), starch synthesis (AGPase) or fatty acid biosynthesis (ACCase). In addition to the regulation of specific reactions within the pathways, fluxes may be controlled by the physical interaction of proteins both within, and between metabolic pathways. Recent research on this emerging concept has been described here in relation to the control of the RPPP and the starch biosynthetic pathway.

Many useful tools are being used in combined approaches to further understand various aspects of plastid metabolism, and the highly compartmentalized nature of plant metabolism in general. For example, the use of genome information with bioinformatics data enable rational predictions of the number of individual isoforms of a particular protein in a pathway, and determine the likely compartmentation of the pathway, based on plastid protein targeting sequences (see section on the OPPP). With cDNA clones now available for many proteins in metabolic pathways of some species, genome arrays for model organisms such as *Arabidopsis* and rice, and reporter-gene technologies, understanding developmental aspects of metabolic compartmentation becomes far more tractable than could be achieved through cell fractionation studies. Metabolic flux analyses using ^{13}C NMR approaches have proved to be extremely valuable non-invasive techniques, which also allow predictions to be made regarding cellular compartmentation (examples included here are for the OPPP and fatty acid synthesis). In addition, metabolic flux measurements are invaluable in determining the effects of genetic lesions or insertion mutants on a given pathway or pathways. The use of whole and partial plant genome sequences in combination with the powerful tool of MS has, and will be, of great value in identifying components of protein complexes and signal transduction cascades within metabolic pathways, and the conditions under which such regulatory mechanisms operate.

References

Aach, H., Hornig, F. and Heise, K.P. (1997) Distribution of lipid radioactivity after fractionation of C-14-labelled zygotic rape embryos. *J. Plant Physiol.*, 151, 323–328.

Abeijon, C., Mandon, E.C. and Hirschberg, C.B. (1997) Transporters of nucleotide sugars, nucleotide sulfate and ATP in the Golgi apparatus. *TIBS*, 22, 203–207.

Abel, G.J.W., Springer, F., Willmitzer, L. and Kossmann, J. (1996) Cloning and functional analysis of a cDNA encoding a novel 139 kDa starch synthase from potato (*Solanum tuberosum* L.). *Plant J.*, 10, 981–991.

Ainsworth, C., Hosein, F., Tarvis, M. *et al.* (1995) Adenosine diphosphate glucose pyrophosphorylase genes in wheat: differential expression and gene mapping. *Planta*, 197, 1–10.

Andrews, T.J. and Whitney, S.M. (2003) Manipulating ribulose bisphosphate carboxylase/oxygenase in the chloroplasts of higher plants. *Arch. Biochem. Biophys.*, 414, 159–169.

Aoki, H. and Ida, S. (1994) Nucleotide sequence of a rice root ferredoxin-NADP$^+$ reductase and its induction by nitrate. *Biochim. Biophys. Acta*, 1183, 553–556.

Aoki, H., Tanaka, K. and Ida, S. (1995) The genomic organisation of the gene encoding a nitrate-inducible ferredoxin-NADP$^+$ oxidoreductase from rice roots. *Biochim. Biophys. Acta*, 1229, 389–392.

ap Rees, T. (1985) The organisation of glycolysis and the pentose phosphate pathway in plants. In *Encyclopedia of Plant Physiology*, Vol. 18 (eds R. Douce and D. Day), Springer-Verlag, Berlin, pp. 391–417.

Bachmann, M., Huber, J.L., Liao, P.-C., Gage, D.A. and Huber, S.C. (1996a) The inhibitor protein of phosphorylated nitrate reductase from spinach (*Spinacia oleracea*) leaves is a 14-3-3 protein. *FEBS Lett.*, 387, 127–131.

Bachmann, M., Shiraishi, N., Campbell, W.H., Yoo, B.-C., Harmon, A. and Huber, S.C. (1996b) Identification of Ser 543 as the major regulatory phosphorylation site in spinach leaf nitrate reductase. *Plant Cell*, 8, 505–517.

Båga, M., Nair, R.B., Repellin, A., Scoles, G.J. and Chibbar, R.N. (2000) Isolation of a cDNA encoding a granule-bound 152-kilodalton starch-branching enzyme in wheat. *Plant Physiol.*, 124, 253–263.

Ball, S.G. and Morell, M.K. (2003) From bacterial glycogen to starch: understanding the biogenesis of the plant starch granule. *Ann. Rev. Plant Biol.*, 54, 207–233.

Ballicora, M.A., Laughlin, M.J., Fu, Y., Okita, T.W., Barry, G.F. and Preiss J. (1995) Adenosine 5'-diphosphate-glucose pyrophosphorylase from potato tuber. Significance of the N-terminus of the small subunit for catalytic properties and heat stability. *Plant Physiol.*, 109, 245–251.

Balmer Y., Koller A., del Val, G., Manieri, W., Schurmann, P. and Buchanan B.B. (2003) Proteomics gives insight into the regulatory function of chloroplast thioredoxins. *Proc. Natl. Acad. Sci. U.S.A.*, 100, 370–375.

Banks, F.M., Driscoll, S.P., Parry, M.A.J. *et al.* (1999) Decrease in phosphoribulokinase activity by antisense RNA in transgenic tobacco. Relationship between photosynthesis, growth and allocation at different nitrogen levels. *Plant Physiol.*, 119, 1125–1136.

Bao, X.M., Focke, M., Pollard, M. and Ohlrogge, J. (2000) Understanding *in vivo* carbon precursor supply for fatty acid synthesis in leaf tissue. *Plant J.*, 22, 39–50.

Bassham, J.A., Kirk, M. and Jensen, R.G. (1968) Photosynthesis by isolated chloroplasts, I: diffusion of labeled photosynthetic intermediates between isolated chloroplasts and suspending medium. *Biochim. Biophys. Acta*, 153, 211–218.

Beatty, M.K., Rahman, A., Cao, H. *et al.* (1999) Purification and molecular genetic characterization of ZPU1, a pullulanase-type starch-debranching enzyme from maize. *Plant Physiol.*, 119, 255–266.

Becker, T.W., Nef-Campa, C., Zehnacker, C. and Hirel, B. (1993) Implication of the phytochrome in light regulation of the tomato gene(s) encoding ferredoxin-dependent glutamate synthase. *Plant Physiol. Biochem.*, 31, 725–729.

Beckles, D.M., Smith, A.M. and apRees, T. (2001) A cytosolic ADP-glucose pyrophosphorylase is a feature of graminaceous endosperms, but not of other starch storing organs. *Plant Physiol.*, 125, 818–827.

Behal, R.H., Lin, M., Back, S.L. and Oliver, D.J. (2002) Role of acetyl-coenzyme A syntheatse in leaves of *Arabidopsis thaliana*. *Arch. Biochem. Biophys.*, 402, 259–267.

Bernacchia, G., Schwall, G., Lottspeich, F., Salamini, F. and Bartels, D. (1995) The transketolase gene family of the resurrection plant *Craterostigma plantagineum*: differential expression during the rehydration phase. *EMBO J.*, 14, 610–618.

Blackwell, R.D., Murray, A.J.S., Lea, P.J. *et al.* (1988) The value of mutants unable to carry out photorespiration. *Photosynth. Res.*, 16, 155–176.

Blauth, S.L., Kim, K.N., Klucinec, J., Shannon, J.C., Thompson, D.B. and Guiltinan, M. (2002) Identification of Mutator insertional mutants of starch-branching enzyme 1 (sbe1) in *Zea mays* L. *Plant Mol. Biol.*, 48, 287–297.

Blauth, S.L., Yao, Y., Klucinec, J.D., Shannon, J.C., Thompson, D.B. and Guiltinan, M. (2001) Identification of Mutator insertional mutants of starch-branching enzyme 2a in corn. *Plant Physiol.*, 125, 1396–1405.

Borchert, S., Grosse, H. and Heldt, H.W. (1989) Specific transport of inorganic phosphate, glucose 6-phosphate, dihydroxyacetone phosphate and 3-phosphoglycerate into amyloplasts. *FEBS Lett.*, 253, 183–186.

Borchert, S., Harborth, J., Schünemann, D., Hoferichter, P. and Heldt, H.W. (1993) Studies of the enzymatic capacities and transport properties of pea root plastids. *Plant Physiol.*, 101, 303–312.

Bouvier, F., Suire, C., Mutterer, J. and Camara, B. (2003) Oxidative remodelling of chromoplasts carotenoids: identification of the carotenoid dioxygenase CsCCD and CsZCD genes involved in crocus secondary metabolite biogenesis. *Plant Cell*, 15, 47–62.

Bowsher, C.G., Boulton, E.L., Rose, J., Nayagam, S. and Emes, M.J. (1992) Reductant for glutamate synthase is generated by the oxidative pentose phosphate pathway in non-photosynthetic root plastids. *Plant J.*, 2, 893–898.

Bowsher, C.G., Hucklesby, D.P. and Emes, M.J. (1989) Nitrite reduction and carbohydrate metabolism in plastids purified from roots of *Pisum sativum* L. *Planta*, 177, 359–366.

Bowsher, C.G., Hucklesby, D.P. and Emes, M.J. (1993) Induction of ferredoxin-NADP$^+$ oxidoreductase and ferredoxin synthesis in pea root plastids during nitrate assimilation. *Plant J.*, 3, 463–467.

Browse, J. and Slack, C.R. (1985) Fatty-acid synthesis in plastids from maturing safflower and linseed cotyledons. *Planta*, 166, 74–80.

Buchanan, B.B. (1980) Role of light in the regulation of chloroplast enzymes. *Annu. Rev. Plant Physiol.*, 31, 341–374.

Buchanan, B.B. (1991) Regulation of CO_2 assimilation in oxygenic photosynthesis – the ferredoxin–thioredoxin system. Perspective on its discovery, present status, and future-development. *Arch. Biochem. Biophys.*, 288, 1–9.

Buléon, A., Colonna, P., Planchot, V. and Ball, S. (1998) Starch granules: structure and biosynthesis. *Int. J. Biol. Macromol.*, 23, 85–112.

Chung, H.J., Sehnke, P.C. and Ferl, R.J. (1999) The 14-3-3 proteins: cellular regulators of plant metabolism. *Trends Plant Sci.*, 4, 367–371.

Clarke, B.R., Denyer, K., Jenner, C.F. and Smith, A.M. (1999) The relationship between the rate of starch synthesis, the adenosine 5'-diphosphoglucose concentration and the amylase content of starch in developing pea embryos. *Planta*, 209, 324–329.

Clasper, S., Easterby, J.S. and Powls, R. (1991) Properties of two high-molecular-mass forms of glyceraldehyde-3-phosphate dehydrogenase from, spinach leaf, one of which also possesses latent phosphoribulokinase activity. *Eur. J. Biochem.*, 202, 1239–1246.

Colleoni, C., Myers, A.M. and James, M.G. (2003) One- and two-dimensional native PAGE activity gel analyses of maize endosperm proteins reveal functional interactions between specific starch metabolizing enzymes. *J. Appl. Glycosci.*, 50, 207–212.

Commuri, P.D. and Keeling, P.L. (2001) Chain-length specificities of maize starch synthase I enzyme: studies of glucan affinity and catalytic properties. *Plant J.*, 25, 475–486.

Cordoba, E., Shishkova, S., Vance, C.P. and Hernández, G. (2003) Antisense inhibition of NADH glutamate synthase impairs carbon/nitrogen assimilation in nodules of alfalfa (*Medicago sativa* L.). *Plant J.*, 33, 1037–1049.

Coschigano, K.T., Melo-Oliveira, R., Lim, J. and Coruzzi, G.M. (1998) *Arabidopsis gls* mutants and distinct Fd-GOGAT genes: Implication for photorespiration and primary nitrogen assimilation. *Plant Cell*, 10, 741–752.

Crete, P., Caboche, M. and Meyer, C. (1997) Nitrite reductase expression is regulated at the post-transcriptional level by the nitrogen source in *Nicotiana plumbaginifolia* and *Arabidopsis thaliana*. *Plant J.*, 11, 625–634.

Critchley, J.H., Zeeman, S.C., Takaha, T., Smith, A.M. and Smith, S.M. (2001) A critical role for disproportionating enzyme in starch breakdown is revealed by a knock-out mutation in *Arabidopsis*. *Plant J.*, 26, 89–100.

Debnam, P.M. and Emes, M.J. (1999) Subcellular distribution of enzymes of the oxidative pentose phosphate pathway in root and leaf tissues. *J. Exp. Bot.*, 50, 1653–1661.

De Fekete, M.A.R., Leloir, L.F. and Cardini, C.E. (1960) Mechanism of starch biosynthesis. *Nature*, 187, 918–919.

Delrue, B., Fontaine, T., Routier, F., Decq, A., Wieruszeski, J.M. and Ball, S. (1992) Waxy *Chlamydomonas reinhardtii*: monocellular algal mutants defective in amylose biosynthesis and granule-bound starch synthase activity accumulate a structurally modified amylopectin. *J. Bacteriol.*, 174, 3612–3620.

Denyer, K., Clarke, B., Hylton, C., Tatge, H. and Smith, A.M. (1996a) The elongation of amylose and amylopectin chains in isolated starch granules. *Plant J.*, 10, 1135–1143.

Denyer, K., Dunlap, F., Thorbjørnsen, T., Keeling, P. and Smith, A.M. (1996b) The major form of ADPglucose pyrophosphorylase in maize endosperm is extraplastidial. *Plant Physiol.*, 112, 779–783.

Denyer, K., Johnson, P., Zeeman, S. and Smith, A.M. (2001) The control of amylose synthesis. *J. Plant Physiol.*, 158, 479–487.

Dieuaide-Noubhani, M., Raffard, G., Canioni, P., Pradet, A. and Raymond, P. (1995). Quantification of compartmented metabolic fluxes in maize root tips using isotope distribution from ^{13}C- or ^{14}C-labeled glucose. *J. Biol. Chem.*, 270, 13147–13159.

Dinges, J.R., Colleoni, C., James, M.G. and Myers, A.M. (2003) Mutational analysis of the pullulanase- type debranching enzyme of maize indicates multiple functions in starch metabolism. *Plant Cell*, 15, 666–680.

Dinges, J.R., Colleoni, C., Myers, A.M. and James, M.G. (2001) Molecular structure of three mutations at the maize *sugary1* locus and their allele-specific phenotypic effects. *Plant Physiol.* 125, 1406– 1418.

Doan, D.N.P., Rudi, H. and Olsen, O.-A. (1999) The allosterically unregulated isoform of ADP-glucose pyrophosphorylase from barley endoserm is the most likely source of ADP-glucose incorporated into endosperm starch. *Plant Physiol.*, 121, 965–975.

Eastmond, P.J., Dennis, D.T. and Rawsthorne, S. (1997) Evidence that a malate/inorganic phosphate exchange translocator imports carbon across the leucoplast envelope for fatty acid synthesis in developing castor seed endosperm. *Plant Physiol.*, 114, 851–856.

Eastmond, P.J. and Rawsthorne, S. (1998) Comparison of the metabolic properties of plastids isolated from developing leaves and embryos of *Brassica napus* L. *J. Exp. Bot.*, 49, 1105–1111.

Eastmond, P.J. and Rawsthorne, S. (2000) Co-ordinate changes in carbon partitioning and plastidial metabolism during the development of oilseed rape (*Brassica napus* L.) embryos. *Plant Physiol.*, 122, 767–774.

Eicks, M., Maurino, V., Knappe, S., Flügge, U.-I. and Fischer, K. (2002) The plastidic pentose phosphate translocator represents a link between the cytosolic and the plastidic pentose phosphate pathways in plants. *Plant Physiol.*, 128, 512–522.

Edwards, A., Fulton, D.C., Hylton, C.M. *et al.* (1999) A combined reduction in activity of starch synthases II and III of potato has novel effects on the starch of tubers. *Plant J.*, 17, 251–261.

Edwards, A., Marshall, J., Sidebottom, C., Visser, R.G.F., Smith, A.M. and Martin, C. (1995) Biochemical and molecular characterisation of a novel starch synthase from potato tubers. *Plant J.*, 8, 283–294.

Emes, M.J., Bowsher, C.G., Hedley, C., Burrell, M.M., Scrase-Field, E.S.F. and Tetlow, I.J. (2003) Starch synthesis and carbon partitioning in developing endosperm. *J. Exp. Bot.*, 54, 569–575.

Emes, M.J. and Fowler, M.W. (1983) The supply of reducing power for nitite reduction in plastids of seedling pea roots (*Pisum sativum* L.). *Planta*, 158, 97–102.

Emes, M.J. and Traska, A. (1987) Uptake of inorganic phosphate by plastids purified from the roots of *Pisum sativum* L. *J. Exp. Bot.*, 38, 1781–1788.

Entus, R., Poling, M. and Herrmann, K. (2002) Redox regulation of *Arabidopsis* 3-deoxy-D-arabino-heptulosonate 7-phosphate synthase. *Plant Physiol.*, 129, 1866–1871.

Entwistle, G. and ap Rees, T. (1988). Enzymic capacities of amyloplasts from wheat endosperm. *Biochem. J.*, 255, 391–396.

Entwistle, G. and ap Rees, T. (1990) Lack of fructose-1,6-bisphosphatase in a range of higher plants that store starch. *Biochem J.*, 271, 467–472.

Fell, D. (1997) *Understanding the Control of Metabolism*, Portland Press, London.

Fernie, A.R., Roessner, U., Trethewey, R.N. and Willmitzer, L. (2001) The contribution of plastidial phosphoglucomutase to the control of starch synthesis within the potato tuber. *Planta*, 213, 418–426.

Fischer, K., Kammerer, B., Gutensohn, M. *et al.* (1997) A new class of plastidic phosphate translocators: a putative link between primary and secondary metabolism by the phosphoenolpyruvate/phosphate antiporter. *Plant Cell*, 9, 453–462.

Fischer, K. and Weber, A. (2002) Transport of carbon in non-green plastids. *Trends Plant Sci.*, 7, 345–351.

Fliege, R., Flügge, U.-I., Werdan, K. and Heldt, H.W. (1978) Specific transport of inorganic phosphate, 3-phosphoglycerate and triosephosphates across the inner membrane of the envelope in spinach chloroplasts. *Biochim. Biophys. Acta*, 502, 232–247.

Flipse, E., Suurs, L., Keetels, C.J.A., Kossmann, J., Jacobsen, E. and Visser, R.G.F. (1996) Introduction of sense and antisense cDNA for branching enzyme in the amylose-free potato mutant leads to physico-chemical changes in the starch. *Planta*, 198, 340–47.

Flügge, U.-I. (1985) Hydrodynamic properties of the Triton X-100 solubilized chloroplast phosphate translocator. *Biochim. Biophys. Acta*, 815, 299–305.

Flügge, U.-I. (1999) Phosphate translocators in plastids. *Annu. Rev. Plant Physiol. Plant Mol. Biol.*, 50, 27–45.

Flügge, U.-I. and Benz, R. (1984) Pore forming activity in the outer membrane of the chloroplast envelope. *FEBS Lett.*, 169, 85–89.

Flügge, U.-I., Fischer, K., Gross, A., Sebald, W., Lottspeich, F. and Eckerskorn, C. (1989) The triose phosphate-3-phosphoglycerate-phosphate translocator from spinach chloroplasts: nucleotide sequence of a full-length cDNA clone and import of the *in vitro* synthesized precursor protein into chloroplasts. *EMBO J.*, 8, 39–46.

Flügge, U.-I. and Heldt, H.W. (1989) The phosphate translocator of the chloroplast envelope. Isolation of the carrier protein and reconstitution of transport. *Biochim. Biophys. Acta*, 638, 296–304.

Fondy, B.R., Geiger, D.R. and Servaites, J.C. (1989) Photosynthesis, carbohydrate metabolism and export in *Beta vulgaris* L. and *Phaseolus vulgaris* L. during square and sinusoidal light regimes. *Plant Physiol.*, 89, 396–402.

Fox, S.R., Hill, L.M., Rawsthorne, S. and Hills, M.J. (2000) Inhibition of the glucose-6-phosphate transporter in oilseed rape (*Brassica napus* L.) plastids by acyl-CoA thioesters reduces fatty acid synthesis. *Biochem. J.*, 352, 525–532.

French, D. (1984) Organization of starch granules. In *Starch: Chemistry and Technology* (eds R.L. Whistler, J.N. BeMiller and E.F. Paschall) Academic Press, Orlando, FL, pp. 183–237.

Frey-Wissling, A. and Kreutzer, E. (1958) Die submikroskopische entwicklung der chromoplasten in den blüten von *Ranunculus repens* L. *Planta (Berlin)*, 51, 104–114.

Fu, Y., Ballicora, M.A., Leykam, J.F. and Preiss, J. (1998) Mechanism of reductive activation of potato tuber ADP-glucose pyrophosphorylase. *J. Biol. Chem.*, 273, 25045–25052.

Fujita, N. and Taira, T. (1998) A 56-kDa protein is a novel granule-bound starch synthase existing in the pericarps, aleurone layers, and embryos of immature seed in diploid wheat (*Triticum monococcum* L.). *Planta*, 207, 125–132.

Galván, A., Rexach, J., Mariscal, V. and Fernandez, E. (2002). Nitrite transport to the chloroplast in *Chlamydomonas reinhardtii*: molecular evidence for a regulated process. *J. Exp. Bot.*, 53, 845–853.

Gao, M., Fisher, D.K., Kim, K-N., Shannon, J.C. and Guiltinan, M.J. (1996) Evolutionary conservation and expression patterns of maize starch branching enzyme I and IIb genes suggests isoform specialization. *Plant Mol. Biol.*, 30, 1223–1232.

Gao, M., Wanat, J., Stinard, P.S., James, M.G. and Myers, A.M. (1998) Characterization of *dull1*, a maize gene coding for a novel starch synthase. *Plant Cell*, 10, 399–412.

Geiger, D.R. and Servaites, J.C. (1994) Diurnal regulation of photosynthetic carbon metabolism in C3 plants. *Annu. Rev. Plant Phys. Plant Mol. Biol.* 45, 235–256.

Ghosh, H.P. and Preiss, J. (1966) Adenosine diphosphate glucose pyrophosphorylase: a regulatory enzyme in the biosynthesis of starch in spinach leaf chloroplasts. *J. Biol. Chem.*, 241, 4491–4504.

Giegé, P., Heazlewood, J.L., Roessner-Tunali, U. *et al.* (2003) Enzymes of glycolysis are functionally associated with the mitochondrion in *Arabidopsis* cells. *Plant Cell*, 15, 2140–2151.

Giroux, M. and Hannah, L.C. (1994) ADPglucose pyrophosphorylase in *shrunken-2* and *brittle-2* mutants of maize. *Mol. Gen. Genet.*, 243, 400–408.

Gómez-Casati, D.F. and Iglesias, A.A. (2002) ADP-glucose pyrophosphorylase from wheat endosperm. Purification and characterisation of an enzyme with novel regulatory properties. *Planta*, 214, 428–434.

Gontero, B., Lebreton, S. and Graciet, E. (2002). Multienzyme complexes involved in the Benson-Calvin cycle and in fatty acid metabolism. In *Annual Reviews, Vol. 7: Protein–Protein Interactions in Pplant Biology* (eds M.T. Mcmanus, W.A. Laing and A. Allan), Sheffield Academic, Sheffield, England, Chapt. 5, pp. 120–144.

Graeve, K., von Schaewen, A. and Scheibe, R. (1994) Purification, characterisation and cDNA sequence of glucose-6-phosphate dehydrogenase from potato (*Solanum tuberosum* L.). *Plant J.*, 5, 353–361.

Gregerson, R.G., Miller, S.S., Twary, S.N., Gantt, J.S. and Vance, C.P. (1993) Molecular characterization of NADH-dependent glutamate synthase from alfalfa nodules. *Plant Cell*, 5, 215–226.

Gross, P. and ap Rees, T. (1986) Alkaline inorganic pyrophosphatase and starch synthesis in amyloplasts. *Planta*, 167, 140–145.

Guan, H.-P. and Preiss, J. (1993) Differentiation of the properties of the branching isozymes from maize (*Zea mays*). *Plant Physiol.*, 102, 1269–1273.

Haake, V., Zrenner, R., Sonnewald, U. and Stitt, M. (1998) A moderate decrease of plastid aldolase activity inhibits photosynthesis, alters the levels of sugars and starch and inhibits growth of potato plants. *Plant J.*, 14, 147–157.

Hall, D.M. and Sayre, J.G. (1973) A comparison of starch granules as seen by both scanning and ordinary light microscopy. *Starch-Stärke*, 25, 292–297.

Harn, C., Knight, M., Ramakrishnan, A., Guan, H.-P., Keeling, P.L. and Wasserman, B.P. (1998) Isolation and characterization of the ZSSIIa and ZSSIIb starch synthase cDNA clones from maize endosperm. *Plant Mol. Biol.*, 37, 639–649.

Harris, G.C. and Koniger, M. (1997) The 'high' concentrations of enzymes within the chloroplast. *Photosynth. Res.* 54, 5–23.

Harrison, E.P., Willingham, N.M., Lloyd, J.C. and Raines, C.A. (1998) Reduced sedoheptulose-1,7-bisphosphatase levels in transgenic tobacco lead to decreased photosynthetic capacity and altered carbohydrate partitioning. *Planta*, 204, 27–36.

Hartman, F.C. and Harpel, M.R. (1994) Structure, function, regulation and assembly of D-ribulose-1,5-bisphosphate carboxylase oxygenase. *Annu. Rev. Biochem.*, 63, 197–234.

Hartwell, J., Bowsher, C.G. and Emes, M.J. (1996) Recycling of carbon in the oxidative pentose phosphate pathway in non-photosynthetic plastids. *Planta*, 200, 107–112.

Hattenbach, B., Müller-Röber, B., Nast, G. and Heineke, D. (1997). Antisense repression of both ADP-glucose pyrophosphorylase and triose phosphate translocator modifies carbohydrate partitioning in potato leaves. *Plant Physiol.*, 115, 471–475.

Häusler, R.E., Schlieben, N.H., Schulz, B. and Flügge, U.-I. (1998). Compensation of decreased triose phosphate/phosphate transport activity by accelerated starch turnover and glucose transport in transgenic tobacco. *Planta*, 204, 366–376.

Hendriks, J.H.M., Kolbe, A., Gibon, Y., Stitt, M. and Geigenberger, P. (2003) ADP-glucose pyrophosphorylase is activated by posttranslational redox-modification in response to light and to sugars in leaves of Arabidopsis and other plant species. *Plant Physiol.*, 133, 1–12.

Henkes, S., Sonnewald, U., Badur, R., Flachmann, R. and Stitt, M. (2001) A small decrease of plastid transketolase activity in antisense tobacco transformants has dramatic effects on photosynthesis and phenylpropanoid metabolism. *Plant Cell*, 13, 535–551.

Herrmann, K.M. and Weaver, L.M. (1999) The shikimate pathway. *Annu. Rev. Plant Physiol. Plant Mol. Biol.*, 50, 473–503.

Hong, Z.Q. and Copeland, L. (1990) Pentose phosphate pathway enzymes in nitrogen-fixing leguminous root nodules. *Phytochemistry*, 29, 2437–2440.

Hsieh, M., Lam, H., Van de loo, F.J. and Coruzzi, G. (1998) A PII-like protein in *Arabidopsis*: putative role in nitrogen sensing. *Proc. Natl. Acad. Sci. U.S.A.*, 95, 13965–13970.

Hylton, C. and Smith, A.M. (1992) The rb mutation of peas causes structural and regulatory changes in ADP glucose pyrophosphorylase from developing embryos. *Plant Physiol.*, 99, 1626–1634.

Ishiyama, K., Hayakawa, T. and Yamaya, T. (1998) Expression of NADH-dependent glutamate synthase protein in the epidermis and exodermis of rice roots in response to the supply of ammonium ions. *Planta*, 204, 288–294.

Island, M.D., Wei, B.Y. and Kadner, J.J. (1992) Structure and function of the *uhp* genes for the sugar phosphate transport system in *E. coli* and *Salmonella typhinurium*. *J. Bacteriol.*, 174, 2754–2762.

Jacquot, J.-P., Lancelin, J.-M. and Meyer, Y. (1997) Thioredoxins: structure and function in plant cells. *New Phytol.*, 136, 543–570.

James, M.G., Robertson, D.S. and Myers, A.M. (1995) Characterization of the maize gene *sugary1*, a determinant of starch composition in kernels. *Plant Cell*, 7, 417–429.

Jarvis, P. (2003) Intracellular signalling: the language of the chloroplast. *Curr. Biol.*, 13, 314–316.

Jebanathirajah, J.A. and Coleman J.R. (1998) Association of carbonic anhydrase with a Calvin cycle enzyme complex in *Nicotiana tabacum. Planta*, 204, 177–182.

Jenkins, P.J., Cameron, R.E. and Donald, A.M. (1993) A universal feature in the starch granules from different botanical sources. *Starke*, 45, 417–420.

Jespersen, H.M., MacGregor, E.A., Henrissat, B., Sierks, M.R. and Svensson, B. (1993) Starch- and glycogen-debranching and branching enzymes: prediction of structural features of the catalytic $(\beta/\alpha)_8$-barrel domain and evolutionary relationship to other amylolytic enzymes. *J. Protein Chem.*, 12, 791–805.

Jobling, S.A., Schwall, G.P., Westcott, R.J. *et al.* (1999) A minor form of starch branching enzyme in potato (*Solanum tuberosum* L.) tubers has a major effect on starch structure: cloning and characterisation of multiple forms of SBE A. *Plant J.*, 18, 163–171.

Johnson, P.E., Fox, S.R., Hills, M.J. and Rawsthorne, S. (2000) Inhibition by long chain acyl-CoAs of glucose-6-phosphate metabolism in plastids isolated from developing embryos of oilseed rape (*Brassica napus* L.). *Biochem. J.*, 348,145–150.

Johnson P.E., Rawsthorne, S. and Hills, M.J. (2002). Export of acyl chains from plastids isolated from embryos of *Brassica napus* L. *Planta*, 215, 515–517.

Journet, E.P. and Douce, R. (1985). Enzymic capacities of purified cauliflower bud plastids for lipid synthesis and carbohydrate metabolism. *Plant Physiol.*, 79, 458–467.

Kakefuda, G. and Preiss, J. (1997). Partial purification and characterization of a diurnally fluctuating novel endoamylase from *Arabidopsis thaliana* leaves. *Plant Physiol. Biochem.*, 35, 907–913.

Kammerer, B., Fisher, K., Hilpert, B. *et al.* (1998) Molecular characterisation of a carbon transporter in plastids from heterotrophic tissues: the glucose 6-phosphate antiporter. *Plant Cell*, 10, 105–117.

Kampfenkel, K., Möhlmann, T., Batz, O., van Montagu, M., Inzé, D. and Neuhaus, H.E. (1995) Molecular characterisation of an *Arabidopsis thaliana* cDNA encoding a novel putative adenylate translocator of higher plants. *FEBS Lett.*, 374, 351–355.

Kang, F. and Rawsthorne, S. (1994) Starch and fatty acid synthesis in plastids from developing embryos of oilseed rape (*Brassica napus* L.). *Plant J.*, 6, 795–805.

Kang, F. and Rawsthorne, S. (1996) Metabolism of glucose-6-phosphate and utilization of multiple metabolites for fatty acid synthesis by plastids from developing oilseed rape embryos. *Planta*, 199, 321–327.

Ke, J., Behal, R.H., Back, S.L., Nikolau, B.J., Wurtele, E.S. and Oliver, D.J. (2000). The role of pyruvate dehydrogenase and acetyl-coenzyme A synthetase in fatty acid synthesis in developing *Arabidopsis* seeds. *Plant Physiol.*, 123, 497–508.

Khoshnoodi, J., Larsson, C.T., Larsson, H. and Rask, L. (1998). Differential accumulation of *Arabidopsis thaliana* SBE2.1 and SBE2.2 transcripts in response to light. *Plant Sci.*, 135, 183–193.

King, S.P., Badger, M.R. and Furbank, R.T. (1998) CO_2 refixation characteristics of developing canola seeds and silique wall. *Aust. J. Plant Physiol.*, 25, 377–386.

Kirk, J.T.O. and Tilney-Bassett, R.A.E. (1978) *The Plastids: Their Chemistry, Structure, Growth and Inheritance*, 2nd edn, Elsevier, Amsterdam/Oxford.

Kleczkowski, L.A. (1994) Glucose activation and metabolism through UDP-glucose pyrophosphorylase in plants. *Phytochemistry*, 37, 1507–1515.

Kleczkowski, L.A., Villand, P., Lüthi, E., Olsen, O.A. and Preiss J. (1993) Insensitivity of barley endosperm ADP-glucose pyrophosphorylase to 3-phosphoglycerate and orthophosphate regulation. *Plant Physiol.*, 101, 179–186.

Knappe, S., Flügge, U.-I. and Fischer, K. (2003a). Analysis of the plastidic phosphate translocator gene family in *Arabidopsis* and identification of new phosphate translocator-homologous transporters, classified by their putative substrate-binding site. *Plant Physiol.* 131, 1178–1190.

Knappe, S., Löttgert, T., Schneider, A., Voll, L., Flügge, U.-I. and Fischer, K. (2003b) Characterization of two functional *phosphoenolpyruvate/phosphate translocator* (*PPT*) genes in *Arabidopsis* – *AtPPT1* may be involved in the provision of signals for correct mesophyll development. *Plant J.*, 36, 411–420.

Knight, J.S., Emes, M.J. and Debnam, P.M. (2001) Isolation and characterisation of a full-length genomic clone encoding a plastidic glucose 6-phosphate dehydrogenase from *Nicotiana tabacum*. *Planta*, 212, 499–507.

Knight, M.E., Harn, C., Lilley, C.E.R. *et al.* (1998) Molecular cloning of starch synthase I from maize (W64) endosperm and expression in *Escherichia coli*. *Plant J.*, 14, 613–622.

Konishi, T., Shinohara, K., Yamada, K. and Sasaki, Y. (1996) Acetyl-CoA carboxylase in higher plants: most plants other than Gramineae have both the prokaryotic and the eukaryotic forms of this enzyme. *Plant Cell Physiol.*, 37, 117–122.

Kossmann, J., Abel, G.J.W., Springer, F., Lloyd, J.R. and Willmitzer, L. (1999) Cloning and functional analysis of a cDNA encoding a starch synthase from potato (*Solanum tuberosum* L.) that is predominantly expressed in leaf tissue. *Planta*, 208, 503–511.

Kossmann, J., Sonnewald, U. and Willmitzer, L. (1994) Reduction of the chloroplastic fructose-16-bisphosphatase in transgenic potato plants impairs photosynthesis and plant growth. *Plant J.*, 6, 637–650

Kozaki, A., Kamada, K., Pagano, Y., Iguchi, H. and Sasaki, Y. (2000) Recombinant carboxyltransferase responsive to redox of pea plastidic acetyl-CoA carboxylase. *J. Biol. Chem.*, 275, 10702–10708.

Kozaki, A. and Sasaki, Y. (1999) Light-dependent changes in redox status of the plastidic acetyl-CoA carboxylase and its regulatory component. *Biochem. J.*, 339, 541–546.

Krapp, A. and Stitt, M. (1994) Influence of high-carbohydrate content on the activity of plastidic and cytosolic isoenzyme pairs in photosynthetic tissues. *Plant Cell Environ.*, 17, 861–866.

Krepinsky, K., Plaumann, M., Martin, W. and Schnarrenberger, C. (2001) Purification and cloning of chloroplast 6-phosphogluconate dehydrogenase from spinach – cyanobacterial genes for chloroplast and cytosolic isoenzymes encoded in eukaryotic chromosomes. *Eur. J. Biochem.*, 268, 2678–2686.

Kruger, N.J. and von Schaewen, A. (2003). The oxidative pentose phosphate pathway: structure and organisation. *Curr. Opin. Plant Biol.*, 6, 236–246.

Kubis, S.E., Pike, M.J., Everett, C.J., Hill, L.M. and Rawsthorne, S. (in press) The import of phosphoenol pyruvate by plastids from developing embryos of oilseed rape *Brassica napus* (L.) and its potential as a substrate for fatty acid synthesis. *J. Exp. Bot.*

Kuipers, A.G.J., Jacobsen, E. and Visser, R.G.F. (1994) Formation and deposition of amylose in the potato tuber are affected by the reduction of granule-bound starch synthase gene expression. *Plant Cell*, 6, 43–52.

La Cognata, U., Willmitzer, L. and Müller-Röber, B. (1995) Molecular cloning and characterisation of novel isoforms of potato ADP-glucose pyrophosphorylase. *Mol. Gen. Genet.*, 246, 538–548.

Lam, H.M., Coschigano, K.T., Oliveira, I.C., Melo-Oliveira, R. and Coruzzi, G.M. (1996) The molecular genetics of nitrogen assimilation into amino acids in higher plants. *Ann. Rev. Plant Physiol. Mol. Biol.*, 47, 569–593.

Lancien, M., Martin, M., Hsieh, M.H., Leustek, T., Goodman, H. and Coruzzi, G.M. (2002) *Arabidopsis* glt1-T mutant defines a role for NADH-GOGAT in the non-photorespiratory ammonium assimilatory pathway. *Plant J.*, 29, 347–358.

Lao, N.T., Schoneveld, O., Mould, R.M., Hibberd, J.M., Gray, J.C. and Kavanagh, T.A. (1999) An *Arabidopsis* gene encoding a chloroplast-targeted beta-amylase. *Plant J.*, 20, 519–527.

Laule, O., Furholz, A., Chang, H.-S. *et al.* (2003) Crosstalk between cytosolic and plastidial pathways of isoprenoid biosynthesis in *Arabidopsis thaliana*. *Proc. Natl. Acad. Sci.*, 100, 6866–6871.

Lawlor, D.W. (2002) The chemistry of photosynthesis. In *Photosythesis*, Bios Scientific, Oxford, Chapt. 7, pp. 139–183.

Lazaro, J.J., Sutton, C.W., Nicholson S. and Powls, R. (1986) Characterization of 2 forms of phosphoribulokinase isolated from the green-alga, *Scenedesmus obliqus*. *Eur. J. Biochem.*, 156 (2): 423–429.

Leegood, R.C., Lea, P.J., Adcock, M.D. and Häusler, R.E. (1995) The regulation and control of photorespiration. *J. Exp. Bot.*, 46, 1397–1414.

Lichtenthaler, H.K. (1999) The 1-deoxy-*d*-xylulose-5-phosphate pathway of isoprenoid biosynthesis in plants. *Annu. Rev. Plant Phys. Plant Mol. Biol.*, 50, 47–65.

Lichtenthaler, H.K., Rohmer, M. and Schwender, J. (1997) Two independent biochemical pathways for isopentenyl diphosphate and isoprenoid biosynthesis in higher plants. *Physiol. Plant.*, 101, 643–652.

Liedvogel, B. and Bäuerle, R. (1986) Fatty acid synthesis in chloroplasts from mustard (*Sinapis alba* L.) cotyledons: formation of acetyl coenzyme A by intraplastidic glycolytic enzymes and a pyruvate dehydrogenase complex. *Planta*, 169, 481–489.

Lin, M., Behal, R. and Oliver, D.J. (2003) Disruption of *plE2*, the gene for the E2 subunit of the plastid pyruvate dehydrogenase complex, in *Arabidopsis* causes and early embryo lethal phenotype. *Plant Mol. Biol.*, 52, 865–872.

Lin, T.P., Caspar, T., Somerville, C. and Preiss, J. (1988) A starch-deficient mutant of *Arabidopsis thaliana* with low ADPglucose pyrophosphorylase activity lacks one of the two subunits of the enzyme. *Plant Physiol.*, 88, 1175–1181.

Maddelein, M.L., Libessart, N., Bellanger, F., Delrue, B., D'Hulst, C. and Ball, S. (1994) Toward an understanding of the biogenesis of the starch granule: Determination of granule-bound and soluble starch synthase functions in amylopectin synthesis. *J. Biol. Chem.*, 269, 25150–25157.

Magasanik, B. (2000) PII: a remarkable regulatory protein. *Trends Microbiol.*, 8, 447–448.

Marshall, J., Sidebottom, C., Debet, M., Martin, C., Smith, A.M. and Edwards, A. (1996) Identification of the major starch synthase in the soluble fraction of potato tubers. *Plant Cell*, 8, 1121–1135.

Matsumara, T., Sakakibara, H., Nakano, R., Kimata, Y., Sugiyama, T. and Hase, T. (1997) A nitrate-inducible ferredoxin in maize roots. Genomic organisation and differential expression of two nonphotosynthetic ferredoxin isoproteins. *Plant Physiol.*, 114, 653–660.

Matt, P., Krapp, A., Haake, V., Mock, H.P. and Stitt, M. (2002) Decreased Rubisco activity leads to dramatic changes of nitrate metabolism, amino acid metabolism and in the levels of phenylpropanoids and nicotine in tobacco antisense RBCS transformants. *Plant J.*, 30, 663–677.

Meyer, Y., Migniac-Maslow, M., Schurmann, P. and Jacquot, J.-P. (2002) Protein–protein interactions in plant thioredoxin dependent systems. In *Annual Reviews, Vol. 7: Protein–Protein Interactions in Plant Biology* (eds M.T. Mcmanus, W.A. Laing and A. Allan), Sheffield Academic, Sheffield, England, Chapt. 1, pp. 1–23.

Miernyk, J.A. and Dennis, D.T. (1992) A developmental analysis of the enolase isoenzymes from *Ricinus communis*. *Plant Physiol.*, 99, 748–750.

Mifflin, B.J. and Lea, P.J. (1980). Ammonia assimilation. In *The Biochemistry of Plants*, Vol. 5 (ed. B.J. Mifflin), Academic Press, New York, pp. 169–202.

Miyawaga, Y., Tamoi, M. and Shigeoka, S. (2001) Overexpression of a cyanobacterial fructose-1,6-/sedoheptulose-1,7-bisphosphatase in tobacco enhances photosynthesis and growth. *Nat. Biotech.*, 19, 965–969.

Möhlmann, T., Scheibe, R. and Neuhaus, H.E. (1994) Interaction between starch synthesis and fatty-acid synthesis in isolated cauliflower-bud amyloplasts. *Planta*, 194, 492–497.

Möhlmann, T., Tjaden, J., Henrichs, G., Quick, W.P., Hausler, R. and Neuhaus, H.E. (1997) ADPglucose drives starch synthesis in isolated maize endosperm amyloplasts: characterisation of starch synthesis and transport properties across the amyloplast envelope. *Biochem. J.*, 324, 503–509.

Moorhead, G.B.G. and Smith, C.S. (2003). Interpreting the plastid carbon, nitrogen, and energy status. A role for PII? *Plant Physiol.*, 133, 492–498.

Morell, M.K., Blennow, A., Kosar-Hashemi, B. and Samuel, M.S. (1997) Differential expression and properties of starch branching enzyme isoforms in developing wheat endosperm. *Plant Physiol.*, 113, 201–208.

Morell, M.K., Kosar-Hashemi, B., Cmiel, M. *et al.* (2003) Barley *sex6* mutants lack starch synthase IIa activity and contain a starch with novel properties. *Plant J.*, 34, 173–185.

Mouille, G., Maddelein, M.-L., Libessart, N. *et al.* (1996) Phytoglycogen processing: a mandatory step for starch biosynthesis in plants. *Plant Cell*, 8, 1353–1366.

Mu-Forster, C., Huang, R., Powers, J.R. *et al.* (1996) Physical association of starch biosynthetic enzymes with starch granules of maize endosperm. Granule-associated forms of starch synthase I and starch branching enzyme II. *Plant Physiol.*, 111, 821–829.

Müller-Rober, B., Kossmann, J., Hannah, L.C., Willmitzer, L. and Sonnewald, U. (1990) Only one of two different ADPglucose pyrophosphorylase genes from potato responds strongly to elevated levels of sucrose. *Mol. Gen. Genet.*, 224, 136–146.

Müller-Rober, B., Sonnewald, U. and Willmitzer, L. (1992) Inhibition of ADP-glucose pyrophosphorylase in transgenic potatoes leads to sugar-storing tubers and influences tuber formation and expression of tuber storage protein genes. *EMBO J.*, 11, 1229–1238.

Myers, A.M., Morell, M.K., James, M.G. and Ball, S.G. (2000) Recent progress toward understanding the biosynthesis of the amylopectin crystal. *Plant Physiol.*, 122, 989–998.

Naeem, M., Tetlow, I.J. and Emes, M.J. (1997) Starch synthesis in amyloplasts purified from developing potato tubers. *Plant J.*, 11, 1095–1103.

Nakamura, T., Vrinten, P., Hayakawa, K. and Ikeda, J. (1998) Characterization of a granule-bound starch synthase isoform found in the pericarp of wheat. *Plant Physiol.*, 118, 451–459.

Nakamura, Y., Fujita, N., Kubo, A., Rahman, S., Morell, M. and Satoh, H.(2003) Engineering amylopectin biosynthesis in rice endosperm. *J. Appl. Glycosci.*, 50, 197–200.

Nakamura, Y. and Kawaguchi, K. (1992) Multiple forms of ADP-glucose pyrophosphorylase of rice endosperm. *Physiol. Plant*, 84, 336–342.

Nakamura, Y., Kubo, A., Shimamune, T., Matsuda, T., Harada, K. and Satoh, H. (1997) Correlation between activities of starch debranching enzymes and α-polyglucan structure in endosperms of *sugary-1* mutants of rice. *Plant J.*, 12, 143–153.

Neilsen, T.H., Krapp, A., Roper-Schwarz, U. and Stitt, M. (1998) The sugar-mediated regulation of genes encoding the small subunit of Rubisoc and the regulatory subunit of ADP glucose pyrophosphorylase is modified by phosphate and nitrogen. *Plant Cell Environ.*, 21, 443–454.

Nelson, O.E. and Rines, H.W. (1962) The enzymatic deficiency in the waxy mutant of maize. *Biochem. Biophys. Res. Commun.*, 9, 297–300.

Neuhaus, H.E. and Maass, U. (1996) Unidirectional transport of orthophosphate across the envelope of isolated cauliflower-bud amyloplasts. *Planta*, 198, 542–548.

Neuhaus, H.E., Thom, E., Batz, O. and Scheibe, R. (1993) Purification of highly intact plastids from various heterotrophic plant tissues. Analysis of enzyme equipment and precursor dependency for starch biosynthesis. *Biochem. J.*, 296, 395–401.

Neuhaus, H.E., Thom, E., Möhlmann, T., Steup, M. and Kampfenkel, K.(1997) Characterization of a novel ATP/ADP transporter from *Arabidopsis thaliana* L. *Plant J.*, 11, 73–82.

Nielsen, T.H., Baunsgaard, L. and Blennow, A. (2002) Intermediary glucan structures formed during starch granule biosynthesis are enriched in short side chains, a dynamic pulse labelling approach. *J. Biol. Chem.*, 277, 20249–20255.

Niittyä, T., Messerli, G., Trevisan, M., Chen, J., Smith, A.M. and Zeeman, S.C. (2004) A previously unknown maltose transporter essential for starch degradation in leaves. *Science*, 303, 87–89.

Nishi, A., Nakamura, Y., Tanaka, N. and Satoh, H. (2001) Biochemical and genetic effects of amylose- extender mutation in rice endosperm. *Plant Physiol.*, 127, 459–472.

Nishimura, M. and Beevers, H. (1979) Subcellular distribution of gluconeogenic enzymes in germinating castor bean endosperm. *Plant Physiol.*, 64, 31–37.

Olcer, H., Lloyd, J.C. and Raines, C.A. (2001) Photosynthetic capacity is differentially affected by reductions in sedoheptulose-1,7-bisphosphatase activity during leaf development in transgenic tobacco plants. *Plant Physiol.*, 125, 982–989.

Olive, M.R., Ellis, R.J. and Schuch, W.W. (1989) Isolation and nucleotide sequences of cDNA clones encoding ADPglucose pyrophosphorylase polypeptides from wheat leaf and endosperm. *Plant Mol. Biol.*, 12, 525–538.

Overlach, S., Diekmann, W. and Raschke, K. (1993) Phosphate translocator of isolated guard-cell chloroplasts from *Pisum sativum* L. transports glucose-6-phosphate. *Plant Physiol.*, 101, 1201–1207.

Parry, M.A.J., Andralojc, P.J., Mitchell, R.A.C., Madgwick, P.J. and Keys, A.J. (2003) Manipulation of Rubisco: the amount, the activity, function and regulation. *J. Exp. Bot.*, 54, 1321–1333.

Paul, M.J., Knight, J.S., Habash, D. *et al.* (1995) Reduction in phosphoribulokinase activity by antisense RNA in transgenic tobacco: effect on CO_2 assimilation and growth at low irradiance. *Plant J.*, 7, 535–542.

Pilling, E. and Smith, A.M. (2003) Growth ring formation in the starch granules of potato tubers. *Plant Physiol.*, 132, 365–371.

Plaxton, W.C. (1996) The organization and regulation of plant glycolysis. *Ann. Rev. Plant Biol.*, 47, 185–214.

Poolman, M., Fell, D. and Raines, C.A. (2003) Elementary modes analysis of photosynthate metabolism in the chloroplast stroma. *Eur. J. Biochem.*, 270, 430–439.

Poolman, M.G., Fell, D.A. and Thomas, S. (2000) Modelling photosynthesis and its control. *J. Exp. Bot.*, 51, 319–328.

Portis, A.R. (2002) The Rubisco activase – Rubisco system: an ATPase-dependent association that regulates photosynthesis. In *Annual Reviews, Vol. 7: Protein–Protein Interactions in Plant Biology* (eds M.T. Mcmanus, W.A. Laing and A. Allan), Sheffield Academic, Sheffield, England, Chapt. 2, pp. 30–52.

Post-Beittenmiller, D., Jaworski, J.G. and Ohlrogge, J.B. (1991) *In vivo* pools of free and acylated acyl carrier proteins in spinach. Evidence for sites of regulation of fatty acid biosynthesis. *J. Biol. Chem.*, 266, 1858–1865.

Post-Beittenmiller, D., Roughan, G. and Ohlrogge, J.B. (1992) Regulation of plant fatty acid biosynthesis. *Plant Physiol.*, 100, 923–930.

Pozueta-Romero, J., Frehner, M., Viale, A.M. and Akazawa, T. (1991) Direct transport of ADP-glucose by adenylate translocator is linked to starch biosynthesis in amyloplasts. *Proc. Natl. Acad. Sci. U.S.A.*, 88, 5769–5773.

Preiss, J. (1991) Biology and molecular biology of starch synthesis and its regulation. In *Oxford Surveys of Cellular and Molecular Biology*, Vol.7 (ed. B.J. Miflin), Oxford University Press, Oxford, UK, pp. 59–114.

Preiss, J. and Sivak, M. (1996) Starch synthesis in sinks and sources. In *Photoassimilate Distribution in Plants and Crops*, Marcel Dekker, New York, pp. 63–69.

Price, G.D., Evans, J.R., Caemmerer, S. von, Yu, J.-W. and Badger, M.R. (1995) Specific reduction of chloroplast glyceraldehyde-3-phosphate dehydrogenase activity by antisense RNA reduces CO_2 assimilation via a reduction in ribulose bisphosphate regeneration in transgenic plants. *Planta*, 195, 369–378.

Prioul, J.-L., Jeanette, E., Reyss, A. *et al.* (1994) Expression of ADPglucose pyrophosphorylase in maize (*Zea mays* L.) grain and source leaf during grain filling. *Plant Physiol.*, 104, 179–187.

Qi, Q., Kleppinger-Sparace, K.F. and Sparace, S.A. (1994). The role of the triose-phosphate shuttle and glycolytic intermediates in fatty-acid and glycerolipid biosynthesis in pea root plastids. *Planta*, 194, 193–199.

Qi, Q., Kleppinger-Sparace, K.F. and Sparace, S.A. (1995) The utilization of glycolytic inter-mediates as precursors for fatty acid biosynthesis by pea root plastids. *Plant Physiol.*, 107, 413–419.

Quick, W.P. and Neuhaus, H.E. (1997) The regulation and control of photosynthetic carbon assimilation. In *A Molecular Approach to Primary Metabolism in Higher Plants* (eds C.H. Foyer and W.P. Quick), Taylor & Francis, London, pp. 41–62.

Quick, W.P., Scheibe, R. and Neuhaus, H.E. (1995) Induction of hexose-phosphate translocator activity in spinach chloroplasts. *Plant Physiol.*, 109, 113–121.

Rahman, S., Regina, A., Li, Z. *et al.* (2001) Comparison of starch-branching enzyme genes reveals evolutionary relationships among isoforms. Characterization of a gene for starch-branching enzyme IIa from wheat D genome donor *Aegilops tauschii*. *Plant Physiol.*, 125, 1314–1324.

Raines, C.A. (2003) The Calvin cycle revisited. *Photosynth. Res.* 75, 1–10.

Raines, C.A., Harrison, E.P., Olcer, H. and Lloyd, J.C. (2000) Investigating the role of the thiol- regulated enzyme sedoheptulose-1,7-bisphosphatase in the control of photosynthesis. *Physiol. Plant*, 110, 303–308.

Raines, C.A., Lloyd, J.C. and Dyer, T.A. (1991) Molecular biology of the C3 – photosynthetic carbon–reduction cycle. *Photosynth. Res.*, 27, 1–14.

Redinbaugh, M.G. and Campbell, W.H. (1998) Nitrate regulation of the oxidative pentose phos-phate pathway in maize (*Zea mays* L.) root plastids: induction of 6-phosphogluconate dehydrogenase activity, protein and transcript levels. *Plant Sci.*, 134, 129–140.

Riesmeier, J.W., Flügge, U.-I., Schulz, B., Heineke, D. and Heldt, H.W. (1993) Antisense repres-sion of the chloroplast triose phosphate translocator affects carbon partitioning in transgenic potato plants. *Proc. Natl. Acad. Sci. U.S.A.*, 90, 6160–6164.

Ritte, G., Lloyd, J.R., Eckermann, N., Rotmann, A., Kossmann, J. and Steup, M. (2002) The starch related R1 protein is an α-glucan, water dikinase. *Proc. Natl. Acad. Sci. U.S.A.*, 99, 1766–1771.

Robinson, S.P. and Wiskich, J.T. (1977) Uptake of ATP analogs by isolated pea chloroplasts and their effect on CO_2 fixation and electron transport. *Biochim. Biophys. Acta*, 461, 131–140.

Roesler, K., Shintani, D., Savage, L., Boddupalli, S. and Ohlrogge, J. (1997) Targetting of the *Arabidopsis* homomeric acetyl-Coenzyme A carboxylase to plastids of rapeseeds. *Plant Physiol.*, 113, 75–81.

Rogers, A., Fischer, B.U., Bryant, J. *et al.* (1998) Acclimation of photosynthesis to elevated CO2 under low-nitrogen nutrition is affected by the capacity for assimilate utilization. Perennial ryegrass under free-air CO_2 enrichment. *Plant Physiol.*, 118, 683–689.

Rontein, D., Dieuaide-Noubhani, M., Dufourc, E.J., Raymond, P. and Rolin, D. (2002) The metabolic architecture of plant cells: stability of central metabolism and flexibility of an-abolic pathways during the growth of tomato cells. *J. Biol. Chem.*, 277, 43948–43960.

Rost, S., Frank, C. and Beck, E. (1996) The chloroplast envelope is permeable for maltose but not for maltodextrins. *Biochim. Biophys. Acta*, 1291, 221–227.

Roughan, P.G., Holland, R., Slack, C.R. and Mudd, J.B. (1979) Acetate is the preferred substrate for long-chain fatty acid synthesis in isolated spinach chloroplasts. *Biochem. J.*, 184, 565–569.

Ruelland, E. and Miginiac-Maslow, M. (1999) Regulation of chloroplast enzyme activities by thioredoxins: activation or relief from inhibition? *Trends Plant Sci.*, 4, 136–141.

Satoh, H., Nishi, A., Fujita, N. *et al.* (2003) Isolation and characterization of starch mutants in rice. *J. Appl. Glycosci.*, 50, 225–230.

Sauer, A. and Heise, K.P. (1983) On the light dependence of fatty-acid synthesis in spinach-chloroplasts. *Plant Physiol.*, 73, 11–15.

Sasaki. Y., Konishi, T. and Nagano, Y. (1995) The compartmentation of acetyl-coenzyme A carboxylase in plants. *Plant Physiol.*, 108, 445–449.

Sasaki, Y., Kozaki, A. and Hatano, M. (1997) Link between light and fatty acid synthesis: thioredoxin- linked reductive activation of plastidic acetyl-CoA carboxylase. *Proc. Natl. Acad. Sci. U.S.A.*, 94, 11096–11101.

Sassenrath-Cole, G.F. and Piercy, R.W. (1992) The role of ribulose-1,5-bisphosphate regeneration in the induction of photosynthetic CO_2 exchange under transient light conditions. *Plant Physiol.*, 99, 227–234.

Sassenrath-Cole, G.F. and Piercy, R.W. (1994) Regulation of photosynthetic induction state by the magnitude and duration of low-light exposure. *Plant Physiol.*, 105, 1115–1123.

Satoh, H., Nishi, A., Fujita, N. *et al.* (2003) Isolation and characterization of starch mutants in rice. *J. Appl. Glycosci.*, 50, 225–230.

Sauer, A. and Heise, K.P. (1983) On the light dependence of fatty-acid synthesis in spinach-chloroplasts. *Plant Physiol.*, 73, 11–15.

Schäfer, G., Heber, U. and Heldt, H.W. (1977) Glucose transport into spinach chloroplasts. *Plant Physiol.*, 60, 286–289.

Scheibe, R. (1991) Redox modulation of chloroplast enzymes. *Plant Physiol.*, 96, 1–3.

Scheibe, R., Wedel, N., Vetter, S., Emmerlich, V. and Sauermann, S.M. (2002) Co-existence of two regulatory NADP-glyceraldehyde 3-P dehydrogenase complexes in higher plant chloroplasts. *Eur. J. Biochem.*, 269, 5617–5624.

Scheidig, A., Frölich, A., Schulze, S., Lloyd, J.R. and Kossmann, J. (2002) Down-regulation of a chloroplast-targeted β-amylase leads to a starch-excess phenotype in leaves. *Plant J.*, 30, 581–591.

Schindler, I., Renz, A., Schmid, F.X. and Beck, E. (2001) Activation of spinach pullulanase by reduction results in a decrease in the number of isomeric forms. *Biochim. Biophys. Acta*, 1548, 175–186.

Schleucher, J., Vanderveer, P.J. and Sharkey, T.D. (1998) Export of carbon from chloroplasts at night. *Plant Physiol.*, 118, 1439–1445.

Schnarrenberger, C., Flechner, A. and Martin, W. (1995) Enzymatic evidence for a complete oxidative pentose phosphate pathway in chloroplasts and an incomplete pathway in the cytosol of spinach leaves. *Plant Physiol.*, 108, 609–614.

Schoenbeck, M.A., Temple, S.J., Trepp, G.B. *et al.* (2000) Decreased NADH glutamate synthase activity in nodules and flowers of alfalfa (*Medicago sativa* L.) transformed with an antisense glutamate synthase transgene. *J. Exp. Bot.*, 51, 29–39.

Schott, K., Borchert, S., Müller-Röber, B. and Heldt, H.W. (1995) Transport of inorganic phosphate and C_3- and C_6-sugar phosphates across the envelope membranes of potato tuber amyloplasts. *Planta*, 196, 647–652.

Schulte, W., Töpfer, R., Stracke, R., Schell, J. and Martini, N. (1997) Multi-functional acetyl-CoA carboxylase from *Brassica napus* is encoded by a multi-gene family: indication for plastidic localization of at least one isoform. *Proc. Natl. Acad. Sci. U.S.A.*, 94, 3456–3470.

Schulz, B., Frommer, W.B., Flügge, U.-I., Hummel, S., Fischer, K. and Willmitzer, L. (1993) Expression of the triose phosphate translocator gene from potato is light dependent and restricted to green tissues. *Mol. Gen. Genet.* 238, 357–361.

Schünemann, D., Borchert, S., Flügge, U.-I. and Heldt, H.W. (1993) ATP/ADP translocator from pea root plastids. Comparison with translocators from spinach chloroplasts and pea leaf mitochondria. *Plant Physiol.*, 103, 131–137.

Schurman, P. and Jacquot, J.-P. (2000) Plant thioredoxin systems revisited. *Annu. Rev. Plant Biol.* 51, 371–400.

Schwender, J. and Ohlrogge, J.B. (2002) Probing in vivo metabolism by stable isotope labelling of storage lipids and proteins in developing *Brassica napus* embryos. *Plant Physiol.*, 130, 347–361.

Schwender, J., Sacher-Hill, Y. and Ohlrogge, J.B. (2003) A flux model of glycolysis and the oxidative pentose phosphate pathway in developing *Brassica napus* embryos. *J. Biol. Chem.*, 278, 29442–29453.

Sehnke, P.C., Chung, H.-J., Wu, K. and Ferl, R.J. (2001) Regulation of starch accumulation by granule-associated plant 14-3-3 proteins. *Proc. Natl. Acad. Sci. U.S.A.*, 98, 765–770.

Sehnke, P.C., Henry, R., Cline, K. and Ferl, R.J. (2000) Interaction of a plant 14-3-3 protein with the signal peptide of a thylakoid-targeted chloroplast precursor protein and the presence of 14-3-3 isoforms in the chloroplast stroma. *Plant Physiol.*, 122, 235–240.

Sellwood, C., Slabas, A.R. and Rawsthorne S. (2000) Effects of manipulating expression of acetyl-CoA carboxylase I in *Brassica napus* L. embryos. *Biochem. Soc. Trans.*, 28, 598–600.

Seo, B.-S., Kim, S., Scott, M.P. *et al.* (2002) Functional interactions between heterologously expressed starch-branching enzymes of maize and glycogen synthases of brewer's yeast. *Plant Physiol.*, 128, 1189–1199.

Shannon, J.C., Pien, F.-M., Cao, H.P. and Lui, K.C. (1998) Brittle-1, an adenylate translocator, facilitates transfer of extraplastidial synthesized ADP-glucose into amyloplasts of maize endosperms. *Plant Physiol.*, 117, 1235–1252.

Shannon, J.C., Pien, F.-M. and Lui, K.C. (1996) Nucleotides and nucleotide sugars in developing maize endosperms: synthesis of ADPglucose in brittle-1. *Plant Physiol.*, 110, 835–843.

Sikka, V.K., Choi, S., Kavakli, I.H. *et al.* (2001) Subcellular compartmentation and allosteric regulation of the rice endosperm ADPglucose pyrophosphorylase. *Plant Sci.*, 161, 461–468.

Slabas, A.R. and Fawcett, T. (1992) The biochemistry and molecular biology of plant lipid biosynthesis. *Plant Mol. Biol.*, 19, 169–191.

Smith, A.M., Denyer, K. and Martin, C. (1997) The synthesis of the starch granule. *Ann. Rev. Plant Physiol. Plant Mol. Biol.*, 48, 67–87.

Smith, A.M., Zeeman, S., Niittylä, T., Kofler, H., Thorneycroft, D. and Smith, S.M. (2003a) Starch degradation in leaves. *J. Appl. Glycosci.*, 50, 173–176.

Smith, C., Weljie, A.M. and Moorhead, G.B.G. (2003b) Molecular properties of the putative nitrogen sensor PII from *Arabidopsis thaliana. Plant J.*, 33, 353–360.

Smith, R.G., Gauthier, D.A., Dennis, D.T. and Turpin, D.H. (1992) Malate- and pyruvate-dependent fatty acid synthesis in leucoplasts from developing castor endosperm. *Plant Physiol.*, 98, 1233–1238.

Spreitzer, R. (1993) Genetic dissection of Rubisco structure and function. *Annu. Rev. Plant Physiol. Plant Mol. Biol.* 44, 411–434.

Spreitzer, R. and Salvucci, M.E. (2002) Rubisco: structure, regulatory interactions, and possibilities for a better enzyme. *Annu. Rev. Plant Biol.* 53, 449–475.

Springer, J. and Heise, K.P. (1989) Comparison of acetate-dependent and pyruvate-dependent fatty-acid synthesis by spinach-chloroplasts. *Planta*, 177, 417–421.

Stark, D.M., Timmerman, K.P., Barry, G.F., Preiss, J. and Kishore, G.M. (1992) Regulation of the amount of starch in plant tissues by ADP glucose pyrophosphorylase. *Science*, 258, 287–292.

Steup, M., Robenek, H. and Melkonian, M. (1983) In vitro degradation of starch granules isolated from spinach chloroplasts. *Planta*, 158, 428–436.

Stitt, M. and ap Rees, T. (1979) Capacities of pea chloroplasts to catalyse the oxidative pentose phosphate pathway and glycolysis. *Phytochemistry*, 18, 1905–1911.

Stitt, M. and Hurry, V. (2002) A plant for all seasons: alterations in the photosynthetic carbon metabolism during cold acclimation in *Arabidopsis. Curr. Opin. Plant Biol.*, 5, 199–206.

Stitt, M. and Krapp, A. (1999) The interaction between elevated carbon dioxide and nitrogen nutrition: the physiological and molecular background. *Plant Cell Environ.*, 22, 583–621.

Stitt, M. and Schulze, E.-D. (1994) Does Rubisco control the rate of photosynthesis and plant growth? An exercise in molecular ecophysiology. *Plant Cell Environ.*, 17, 465–487.

Stitt, M. and Steup, M. (1985) Starch and sucrose degradation. In *Encyclopedia of Plant Physiology*, Vol. 18 (eds R. Douce and D.A. Day), Springer-Verlag, Heidelberg, pp. 347–390.

Strand, A., Asami, T., Alonso, J., Ecker, J.R. and Chory, J. (2002) Chloroplast to nucleus communication triggered by accumulation of Mg- protoporphoryn IX. *Nature*, 421, 79–83.

Streatfield, S.J., Weber, A., Kinsman, E.A. *et al.* (1999) The phosphoenolpyruvate/phosphate translocator is required for phenolic metabolism, palisade cell development, and plastid-dependent nuclear gene expression. *Plant Cell*, 11, 1609–1622.

Sullivan, T.D. and Kaneko, Y. (1995) The maize brittle1 gene encodes amyloplast membrane polypeptides. *Planta*, 196, 477–484.

Suzuki, A., Rioual, S., Lemarchand, S. *et al.* (2001) Regulation by light and metabolites of ferredoxin-dependent glutamate synthase in maize. *Physiol Plant*, 112, 524–530.

Takeda, Y., Guan, H.-P. and Preiss, J. (1993) Branching of amylose by the branching isoenzymes of maize endosperm. *Carbohydr. Res.*, 240, 253–263.

Tanaka, T., Ida, A., Irifune, K., Oeda, K. and Morikawa, H. (1994) Nucleotide sequence of a gene for nitrite reductase from *Arabidopsis thaliana*. *J. DNA Seq. Mapp.*, 5, 57–61.

Tatge, H., Marshall, J., Martin, C., Edwards, E.A. and Smith, A.M. (1999) Evidence that amylase synthesis occurs within the matrix of the starch granule in potato tubers. *Plant Cell Environ.*, 22, 543–550.

Temple, S.J., Vance, C.J. and Gantt, J.S. (1998) Glutamate synthase and nitrogen assimilation. *Trends Plant Sci.*, 3, 51–56.

Tetlow, I.J., Blissett, K.J. and Emes, M.J. (1994) Starch synthesis and carbohydrate oxidation in amyloplasts from developing wheat endosperm. *Planta*, 194, 454–460.

Tetlow, I.J., Blissett, K.J. and Emes, M.J. (1998) Metabolite pools during starch synthesis and carbohydrate oxidation in amyloplasts isolated from wheat endosperm. *Planta*, 204, 100–108.

Tetlow, I.J., Bowsher, C.G. and Emes, M.J. (1996) Reconstitution of the hexose phosphate translocator from the envelope membranes of wheat endosperm amyloplasts. *Biochem. J.*, 319, 717–723.

Tetlow, I.J., Bowsher, C.G. and Emes, M.J. (2003a) Biochemical properties and enzymic capacities of chromoplasts isolated from wild buttercup (*Ranunculus acris* L.). *Plant Sci.*, 165, 383–394.

Tetlow, I.J., Bowsher, C.G., Scrase-Field, E.F.A.L., Davies, E.J. and Emes, M.J. (2003b) The synthesis and transport of ADPglucose in cereal endosperms. *J. Appl. Glycosci.*, 50, 231–236.

Tetlow, I.J., Davies, E.J., Vardy, K.A., Bowsher, C.G., Burrell, M.M. and Emes, M.J. (2003c) Subcellular localization of ADPglucose pyrophosphorylase in developing wheat endosperm and analysis of a plastidial isoform. *J. Exp. Bot.*, 54, 715–725.

Tetlow, I.J., Wait, R., Lu, Z. *et al.* (2004) Protein phosphorylation in amyloplasts regulates starch branching enzyme activity and protein–protein interactions. *Plant Cell*, 16, 694–708.

Thom, E., Möhlmann, T., Quick, W.P., Camara, B. and Neuhaus, H.E. (1998) Sweet pepper plastids: enzymic equipment, characterisation of the plastidic oxidative pentose-phosphate pathway, and transport of phosphorylated intermediates across the envelope membrane. *Planta*, 204, 226–233.

Thompson, D.B. (2000) On the non-random nature of amylopectin branching. *Carbohydr. Polym.*, 43, 223–239.

Thorbjørnsen, T., Villand, P., Denyer, K., Olsen, O.A. and Smith, A.M. (1996) Distinct isoforms of ADPglucose pyrophosphorylase occur inside and outside the amyloplasts in barley endosperm. *Plant J.*, 10, 243–250.

Tiessen, A., Hendriks, J.H.M., Stitt, M. *et al.* (2002) Starch synthesis in potato tuber is regulated by post-translational redox modification of ADP-glucose pyrophosphorylase. *Plant Cell*, 14, 2191–2213.

Tjaden, J., Möhlmann, T., Kampfenkel, K., Henrichs, G. and Neuhaus, H.E. (1998a) Altered plastidic ATP/ADP-transporter activity influences potato (*Solanum tuberosum* L.) tuber morphology, yield and composition of tuber starch. *Plant J.*, 16, 531–540.

Tjaden, J., Schwöppe, C., Möhlmann, T. and Neuhaus, H.E. (1998b) Expression of the plastidic ATP/ADP transporter gene in *Escherichia coli* leads to the presence of a functional adenine nucleotide transport system in the bacterial cytosolic membrane. *J. Biol. Chem.*, 273, 9630–9636.

Trepp, G.B., Plank, D.W., Gantt, J.S. and Vance, C.P. (1999) NADH-glutamate synthase in alfalfa root nodules. Immunocytochemical localization. *Plant Physiol.*, 119, 829–837.

Trimming, B.A. and Emes, M.J. (1993) Glycolytic enzymes in non-photosynthetic plastids of pea (*Pisum sativum* L.) roots. *Planta*, 190, 439–445.

Van de Wal, M., D'Hulst, C., Vincken, J.-P., Buléon, A., Visser, R. and Ball, S. (1998) Amylose is synthesized *in vitro* by extension of and cleavage from amylopectin. *J. Biol. Chem.*, 273, 22232–22240.

Villand, P., Aalen, R., Olsen, O.-A., Lonneborg, A., Lüthi, E. and Kleczkowski, L.A. (1992a). PCR-amplification and sequence of cDNA clones for the small and large subunits of ADP-glucose pyrophosphorylase from barley tissues. *Plant Mol. Biol.*, 19, 381–389.

Villand, P., Olsen, O.-A., Killan, A. and Kleczkowski, L.A. (1992b). ADPglucose pyrophosphorylase large subunit cDNA from barley endosperm. *Plant Physiol.*, 100, 1617–1618.

Vincentz, M., Moureaux, T., Leydecker, M.T., Vaucheret, H. and Caboche, M. (1993) The regulation of nitrate and nitrite reductase expression in *Nicotiana plumbaginofolia* leaves by carbon and nitrogen metabolites. *Plant J.*, 3, 315–324.

Visser, R.G.F., Somhorst, I., Kuipers, G.J., Ruys, N.J., Feenstra, W.J. and Jacobsen, E. (1991) Inhibition of expression of the gene for granule-bound starch synthase in potato by antisense constructs. *Mol. Gen. Genet.*, 225, 289–296.

Von Caemmerer, S. (2000) *Biochemical Models of Leaf Photosynthesis*, CSIRO Publishing, Collingwood, Ontario.

von Schaewen, A., Langenkämper, G., Graeve, K., Wenderoth, I. and Scheibe, R. (1995) Molecular characterisation of the plastidic glucose-6-phosphate dehydrogenase from potato in comparison to its cytosolic counterpart. *Plant Physiol.*, 109, 1327–1335.

Vrinten, P. and Nakamura, T. (2000) Wheat granule-bound starch synthase I and II are encoded by separate genes that are expressed in different tissues. *Plant Physiol.*, 122, 255–263.

Wang, S.-J., Yeh, K.-W. and Tsai, C.-Y. (2001) Regulation of starch granule-bound starch synthase I gene expression by circadian clock and sucrose in the source tissue of sweet potato. *Plant Sci.*, 161, 635–644.

Weaire, B.P. and Kekwick, R.G.O. (1975) The synthesis of fatty acids in avocado mesocarp and cauliflower bud tissue. *Biochem. J.*, 146, 425–437.

Weber, A., Menzlaff, E., Arbinger, B., Gutensohn, M., Eckerskorn, C. and Flügge, U.-I. (1995) The 2-oxoglutarate/malate translocator of chloroplast envelope membranes: molecular cloning of a transporter containing a 12-helix motif and expression of the functional protein in yeast cells. *Biochemistry*, 34, 2621–2627.

Weber, A., Servaites, J.C., Geiger, D.R. *et al.* (2000) Identification, purification, and molecular cloning of a putative plastidic glucose translocator. *Plant Cell*, 12, 787–801.

Weber, H., Heim, U., Borisjuk, L. and Wobus, U. (1995) Cell-type specific, coordinate expression of two ADPglucose pyrophosphorylase genes in relation to starch biosynthesis during seed development in *Vicia faba* L. *Planta*, 195, 352–361.

Wedel, N. and Soll, J. (1998) Evolutionary conserved light regulation of Calvin cycle activity by NADPH-mediated reversible phosphoribulokinase/CP12/glyceraldehyde-3-phosphate dehydrogenase complex dissociation. *Proc. Natl. Acad. Sci. U.S.A.*, 95, 9699–9704.

Wedel, N., Soll, J. and Paap, B.K. (1997) CP12 provides a new mode of light regulation of Calvin cycle activity in higher plants. *Proc. Natl. Acad. Sci. U.S.A.*, 94, 10479–10484.

Weiner, H., Stitt, M. and Heldt, H.W. (1987) Subcellular compartmentation of pyrophosphate and alkaline pyrophosphatase in leaves. *Biochim. Biophys. Acta*, 893, 13–21.

Wenderoth, I., Scheibe, R. and von Schaewen, A. (1997) Identification of the cysteine residues involved in redox modification of plant plastidic glucose-6-phosphate dehydrogenase. *J. Biol. Chem.*, 272, 26985–26990.

Wendt, U.A., Wenderoth, I., Tegeler, A. and von Schaewen, A. (2000) Molecular characterisation of a novel glucose-6-phosphate dehydrogenase from potato (*Solanum tuberosum* L.). *Plant J.*, 23, 723–733.

Weier, T.E. (1942) A cytological study of the carotene in the root of *Daucus carota* under various experimental treatments. *Am. J. Bot.*, 29, 35–44.

Wiese, A., Gröner, F., Sonnewald, U. *et al.* (1999) Spinach hexokinase I is located in the outer envelope membrane of plastids. *FEBS Lett.*, 461, 13–18.

Willms, J.R., Salon, C. and Layzell, D.B. (2000) Evidence for light-stimulated fatty acid synthesis in soybean fruit. *Plant Physiol.*, 120, 1117–1127.

Winkler, H.H. (1991) Molecular biology of Rickettsia. *Eur. J. Epidemiol.*, 7, 207–212.

Woodrow, I.E. and Berry, J.A. (1988) Enzymic regulation of photosynthetic CO_2 fixation in C_3 plants. *Annu. Rev. Plant Physiol. Plant Mol. Biol.*, 39, 533–594.

Wright, D.P., Huppe, H.C. and Turpin, D.H. (1997) *In vivo* and *in vitro* studies of glucose 6-phosphate dehydrogenase from barley root plastids in relation to reductant supply for NO_2^- assimilation. *Plant Physiol.*, 114, 1413–1419.

Yu, T.S., Kofler, H., Häusler, R.E. *et al.* (2001) The Arabidopsis *sex*1 mutant is defective in the R1 protein, a general regulator of starch degradation, and not in the chloroplastic hexose transporter. *Plant Cell*, 13, 1907–1918.

Zeeman, S.C., Northrop, F., Smith, A.M. and ap Rees, T. (1998a) A starch-accumulating mutant of *Arabidopsis thaliana* deficient in a chloroplastic starch-hydrolysing enzyme. *Plant J.*, 15, 357–365.

Zeeman, S.C., Tiessen, A., Pilling, E., Kato, L., Donald, A.M. and Smith, A.M. (2002). Starch synthesis in *Arabidopsis*; granule synthesis, composition, and structure. *Plant Physiol.*, 129, 516–529.

Zeeman, S.C., Umemoto, T., Lue, W.L. *et al.* (1998b). A mutant of *Arabidopsis* lacking a chloroplastic isoamylase accumulates both starch and phytoglycogen. *Plant Cell*, 10, 1699–1712.

Ziegler, P. and Beck, E. (1986) Exoamylase activity in vacuoles isolated from pea and wheat leaf protoplasts. *Plant Physiol.*, 82, 1119–1121.

4 Plastid division in higher plants

Simon Geir Møller

4.1 Introduction

All plant cells (except for pollen) have plastids, which are derived from undifferentiated proplastids found in dividing meristematic cells, and during cell differentiation, proplastids differentiate into a spectrum of plastid types, depending on the cell type (Pyke, 1999). Plastids are not created *de novo*, but arise by division from existing plastids in the cytoplasm, and the division process is essential for the maintenance of plastid populations in dividing cells and, for instance, in the accumulation of large numbers of chloroplasts in photosynthetic cells. The division process itself comprises an elaborate pathway of coordinated events, including assembly of the division machinery at the division site, the constriction of inner and outer envelope membranes, membrane envelope fusion at late stages of constriction and ultimately the separation of the two new organelles.

Because of their prokaryotic origin, plastid division, as for many plastid processes, share common features with bacterial division. Plastid division is initiated by the polymerisation of FtsZ proteins which form a contractile Z-ring at the site of division (Osteryoung *et al.*, 1998; Strepp *et al.*, 1998; Miyagishima *et al.*, 2001c; Mori *et al.*, 2001; Vitha *et al.*, 2001; Kuroiwa *et al.*, 2002). Plant FtsZs were identified based on their similarity to the bacterial FtsZ protein involved in septum formation during cell division (Lutkenhaus *et al.*, 1980). In contrast to the one FtsZ protein found in bacteria, plants harbour at least two types of FtsZ proteins (Mandrel, *et al.*, 2001) acting together at the division site. The correct placement of the Z-ring during initiation is mediated by the MinD and MinE proteins. As for FtsZ, MinD and MinE were identified based on their similarity to their bacterial counterparts (Colletti *et al.*, 2000; Itoh *et al.*, 2001; Maple *et al.*, 2002). In bacteria, MinD acts together with the topological specificity factor MinE, ensuring that FtsZ polymerisation occurs only at midcell (Hu and Lutkenhaus, 1999; Raskin and de Boer, 1999; Rowland *et al.*, 2000; Fu *et al.*, 2001; Shih *et al.*, 2003). Similarly, during chloroplast division, MinD acts together with the topological specificity factor MinE, ensuring FtsZ polymerisation occurs only at the central division site (Maple and Møller, unpublished results, 2004). In contrast to FtsZ, MinD and MinE localise to discrete polar regions inside chloroplasts (Maple *et al.*, 2002).

Following division site placement, constriction takes place. The constriction event is driven by electron-dense structures termed *plastid-dividing (PD) rings* (Hashimoto, 1986; Mita *et al.*, 1986; Tewinkel and Volkmann, 1987; Oross and

Possingham, 1989; Duckett and Ligrone, 1993; Kuroiwa *et al.*, 1998), which are separate from the Z-ring (Miyagishima *et al.*, 2001c; Kuroiwa *et al.*, 2002). The cytoplasmic outer PD ring and the stromal inner PD ring act in concert during constriction but appear to have distinct roles. The inner PD ring acts as a transient constriction collar that disassembles prior to completed constriction, whilst the outer PD ring remains attached to the cytosolic surface until after completed division (Miyagishima *et al.*, 2001a). The PD ring composition in higher plants remains unknown; however, in red alga the outer PD ring consists of 5-nm bundles comprising globular proteins (Miyagishima *et al.*, 2001b). The involvement of a cytosolic dynamin-like protein during plastid division in *Arabidopsis* (Gao *et al.*, 2003) and in red alga (Miyagishima *et al.*, 2003) shows that dynamins play an important role during plastid constriction.

The *accumulation and replication of chloroplasts* (*arc*) mutants represent an invaluable source of new plastid division components (Pyke, 1997, 1999; Marrison *et al.*, 1999), and the recent cloning of several *arc* loci have identified a dynamin-like protein (Gao *et al.*, 2003) and a J-domain protein (Vitha *et al.*, 2003) involved in *Arabidopsis* plastid division, in addition to shedding light on the mode of action of the *Arabidopsis* MinD protein (Fujiwara *et al.*, 2004). The continued cloning of the remaining nine *arc* loci will undoubtedly add to our knowledge of plastid division in higher plants.

ARTEMIS, a GTPase involved in late stages of plastid division (Fulgosi *et al.*, 2002), and GIANT CHLOROPLAST 1, involved in early stages of the division process (Maple *et al.*, 2004), are cyanobacterial cell division descendants. Both proteins are inner envelope associated and represent yet another added complexity to the process of plastid division in higher plants.

Plastid division clearly represents a complex but fundamental biological process involving a spectrum of different protein components. During the last 5 years, our understanding of plastid division in higher plants has increased dramatically largely because of the variety of approaches taken to dissect the process. This chapter summarises recent advances in the field and attempts to bring together the various findings into a coherent pathway of events.

4.2 The morphology of plastid division

During recent years, numerous protein components involved in plastid division have been isolated and our knowledge of the molecular complexity of the division process is rapidly increasing. However, it is important to view the individual protein components in the context of the morphological and ultrastructural changes that take place during the division process. Although the changes that take place are complex, the pathway of events can be broken down into a basic series of steps, as shown in Figure 4.1 (Leech *et al.*, 1981; Tewinkel and Volkmann, 1987; Oross and Possingham, 1989; Robertson *et al.*, 1996). Plastid division is initiated by slight plastid elongation, followed by a constriction at the centre of the plastid.

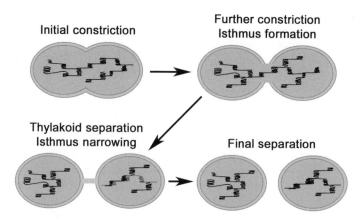

Figure 4.1 Schematic overview of the plastid division pathway in plants. Plastid division is initiated by slight elongation, followed by further constriction and isthmus formation. Later stages of constriction involve isthmus narrowing and separation of the thylakoid membranes. The isthmus then breaks, followed by separation of the two daughter plastids.

Constriction then continues, leading to the formation of a thin isthmus joining the two daughter plastids. During later stages of isthmus narrowing, the thylakoid membranes become separated into the two daughter plastids; the isthmus then breaks, followed by envelope membrane resealing. Recent research has shed light on the individual steps in this process and this is described in the following sections.

4.2.1   Early observations

Although it is becoming evident that the division process is probably similar for all plastid types, the majority of information regarding morphological division events is based on observations of chloroplasts. This is mainly because they are easily viewed using conventional light microscopy. The notion that chloroplasts could divide inside plant cells came from an early study on spinach where elongated and constricted dumbbell-shaped chloroplasts could be observed (Possingham and Saurer, 1969). With the advancement of microscopic techniques, it became increasingly evident that chloroplasts from a number of species did indeed have central constrictions (Boasson *et al.*, 1972; Cran and Possingham, 1972; Platt-Aloia and Thomson, 1977) and these were generally accepted as being dividing chloroplasts. Detailed ultrastructural analyses of the isthmus region and of membrane changes at different stages of division confirmed these observations and showed that dividing chloroplasts can be observed in a variety of tissues in a number of species (Leech *et al.*, 1981). Although division events have been captured both *in vitro* (Ridley and Leech, 1970) and *in vivo* (Honda et al., 1971), it appears to be inherently difficult to follow the actual division process in higher plants using conventional methods.

4.2.2   What drives the constriction event?

The idea that a central constriction event initiates division and presumably drives the division forward suggested the presence of a motive force. Using transmission electron microscopy and homing in on the constricted region of dividing chloroplast in various species, an electron-dense ring was observed at later stages of division in alga (Mita *et al.*, 1986), in moss (Tewinkel and Volkmann, 1987), in higher plants (Hashimoto, 1986; Oross and Possingham, 1989; Robertson *et al.*, 1996) and in ferns (Duckett and Ligorne, 1993). This ring structure was presumed to be contractile and is most often referred to as the PD ring. Fine-section ultrastructural studies in *Avena sativa* (pea) further revealed that the electron-dense ring is in fact a PD doublet consisting of an inner PD ring on the stromal side of the inner envelope and an outer PD ring on the cytoplasmic side of the outer envelope (Hashimoto, 1986). Subsequent studies in alga and in higher plants, showing the presence of an inner and outer PD ring, suggested that this doublet PD structure is most probably ubiquitous in plant cells (Tewinkel and Volkmann, 1987; Oross and Possingham, 1989; Duckett and Ligrone, 1993). Although no electron-dense structures were observed in the lumen between the outer and the inner envelope membranes in higher plants (Hashimoto, 1986), studies on the red alga *Cyanidioschyzon merolae* provided evidence that a middle PD ring exists in the intermembrane space (Miyagishima *et al.*, 1998a). PD rings are small structures and can only be observed, using high-quality fixation techniques, at late stages of constriction. Together with the fact that plastids in higher plant show non-synchronised division characteristics, it is possible that a middle PD ring does exist in the intermembrane space of higher plants.

4.2.3   PD rings and FtsZ

Using synchronised *C. merolae* cultures, insight into PD ring formation has been revealed (Miyagishima *et al.*, 1998b). Constriction follows a coordinated pathway where the inner PD ring forms first, followed by the middle and outer PD rings. As constriction proceeds the inner PD ring continues to be of constant thickness (the volume decreases at a constant rate with constriction), indicating that inner PD ring components are lost during the process. In contrast, the outer PD ring becomes thicker and maintains a constant volume, suggesting that components of the cytosolic PD ring are retained (Miyagishima *et al.*, 1999). These differences suggest that the two PD rings play divergent roles during constriction. This notion has been further strengthened by the finding that in *C. merolae* chloroplasts, the inner PD ring disassembles prior to completion of division whilst the outer PD ring remains on the cytosolic surface until after division has been completed (Miyagishima *et al.*, 2001a). These results are intriguing and suggest a complex dynamic interplay between the PD ring structures: the inner PD ring acts as a partially transient constricting collar whilst the outer PD ring functions throughout the division cycle.

Plastids have arisen from a bacterial endosymbiont (Gray, 1999), and bacterial cell division is initiated by the formation of a central Z-ring (Lutkenhaus and

Addinall, 1997) composed of polymerised FtsZ proteins (see Section 4.3). The identification of FtsZ homologues in plants (cf. Osteryoung and McAndrew, 2001) and recent evidence showing that FtsZ forms ring structures in *Arabidopsis*, which appear similar to PD rings (Vitha *et al.*, 2001), prompted the notion that FtsZ proteins are components of the PD rings. Detailed analysis using immunofluorescence in combination with electron microscopy revealed, however, that this is not the case.

In *C. merolae*, a series of elegant experiments by Miyagishima *et al.* (2001a–c) conclusively showed the relationship between the Z-ring and the PD rings. Using immunofluorescence, electron microscopy and biochemical approaches, a coordinated pathway of events has been constructed (Figure 4.2). The Z-ring forms initially in the stroma, followed by inner PD ring formation and then by outer PD ring formation (Miyagishima *et al.*, 2001c). At late stages of constriction the Z-ring disappears first from the constriction site and disperses towards the two daughter chloroplasts. Following this, the inner PD ring (and the middle PD ring) disassembles whilst the outer PD ring remains in the cytosol until completed division.

In the higher plant *Pelargonium zonale*, immunogold particles from anti-FtsZ antibodies do not co-localise with the PD rings but are found in the stromal region of chloroplasts (Kuroiwa *et al.*, 2002). In addition, the Z-ring forms prior to the initial constriction event, followed by formation of the inner and outer PD rings. The Z-ring also appears wider (80 nm) than the inner PD ring (40 nm) and the outer PD ring (20 nm), although the outer PD ring increases in thickness as constriction proceeds.

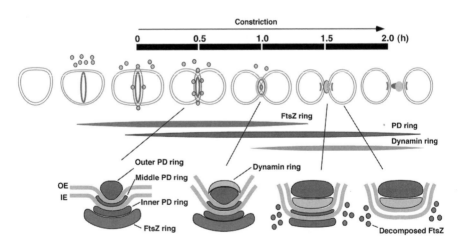

Figure 4.2 Schematic diagram showing chloroplast division in the red alga *Cyanidioschyzon merolae* mediated by FtsZ, PD rings and dynamin rings. The time-scale of division is shown representing division from initial trapezoidal chloroplasts to just completed division. The FtsZ ring, the PD rings and the dynamin ring are indicated. Reprinted with permission from Miyagishima *et al.* (2003). Copyright (2003) The American Society of Plant Biologists.

4.2.4   PD ring composition

Although the pathway of constriction events has been described at the ultrastructural level, the isolation of the different PD ring components will certainly enhance our understanding of the process. The first step towards this has been taken, where it has been shown that a bundle of 5-nm filaments consisting of globular proteins, one of which may be a highly stable 56-kDa protein, is part of the outer PD ring in *C. merolae* chloroplasts (Miyagishima *et al.*, 2001b).

The recent cloning of the disrupted gene in the chloroplast division mutant *accumulation and replication of chloroplasts 5* (*arc5*) (see Section 4.5.2) has revealed that *ARC5* encodes a dynamin-like protein that localises to a cytosolic ring structure (Gao *et al.*, 2003). Since the topology of the outer PD ring is similar to that of the ARC5 ring and since the diameter of dynamin strands (Klockow et al., 2002) are similar to the outer PD ring filament diameter (Miyagishima *et al.*, 2001b), the possibility exists that ARC5 is a component of the outer PD ring in *Arabidopsis* (Gao *et al.*, 2003). However, this remains to be shown.

The involvement of dynamin-like proteins in chloroplast division has been further verified by recent findings showing that a dynamin-related protein (CmDnm2) from *C. merolae* chloroplasts forms a cytosolic ring at the chloroplasts division site at late stages of division (Miyagishima *et al.*, 2003). On the basis of immunoelectron microscopy, CmDnm2 is proposed to be recruited from cytosolic patches to the cytosolic side of the outer PD ring after outer PD formation (Figure 4.2). The recent characterisation of CmDnm2 has clearly raised the complexity level of the coordinated interplay by multiple protein rings during chloroplast division.

Although most research into PD ring structures has focused on alga, the presence of multiple ring structures (Z-ring and PD rings) in higher plant chloroplasts (Kuroiwa *et al.*, 2002) suggests that the mechanism is most probably conserved between species. To date, the structure and composition of the inner and middle PD rings remain unknown, but as the components of the individual ring structures are identified, we can start to assemble the mechanisms that the host eukaryotic cell imposed on the inherited prokaryotic-derived division machinery in order to generate a functional chloroplast division apparatus.

4.3   Plastid division initiation by FtsZ

Plastids have arisen from a cyanobacterial endosymbiont (Gray, 1999), and in the early 1980s, it was suggested that plastid division components may share common features with proteins involved in prokaryotic cell division (Possingham and Lawrence, 1983). Further evidence implying the involvement of prokaryotic cell-division-like proteins in plastid division came from the identification of a nuclear gene in *Arabidopsis*, encoding a protein with over 40% amino acid similarity to the bacterial FtsZ cell division protein (Osteryoung and Vierling, 1995). Despite this, it took almost 15 years from the initial concept to a conclusive demonstration that nuclear plant genes encoding plastid-targeted prokaryotic-like cell division proteins are indeed involved in plastid division in plants (Osteryoung *et al.*, 1998; Strepp *et al.*, 1998). The similarity between prokaryotic cell division and plastid division has now been demonstrated unequivocally and bacterial cell division is often used as a paradigm for plastid division.

4.3.1   Bacterial FtsZ

In a genetic screen for temperature-sensitive cell division mutants of *Escherichia coli*, a number of mutants that failed to divide at the restrictive temperature were identified and these were named *filamentous temperature sensitive* (*fts*) mutants (Hirota *et al.*, 1968). One of these mutants, originally labelled as PAT84, was called *ftsZ*, and the extreme filamentation phenotype of the *ftsZ* mutant was due to loss of septum formation, indicating that FtsZ is important during initiation of bacterial cell division (Lutkenhaus *et al.*, 1980). FtsZ was shown to form a contractile ring (Z-ring) at the division site (Bi and Lutkenhaus, 1991; Lutkenhaus and Addinall, 1997) on the cytosolic face of the cell membrane, acting as a structural cytoskeletal division component (Ward and Lutkenhaus, 1985; Addinall and Lutkenhaus, 1996; Baumann and Jackson, 1996; Margolin *et al.*, 1996; Wang and Lutkenhaus, 1996). In contrast to the actin-based contractile ring formed in eukaryotic cells, FtsZ is most probably an ancient tubulin: FtsZ shows the presence of the tubulin signature motif GGGTGS/TG, forms GTP-dependent polymers *in vitro* and shows similarity at the structural level to α- and β-tubulin (de Boer *et al.*, 1992; RayChaudhuri and Park, 1992; Mukherjee *et al.*, 1993; Bramhill and Thompson, 1994; Mukherjee and Lutkenhaus, 1994; Erickson, 1995, 1998; de Pereda *et al.*, 1996; Bramhill, 1997; Yu and Margolin, 1997; Löwe and Amos, 1998). During cytokinesis, the Z-ring (FtsZ) stays associated with the leading edge of the division septum but disassembles after cell separation before reappearing again at midcell in the resulting daughter cells (Bi and Lutkenhaus, 1991; Addinall and Lutkenhaus, 1996; Sun and Margolin, 1998). The role of FtsZ in septum constriction is not yet known, but it is thought that polymerisation-induced GTP hydrolysis may provide the force and induce Z-ring curvature (Lu *et al.*, 2000).

4.3.2   Plant FtsZ proteins

AtFtsZ1-1 shows over 40% similarity at the amino acid level to the bacterial FtsZ protein and 50% similarity to the cyanobacterial FtsZ protein, and AtFtsZ1-1 is post-translationally targeted to chloroplasts (Osteryoung and Vierling, 1995). In contrast to the majority of bacterial species, containing only a single *FtsZ* gene, plants contain two FtsZ gene families, FtsZ1 and FtsZ2 (Osteryoung *et al.*, 1998; Osteryoung and McAndrew, 2001; Wang *et al.*, 2003). The role of FtsZ proteins in plastid division was first shown in the moss *Physcomitrella patens*, where a knockout mutation (homologous recombination) in one of the *FtsZ2* genes results in cells containing one giant chloroplast in contrast to wild-type cells containing multiple smaller chloroplasts (Strepp *et al.*, 1998). In higher plants the role of FtsZ proteins in plastid division came from studies using antisense technology (Osteryoung *et al.*, 1998). In contrast to wild-type mesophyll cells containing over 100 chloroplasts, cells with reduced levels of either FtsZ1-1 or FtsZ2-1 show, as observed in *P. patens*, the presence of one giant chloroplast. In *E. coli*, over-expression of FtsZ results in either asymmetrical division (modest over-expression) or filamentation (high over-expression), implying a delicate stoichiometric balance between FtsZ proteins and other division components (Ward and Lutkenhaus, 1985). FtsZ1-1 over-expression in *Arabidopsis* has a similar effect, resulting in one or two giant chloroplasts in the most severe cases (Stokes *et al.*, 2000). In contrast, over-expression of FtsZ2-1 in *Arabidopsis* has no effect on chloroplast size or number, implying that AtFtsZ1-1 and FtsZ2-1 have different roles during chloroplast division (Stokes *et al.*, 2000). Combined with the identification of a third FtsZ gene in *Arabidopsis*, FtsZ2-2, this suggests a complex interplay between these ancient tubulin proteins during division (McAndrew *et al.*, 2001).

Early studies showed that FtsZ1-1 was imported into chloroplasts (Osteryoung and Vierling, 1995) whilst FtsZ2-1 seemed to be present on the cytosolic surface (Osteryoung *et al.*, 1998). However, it is now clear that FtsZ1-1, FtsZ2-1 and FtsZ2-2 have N-terminal extensions and are all imported into chloroplasts (Fujiwara and Yoshida, 2001; McAndrew *et al.*, 2001). The burning question at this time was whether plant FtsZ proteins form ring-like structures inside plastids. The first report of FtsZ localisation came from studies in *P. patens*, showing that an FtsZ–GFP fusion protein predominantly localises to organised networks inside plastids (Kiessling *et al.*, 2000). This was surprising because it did not mirror the situation observed in bacteria. Subsequent studies however, using FtsZ-specific antibodies, demonstrated that the FtsZ network was most probably an artifact and that in *Lilium longiflorum* and in *Arabidopsis*, FtsZ forms ring structures at the chloroplast midpoint (Mori *et al.*, 2001; Vitha *et al.*, 2001). In *Arabidopsis*, both FtsZ1-1 and FtsZ2-1 localise to ring structures at the chloroplast midpoint and have been shown to co-localise through double immunofluorescence labelling approaches (McAndrew *et al.*, 2001; Vitha *et al.*, 2001). This suggests that either FtsZ1-1 and FtsZ2-1 form homopolymers, which then subsequently assemble laterally to form the Z-ring, or that FtsZ1-1 and FtsZ2-1 form heteropolymers similar to the association of α- and β-tubulin in eukaryotes (Nogales *et al.*, 1998). A third possibility exists where FtsZ1 and FtsZ2 can form both homo- and heteropolymers in any given Z-ring, although such a seemingly disorganised model is less likely. Direct protein–protein interaction studies would address this question. In addition, it would be interesting to assess whether FtsZ1 can functionally substitute for FtsZ2 and vice versa.

4.3.3   The domains of FtsZ

The structural features of prokaryotic FtsZ proteins have given clues towards their function during cell division, and the structural similarities between FtsZ proteins across species are high (Löwe, 1998; Löwe and Amos, 1998). FtsZ proteins can be broadly divided into an N- and C-terminal domain. The N-terminal domain is highly conserved (when excluding chloroplast transit peptides) and contains a Rossman fold, found in proteins such as p21ras and EF-Tu, responsible for GTP binding and hydrolysis (Löwe and Amos, 1998). The N-terminal region has also been shown to be sufficient for polymerisation (Wang *et al.*, 1997). The overall secondary and predicted tertiary structures in this region between FtsZ1, FtsZ2 and bacterial FtsZs are almost indistinguishable, making the bacterial FtsZ structural features valuable tools for further dissection of plant FtsZ proteins.

In contrast, the C-terminal domain of FtsZ proteins is highly variable. Despite this there are two main recognisable features: Firstly, C-terminal loop structures involved in Ca^{2+} binding are present in bacterial FtsZs and probably in FtsZ1 and FtsZ2, and these are thought to stabilise FtsZ polymers (Erickson *et al.*, 1996; Yu and Margolin, 1997; Löwe and Amos, 1999; Mukherjee and Lutkenhaus, 1999). Secondly, a surface-exposed hydrophilic domain at the extreme C-terminal end, similar to the MAP-binding domain of tubulin (Desai and Mitchison, 1997, 1998), contains a highly conserved sequence (D/E-I/V-P-X-F/Y-L) named the *core domain* (Ma and Margolin, 1999). This core domain is responsible for FtsZ interaction with the cell division proteins ZipA and FtsA (Wang *et al.*, 1997; Din *et al.*, 1998; Liu *et al.*, 1999; Hale *et al.*, 2000; Mosyak *et al.*, 2000; Yan *et al.*, 2000). Although ZipA or FtsA homologues have not been identified in plants to date, FtsZ2 seems to contain the core domain (Osteryoung and McAndrew, 2001). Interestingly, FtsZ1 does not contain this domain, suggesting a possible functional distinction between FtsZ1 and FtsZ2. However, because neither FtsA nor ZipA has been identified in plants, the FtsZ2 core domain may merely represent an evolutionary relic.

4.4   Division site placement

The entire cell membrane in *E. coli* is competent for Z-ring formation, meaning that FtsZ polymerisation can occur throughout the bacterial cell. Clearly this is not the case, and several bacterial species contain three proteins that in combination ensure the accurate placement of the Z-ring at midcell. Recent evidence has demonstrated that plants have at least two of these proteins and that they act together during plastid division in a similar fashion to their bacterial counterparts to govern division site placement at the centre of plastids. Z-ring placement appears to be highly conserved between prokaryotic cell division and plastid division in plants, and it is therefore appropriate to draw parallels between the two systems.

4.4.1   Division site placement in bacteria

In *E. coli*, Z-ring placement is negatively regulated by the *minB* operon that encodes the proteins MinC, MinD and MinE (de Boer *et al.*, 1989; Bi and Lutkenhaus, 1993; Rothfield *et al.*, 1999). The importance of the *minB* operon in division site place-ment was recognised in the late 1960s, where mutations in the *minB* locus were shown to lead to minicell formation (Adler *et al.*, 1967). Minicells are the result of asymmetric division and represent anucleate 'mini-bacteria'. The *minB* locus was dissected and shown to encode MinC, MinD and MinE, which are coordinately expressed, resulting in the proper placement of the division septum (de Boer *et al.*, 1988, 1989). MinC acts directly on FtsZ, preventing polymerisation and the formation of a stable cytokinetic ring (Hu *et al.*, 1999; Pichoff and Lutkenhaus, 2001). MinC does however lack site specificity and can therefore inhibit FtsZ polymerisation throughout the entire cell, resulting in filamentation (de Boer *et al.*, 1992). MinC site specificity is governed by MinD and MinE, and during division, MinC and MinD forms a division inhibitor complex. MinD, a peripheral membrane ATPase, forms dimers/polymers in the presence of ATP, and ATP hydrolysis is essential for its function (de Boer *et al.*, 1991; Hu *et al.*, 2002; Suefuji *et al.*, 2003). ATP-bound MinD can interact with MinC, directing it to the membrane forming a stable protein complex (Hu *et al.*, 2003; Hu and Lutkenhaus, 2003). Binding of MinE to the MinD/C complex stimulates ATP hydrolysis (Hu and Lutkenhaus, 2001), leading to MinD membrane release (Hu *et al.*, 2002). It is interesting to note that MinC can be directly released from the MinD/membrane complex by MinE and that this step is independent of ATP hydrolysis. These sequential protein interactions and the MinD-dependent ATP hydrolysis are important for the observed oscillatory behaviour of the Min proteins during division (Lutkenhaus and Sundaramoorthy, 2003). The oscillatory behaviour of the Min proteins between the cell poles ensures that the concentration of the MinC/MinD complex is lowest at midcell, allowing FtsZ polymerisation only at this site (Hu and Lutkenhaus, 1999; Raskin and de Boer, 1999; Rowland *et al.*, 2000; Fu *et al.*, 2001; Shih *et al.*, 2003). MinE acts as a topological specificity factor during division and induces the redistribution of MinD and MinC into polar zones at one end of cells (Hu and Lutkenhaus, 1999; Raskin and de Boer, 1999; Rowland *et al.*, 2000). The majority of MinE forms a ring structure (MinE ring), and together with the MinC/MinD/MinE polar zone they undergo rapid and repeated oscillation from pole to pole (Raskin and de Boer, 1997; Hu and Lutkenhaus, 1999; Rowland *et al.*, 2000; Fu *et al.*, 2001; Hale *et al.*, 2001). The distribution of the Min proteins at the polar zones is not random but rather organised into membrane-associated coiled structures (Shih *et al.*, 2003).

4.4.2 *Plastid division site placement*

The discovery that MinD- and MinE-like sequences are present in the chloroplast genome of the alga *Chlorella vulgaris* and *Guillardia theta* (Wakasugi *et al.*, 1997; Douglas and Penny, 1999) prompted the notion that division site placement in plastids most probably shares similarities to the bacterial cell division site placement mechanism. In plants, genes encoding homologues to MinD have been identified in *Arabidopsis*, rice and Marigold (Colletti *et al.*, 2000; Kanamaru *et al.*, 2000; Moehs *et al.*, 2001), whilst to date MinE has only been identified in *Arabidopsis* (Itoh *et al.*, 2001; Maple *et al.*, 2002). It is interesting to note that to date no MinC-like protein has been identified in any plant species.

The involvement of MinD in division site placement in plants came from studies using the *Arabidopsis* MinD homologue AtMinD1 (Colletti *et al.*, 2000). *AtMinD1* was identified on chromosome 5 in *Arabidopsis* by homology searches using the *C. vulgaris* MinD as the query input sequence, and *AtMinD1* encodes a protein of 326 amino acids containing an N-terminal putative chloroplast targeting transit peptide. Not unexpectedly, AtMinD1 localises to chloroplasts as shown by *in vitro* chloroplast import assays. Firm proof that AtMinD1 plays a role in plastid division came from subsequent studies of transgenic *Arabidopsis* plants with reduced levels of AtMinD1. Mesophyll cells from these plants showed a reduced number of large chloroplasts, indicating that division events are less frequent than in wild-type plants. In addition, the chloroplast size was highly variable, showing a heterogeneous population of chloroplasts within single cells. This situation mirrors the asymmetric division phenotype observed in *E. coli* mutants deficient for MinD, and using *Arabidopsis* petals, it was shown that a reduction in AtMinD1 levels results in asymmetric division of chloroplasts. A more severe phenotype is observed upon AtMinD1 over-expression in *Arabidopsis* where mesophyll and palisade cells contain five or fewer chloroplasts per cell (Colletti *et al.*, 2000; Kanamaru *et al.*, 2000). This phenotype resembles the *E. coli* filamentation phenotype observed upon MinD over-expression indicative of a loss of FtsZ polymerisation (de Boer *et al.*, 1989). The role of AtMinD1 in plastid division is most probably conserved amongst different plant species since over-expression of AtMinD1 in transgenic tobacco results in fewer but larger chloroplasts (Dinkins *et al.*, 2001).

In *E. coli*, MinD shows polar localisation, and detailed localisation analysis in transgenic plants harbouring an AtMinD1–GFP fusion protein shows that AtMinD1 localises to distinct regions inside chloroplasts (Maple *et al.*, 2002). AtMinD1 localises in most cases as two spots at each pole of chloroplasts but does also in some cases localise to a single spot at one end of chloroplasts (Figure 4.3A). These observations suggest that AtMinD1 has a similar localisation pattern to MinD in *E. coli*; however, whether AtMinD1 shows dynamic behaviour remains to be shown.

Membrane localisation of MinD in bacteria is mediated through a direct interaction between a C-terminal amphipathic helix and membrane phospholipids (Szeto *et al.*, 2002; Hu and Lutkenhaus, 2003). This helix is highly conserved and is present in AtMinD1. Deletion of this putative amphipathic helix in AtMinD1 results in mislocalisation of the protein in transgenic *Arabidopsis* plants (J. Maple and S.G. Møller, unpublished data, 2004). However, because AtMinD1 can form homodimers and dimerisation is mediated by the C-terminal domain (Section 4.5.4), it is possible that the mislocalisation of the C-terminal-truncated AtMinD1 protein is indirectly due to loss of dimerisation capacity.

The demonstration that MinD plays a crucial role in plastid division suggested the presence of other Min protein homologues in plants. AtMinE1 was identified from *Arabidopsis* on chromosome 1, based on its similarity to the chloroplast-encoded and bacterial MinE proteins (Itoh *et al.*, 2001; Maple *et al.*, 2002). *AtMinE1* encodes a protein of 229 amino acids and contains an N-terminal chloroplast targeting transit

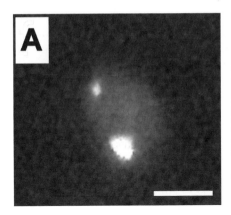
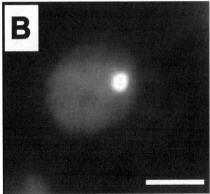

Figure 4.3 Intraplastidic localisation of AtMinD1 (A) and AtMinE1 (B) in transgenic *Arabidopsis* plants. Reproduced from Maple *et al.* (2002), with permission from Blackwell Publishing.

peptide absent in prokaryotic MinE proteins. The *E. coli* MinE has two functional domains: the N-terminal anti-MinCD (AMD) domain, which is necessary and sufficient for counteracting MinC/MinD activity; and the C-terminal domain (TSD), which imparts topological specificity (Zhao *et al.*, 1995). Sequence alignments suggest that AtMinE1 harbours an N-terminal AMD domain; however, the C-terminal TSD domain appears less conserved. TSD domains from various species show limited similarity, suggesting evolutionary divergence of the TSD function. This notion is strengthened by the fact that MinE is absent in *Bacillus subtilis* and that the anti-MinCD function is performed by DivIVA through a mechanism different to that observed in *E. coli* (Cha and Stewart, 1997; Edwards and Errington, 1997; Marston *et al.*, 1998).

The functional role of AtMinE1 in plastid division came from studies using transgenic *Arabidopsis* and tobacco plants over-expressing AtMinE1 (Itoh *et al.*, 2001; Maple *et al.*, 2002; Reddy *et al.*, 2002). In these plants, mesophyll cells contain a heterogeneous population of fewer but larger chloroplasts. As for AtMinD1 antisense plants (Colletti *et al.*, 2000) these data suggest that AtMinE1 over-expression results in asymmetric plastid division. Indeed, detailed observations in *Arabidopsis* hypocotyl cells show that AtMinE1 over-expression results in asymmetric division and misplacement of division sites away from the midpoint, giving rise to a 'mini-cell' phenotype (Maple *et al.*, 2002). In extreme cases, elongated chloroplasts are observed showing the presence of multiple constriction sites, indicating severe loss of topological specificity. The misplacement of division sites and the apparent minicell phenotype in these chloroplasts suggest that MinE mode of action represents an evolutionary conserved process. MinE over-expression in *E. coli* leads to a classical minicell phenotype (de Boer *et al.*, 1989) and the evolutionary conservation of AtMinE1 mode of action was demonstrated by AtMinE1 over-expression in

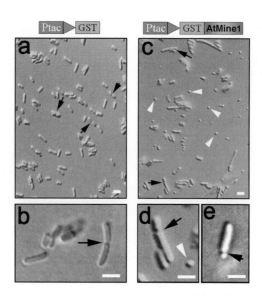

Figure 4.4 Over-expression of AtMinE1 in *E. coli*. Control *E. coli* expressing the GST tag (a, b) showing normal division at midcell, and *E. coli* over-expressing AtMinE1 (c–e) showing asymmetric division. Reproduced from Maple *et al.* (2002), with permission from Blackwell Publishing.

E. coli (Maple *et al.*, 2002). In contrast to wild-type *E. coli* dividing at midcell, AtMinE1 over-expression leads to asymmetric division and minicell formation in *E. coli* (Figure 4.4).

Intraplastidic localisation analysis shows that in a similar fashion to AtMinD1, AtMinE1 exhibits polar localisation (Maple *et al.*, 2002). However, in slight contrast to AtMinD1 appearing as two spots at either pole of chloroplasts, AtMinE1 localises either as one spot or as two spots in close proximity at one pole (Figure 4.3B). The similarity in localisation patterns of AtMinD1 and AtMinE1 suggests that these two proteins act in concert during division. Indeed in *E. coli*, MinE interacts with MinD, stimulating ATP hydrolysis and ensuring release from the cell membrane, leading to dynamic oscillations during the division cycle (Hu and Lutkenhaus, 2001; Hu *et al.*, 2002). Similarly, AtMinD1 shows direct protein–protein interactions with AtMinE1 in yeast two-hybrid assays (J. Maple and S.G. Møller, unpublished data, 2004).

Based on the evolutionary conservation of MinD and MinE proteins, it is somewhat puzzling that higher plants do not appear to have a classical MinC-like protein. However, based on the low similarity between MinC proteins from prokaryotes (Hu and Lutkenhaus, 2000) the possibility still remains that a MinC-like protein is present in plants but avoids detection by homology searches. A more likely scenario is that plants have substituted the prokaryotic classical MinC-like protein function with an acquired eukaryotic FtsZ polymerisation inhibitor protein.

4.5 *arc* mutants

It was realised during the early 1990s that in order to gain a non-bias molecular handle on plastid division in higher plants, mutants defective in plastid division were needed. A rapid genetic screen was therefore developed based on visually identifying altered chloroplast number and size in ethyl methane sulfonate (EMS) mutagenised *Arabidopsis* seedlings (Pyke and Leech, 1991). This screen has subsequently been expanded and used on T-DNA-mutagenised seedling populations (Rutherford, 1996), and in combination 12 *arc* mutants have now been identified and characterised and the main features of 11 of these are summarised in Table 4.1 (Pyke and Leech, 1991, 1992, 1994; Pyke *et al.*, 1994; Robertson *et al.*, 1995, 1996; Rutherford, 1996; Marrison *et al.*, 1999; Pyke, 1999; Yamamoto *et al.*, 2002; Fujiwara *et al.*, in press).

4.5.1 *arc* mutant physiology

In both *arc1* and *arc7* plants there appears to be an increase in the rate of chloroplast accumulation during cell expansion, leading to an increase in chloroplast number per cell (Pyke and Leech, 1992; Rutherford, 1996; Marrison *et al.*, 1999; Pyke, 1999). *arc1* and *arc7* are both recessive mutations and have pale leaves showing reduced rates of greening, suggesting that the *ARC1* and *ARC7* loci are not involved in chloroplast division but rather in chloroplast development. The increased chloroplast number may actually be a compensatory mechanism for the reduced chloroplast size.

The most striking *arc* mutant is *arc6*, showing the presence of one or two giant chloroplasts (Pyke *et al.*, 1994; Robertson *et al.*, 1995; Vitha *et al.*, 2003). *arc6* seedlings also show a reduced number of proplastids in meristems, with only two enlarged proplastids in the apical meristem. In addition, *arc6* stomatal guard cells show a perturbation in proplastid populations, resulting in abnormal plastid segregation and plastid-less guard cells (Robertson *et al.*, 1995). All cell types in *arc6* plants studied to date show altered plastid phenotypes, indicating that ARC6 plays a global role in both proplastid and chloroplast division initiation (Robertson *et al.*, 1995; see Section 4.5.3). *arc12* is not allelic to *arc6* but shows a similar phenotype (Pyke, 1999; Yamamoto *et al.*, 2002).

arc3 and *arc5* mutants both show a similar phenotype in that they contain between 13 and 15 enlarged chloroplasts per cell (Pyke and Leech, 1992, 1994; Robertson *et al.*, 1996; Marrison *et al.*, 1999). Recent studies have, however, suggested that *arc5* cells can harbour between 3 and 15 chloroplasts, depending on the growth conditions (Gao *et al.*, 2003). In contrast to the *arc5* dumb-bell-shaped chloroplasts, *arc3* chloroplasts do not show a central constriction, indicating that ARC3 is involved in division initiation (Pyke and Leech, 1992, 1994; Marrison *et al.*, 1999). In *arc5* mesophyll cells it appears that chloroplasts enter division but arrest when they have become centrally restricted, suggesting that ARC5 is required to complete the separation process (Roberston *et al.*, 1996; see Section 4.5.2).

Table 4.1 A summary of 11 *Arabidopsis arc* mutants, indicating the phenotype, chloroplast size and chloroplast number

Geno type	Eco type	Chloroplast size (μm^2)	Chloroplast number/cell	Chloroplasts/ 1000 μm^2 mesophyll cell plan area	Notes	Reference
WT	Ler	50	120	25	Spherical	Pyke and Leech, 1992
WT	Ws	50	80–90	20–23	Spherical	Pyke et al., 1994; Rutherford, 1996
WT	Col	50	100	23	Spherical	Osteryoung et al., 1998
arc1	Ler	25	108	32	Increased number of smaller chloroplasts and pale leaves	Pyke and Leech, 1992; Marrison et al., 1999
arc2	Ler	110	40	9	Fewer chloroplasts/cell than WT	Pyke and Leech, 1992
arc3	Ler	200–300	18	4–5	Heterogeneous chloroplast size	Pyke and Leech, 1992; Pyke and Leech, 1994; Marrison et al., 1999
arc5	Ler	300–900	3–15	1–4	Dumb-bell-shaped chloroplasts	Pyke and Leech, 1994; Robertson et al.,1996; Marrison et al., 1999; Gao et al., 2003
arc6	Ws	1000	2	0.5	One or two large chloroplasts	Pyke et al., 1994; Robertson et al., 1995; Vitha et al., 2003
arc7	Ws	40	80	26	Pale first leaves	Rutherford, 1996; Pyke, 1999
arc8	Ws	110	45	10	Moderatly enlarged chloroplasts	Rutherford, 1996
arc9	Ws	140	34	12	Moderatly enlarged chloroplasts	Rutherford, 1996
arc10	Ws	170	38	6	Highly variable in size	Rutherford, 1996; Pyke, 1999
arc11	Ler	110	30	7	Heterogeneous chloroplast size	Marrison et al., 1999; Colletti et al., 2000; Fujiwara et al., 2004
arc12	Col	ND	1–2	ND	Similar to arc6	Pyke, 1999; Yamamoto et al., 2002

Note: WT: wild type; ND: no data.

arc11 chloroplasts, as observed for *arc10* chloroplasts (Rutherford, 1996; Pyke, 1999), show a highly heterogeneous population of chloroplasts (Marrison *et al.*, 1999). Approximately 50% of the chloroplast population in *arc11* mesophyll cells are within wild-type size whilst the other half are larger than wild-type. Division of *arc11* chloroplasts is clearly asymmetric since the appearance of 'budding' chloroplasts (Marrison *et al.*, 1999), multiple arrayed chloroplasts and spherical minichloroplasts (Fujiwara *et al.*, in press) can be observed, indicating that ARC11 is indeed involved in placement of the division site.

Through a series of double mutant studies using five *arc* mutants (*arc1*, *arc3*, *arc5*, *arc6*, *arc11*) the hierarchy of some of the ARC gene products in chloroplast division has been established (Marrison *et al.*, 1999). ARC1 is in a separate pathway to ARC3, ARC5, ARC6 and ARC11 and down-regulates proplastid division, whilst ARC6 initiates proplastid and chloroplast division. Next, ARC3 seems to control the rate of chloroplast expansion whilst ARC11 is clearly involved in controlling division site placement. Finally, ARC5 assists the separation of the two daughter chloroplasts. Although not complete, these experiments have generated a framework for further study of the ARC gene products and their place in the division process.

4.5.2 arc5

ARC5 represents the first *ARC* gene to be identified and characterised from *Arabidopsis*. The ARC5 gene was previously mapped to chromosome 3 (Marrison *et al.*, 1999), and by using a combination of fine mapping and a novel antisense strategy, a candidate gene for *ARC5* was identified from a BAC clone (MMB12) showing a G-to-A mutation, changing a tryptophan codon to a stop codon (Gao *et al.*, 2003). ARC5 encodes a 777-amino acid protein and shows similarity to the dynamin protein family containing conserved domains found in other dynamin-like proteins. ARC5 contains an N-terminal GTPase domain, a pleckstrin homology (PH) domain and a C-terminal GTPase effector domain. PH domains have been shown to be involved in membrane association whilst GTPase effector domains have been implicated in GTPase domain interaction and self-assembly (Danino and Hinshaw, 2001). Further phylogenetic analysis reveals that ARC5 is clustered distantly to dynamin-like proteins involved in cell-plate formation (Gu and Verma, 1996) and mitochondrial division such as ADL2b (Arimura and Tsutsumi, 2002), suggesting that ARC5 represents a new class of dynamin-like protein involved in chloroplast division.

The expression of a GFP–ARC5 fusion protein in transgenic plants revealed that ARC5 localises to a ring-like structure at the site of constriction. Although the ring structure is visible at early stages of chloroplast constriction, it becomes more noticeable at later stages. This indicates that ARC5 acts both during early and late stages of constriction. Interestingly, ARC5 was not predicted to contain a chloroplast targeting transit peptide, suggesting that ARC5 forms a cytosolic ring structure on the outside of chloroplasts and this was confirmed by *in vitro* chloroplast import/protease protection assays. The localisation of ARC5 resembles that of the dynamin-like protein Dnm1p involved in yeast mitochondrial division (Bleazard *et al.*, 1999). The firm proof that dynamins are involved in chloroplast division in higher plants is exciting. Dynamins participate in budding of endocytic and Golgi-derived vesicles, mitochondrial fission and fusion and cell plate formation (Hinshaw, 2000; Danino and Hinshaw, 2001); however, the precise mode of dynamin action remains unknown. Based on structural and cell biological studies (Hinshaw and Schmid, 1995; Niemann *et al.*, 2001) it has been proposed that one role for dynamins could be to form a GTP-stimulated collar driving, for example the budding of vesicles during endocytosis. ARC5 may be performing this role during chloroplast constriction (Gao *et al.*, 2003). The finding that a *bona fide* cytosolic protein is involved in chloroplast constriction in higher plants has triggered curiosity into how protein components separated by two envelope membranes are coordinated to ensure correct division.

4.5.3   arc6

arc6 has the most striking phenotype out of the *arc* mutants, showing the presence of one or two giant chloroplasts per cell (Pyke *et al.*, 1994; Robertson *et al.*, 1995; Vitha *et al.*, 2003). The *arc6* mutation was mapped to chromosome 5 (Marrison *et al.*, 1999), close to a gene showing significant similarity to the cyanobacterial cell division gene *Ftn2* (Koksharova and Wolk, 2002). The *Ftn2*-like gene in *Arabidopsis* encodes a protein of 801 amino acids, and in *arc6* plants this gene has a mutation at nucleotide 1141 of its open reading frame, resulting in a premature stop codon. Sequence alignments revealed further that *ARC6*-like sequences are present in fern (*Ceratopteris richardii*), moss (*Physcomitrella patens*) and green alga (*Chlamydomonas reinhardtii*) but not in non-cyanobacterial prokaryotes, indicating that *ARC6* is a descendant of the cyanobacterial *Ftn2* gene.

   At the amino acid level ARC6-like proteins contain a conserved N-terminus, harbouring a putative J-domain, a conserved C-terminal domain and a transmembrane domain. The ARC6 J-domain is similar to the J-domain found in DnaJ cochaperones (Cheetham and Caplan, 1998). DnaJ serves a dual role delivering polypeptides to chaperones such as Hsp70 and at the same time regulating Hsp70 activity by direct J-domain interaction (Bukau and Horwich, 1998). In *E. coli*, Hsp70 interacts with FtsZ, possibly playing a role during cell division (Uehara *et al.*, 2001), and it is tempting to speculate that ARC6 may act as an Hsp70 cochaperone during chloroplast division in *Arabidopsis*.

   ARC6 has been shown to be an integral inner envelope membrane protein, with the N-terminal J-domain extending into the chloroplast stroma whilst the C-terminal region is present in the intermembrane space. Moreover, an ARC6–GFP fusion protein localises to a ring structure at the centre of chloroplasts prior to and during constriction similar to that of FtsZ1 and FtsZ2 (Vitha *et al.*, 2001). In contrast to wild-type chloroplasts (Figure 4.5A), *arc6* chloroplasts do not harbour intact FtsZ rings but rather numerous short FtsZ filaments (Figure 4.5B), suggesting that ARC6 is involved in FtsZ ring assembly and maintenance. Consistent with this idea is the finding that FtsZ filaments are numerous and appear long and spiral in large chloroplasts from plants over-expressing ARC6, suggesting ARC6-governed excessive FtsZ polymerisation and/or stabilisation (Figure 4.5D).

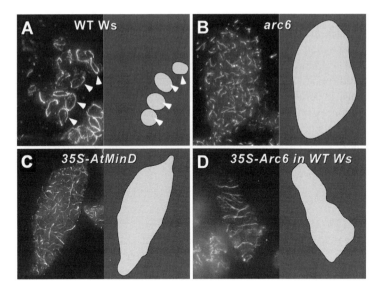

Figure 4.5 Localisation of FtsZ in leaf mesophyll chloroplasts in (A) wild type (WT) showing a single FtsZ ring (arrow head), (B) *arc6* showing the presence of numerous short FtsZ filaments, (C) AtMinD1 over-expressing plants showing the presence of FtsZ fragmentation and (D) ARC6 over-expressing plants showing the presence of excessive FtsZ polymerisation. Reprinted with permission from Vitha *et al.* (2003). Copyright (2003) The American Society of Plant Biologists.

The FtsZ fragmentation phenotype observed in *arc6* is also observed in plants with elevated AtMinD1 levels (Figure 4.5C). It has been shown that *AtMinD1* transcript levels in *arc6* seedlings is elevated compared to wild-type (Kanamaru *et al.*, 2000); however, in contrast, Vitha *et al.* (2003) show evidence that *AtMinD1* transcript levels are not increased in *arc6*. This discrepancy could be due to the seedling stage during analysis since *AtMinD1* transcript levels fluctuate during seedling development (Kanamaru *et al.*, 2000).

During bacterial division, FtsA and ZipA are thought to stabilise assembled FtsZ (Errington *et al.*, 2003). However, no obvious FtsA or ZipA homologues are present in plants and it is possible that ARC6 performs a function analogous to FtsA and ZipA, stabilising and/or anchoring FtsZ during chloroplast division. In contrast, AtMinD1 may destabilise FtsZ ring formation acting in the opposite direction to ARC6. Together, this implies a complex interplay between ARC6, AtMinD1 and most probably other to-date uncharacterised components, ensuring correct FtsZ ring formation at central constriction sites during chloroplast division.

4.5.4   arc11

Division of *arc11* chloroplasts is asymmetric, showing either single asymmetric constriction sites (Marrison *et al.*, 1999) or elongated multiple-arrayed chloroplasts (Fujiwara *et al.*, 2004), ultimately leading to a heterogeneous chloroplast population. This chloroplast morphology is similar to that observed in *Arabidopsis* AtMinD1 antisense plants (Colletti *et al.*, 2000), and genetic mapping data placed the *arc11* mutation in close proximity to *AtMinD1* on chromosome 5 (Marrison *et al.*, 1999; Colletti *et al.*, 2000; Kanamaru *et al.*, 2000). *AtMinD1* in *arc11* has a single-point mutation at nucleotide 887 of its open reading frame, resulting in an alanine-to-glycine substitution at amino acid residue 296 in a predicted helical region towards the extreme C-terminus (Fujiwara *et al.*, 2004). Although this single-point mutation does not alter endogenous *AtMinD1* transcript levels in *arc11*, complementation analysis demonstrates that it is the cause of the *arc11* chloroplast phenotype. The substitution of an α-helix-favourable alanine residue to an α-helix-unfavourable glycine residue probably distorts the overall structure of the extreme C-terminal domain of AtMinD1 in *arc11*. Despite this, AtMinD1(A296G) has retained its ability to inhibit chloroplast division in *Arabidopsis* as demonstrated by over-expression studies in transgenic plants, presumably through lack of FtsZ ring formation as observed by Vitha *et al.* (2003).

The single-point mutation in AtMinD1 disrupts normal intraplastidic localisation patterns where expression of an AtMinD1(A296G)–YFP fusion protein results in distorted fluorescent aggregates and/or multiple fluorescent spots. This is in sharp contrast to the defined punctate single/double spot localisation of wild-type AtminD1 (Figure 4.3). In *E. coli*, membrane localisation of MinD is mediated by an amphipathic C-terminal α-helix (Szeto *et al.*, 2002; Hu and Lutkenhaus, 2003) and it appears that correct AtMinD1 localisation in *Arabidopsis* is governed by the extreme C-terminal domain. In *E. coli* it has been further demonstrated that MinD membrane localisation is mediated by an ATP-driven dimerisation/polymerisation reaction (Hu *et al.*, 2002; Suefuji *et al.*, 2003). Interestingly, AtMinD1 is capable of forming homodimers as shown by yeast two-hybrid assays and this dimerisation capacity is abolished by the single-point mutation in AtMinD1 (Fujiwara *et al.*, 2004). This suggests further that the C-terminal end of AtMinD1 is involved in dimerisation; however, whether dimerisation occurs prior to localisation or vice versa is currently unknown. Further studies have verified that AtMinD1 does indeed dimerise inside living chloroplasts. Using fluorescence resonance energy transfer (FRET) assays in living plant cells, we have shown that energy transfer occurs between an AtMinD1–CFP donor and an AtMinD1–YFP acceptor, demonstrating dimerisation *in planta* (Fujiwara *et al.*, 2004; Plate 1). Together, these data demonstrate that AtMinD1 forms dimers *in vivo* and that dimerisation is important for correct AtMinD1 localisation, ultimately ensuring appropriate division site placement during the division process.

4.6   Non-*arc*-related chloroplast division components

Although the *arc* mutant collection provides an invaluable source for the isolation of plastid division genes, new division components are also being identified based on their similarity to cell division proteins in cyanobacteria and on their involvement in general chloroplast biogenesis. The isolation of additional non-*arc*-related plastid division proteins is probably due to the fact that the *arc* mutants were identified based on a visual screen and moreover that the screen may not have been saturated.

4.6.1 ARTEMIS

ARTEMIS (*Arabidopsis thaliana* envelope membrane integrase) was identified in a search for proteins involved in chloroplast biogenesis (Fulgosi *et al.*, 2002). The 1013-amino acid protein, encoded by a gene on chromosome 1, has a unique molecular structure containing a C-terminal domain similar to the Alb3 protein with conserved YidC translocase motifs, an N-terminal domain containing similarities to receptor protein kinases and a middle portion that contains an ATP/GTP-binding domain. ARTEMIS localises to the inner envelope membrane of chloroplasts and fractionation experiments demonstrate that ARTEMIS is an integral membrane protein. Based on the molecular architecture of ARTEMIS, the middle domain is predicted to bind nucleotides. GTP–agarose matrix and labelling experiments confirm this, showing that ARTEMIS can bind GTP, but ATP only weakly, implying that ARTEMIS function may be regulated by GTP hydrolysis. The role of ARTEMIS in chloroplast division came from studies on mutant *Arabidopsis* plants with highly reduced levels of the protein. Although these plants grow as normal, they show extended duplicated and tripolar undividing chloroplasts. The thylakoid network in these chloroplasts however appears normal, extending uninterruptedly between the two chloroplast halves, suggesting that ARTEMIS is most probably not involved in general chloroplast protein translocation. ARTEMIS seems to function late in the division process through proper placement of the envelope division constriction site and/or through insertion of plastid division components into the envelope membrane by the YidC/Alb3-like translocase motif.

A plasma membrane protein showing similarity to the YidC/Alb3-like domain of ARTEMIS has been identified in *Synechocystis* PCC6803, and a deletion mutant cell line for this gene (slr1471) shows the formation of tetrameric and hexameric clusters of cells indicative of late cell division arrest (Fulgosi *et al.*, 2002). Moreover, the fission events are unevenly distributed, leading to irregularly shaped cells. The evolutionary conservation of ARTEMIS has been further shown by rescue experiments where the YidC/Alb3-like domain of the *Arabidopsis* ARTEMIS can restore wild-type division characteristics in the slr1471 mutant cell line. It is interesting to note that slr1471 does not contain the N-terminal receptor-like domain nor the GTP-binding domain, suggesting that ARTEMIS represents an evolutionary protein hybrid: the eukaryotic receptor domain may be involved in the nuclear control of chloroplast division whilst the prokaryotic YidC/Alb3-like domain might aid in the integration/positioning of the division machinery.

4.6.2 GIANT CHLOROPLAST 1

The protein GIANT CHLOROPLAST 1 (GC1) was identified based on its similarity to a putative cyanobacterial (*Anabaena* sp. PCC 7120) cell division protein, although no actual function for the cyanobacterial protein has been reported (All2390). *GC1* is located on *Arabidopsis* chromosome 2 and encodes a 347-amino acid protein with similarity to nucleotide-sugar epimerases (Maple *et al.*, 2004). The conjugation of uridine-diphosphates (UDP) to sugars and the subsequent epimerase interconversion is important in prokaryotes for sugar activation to form polymers for a variety of functions, including cell envelope biogenesis (Baker *et al.*, 1998).

GC1 is expressed ubiquitously in *Arabidopsis* but shows highest transcript levels in photosynthetic tissues. Although GC1 has no transmembrane domains, a GC1–YFP fusion protein localises uniformly to the inner chloroplast envelope in transgenic plants, suggesting that the entire envelope membrane is equally competent for GC1. Detailed secondary structure predictions of GC1 shows the presence of an amphipathic helix at the extreme C-terminal end, and through protein domain deletion experiments it was established that GC1 is anchored to the stromal side of the inner envelope membrane through this C-terminal amphipathic helix. The role of GC1 in chloroplast division came from analysis of transgenic *Arabidopsis* plant with elevated and reduced levels of GC1. Highly elevated levels of GC1 have no effect on chloroplast division, which might be consistent with the idea that *GC1* encodes an enzyme. In contrast, GC1 deficiency by co-suppression, but not ~70% reduction by antisense expression, leads to severe division inhibition, with mesophyll and hypocotyl cells containing only one or two giant chloroplasts (Figure 4.6).

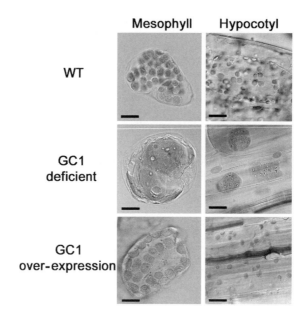

Figure 4.6   GC1-deficient and over-expressing mesophyll cells and hypocotyl cells. GC1 deficiency by co-suppression results in inhibition of chloroplast division, with the presence of one to two giant chloroplast, whilst GC1 over-expression has no effect of chloroplast size or number as compared to wild type (WT). Scale bar: 25 μm.

Ultrastructural analysis of GC1-deficient chloroplasts indicates that thylakoid biogenesis is normal but that grana are more closely stacked in addition to a reduction in starch grains.

4.7   DNA segregation during division

Meristematic proplastids contain ~50 genomes whilst mature chloroplasts contain in excess of 100 copies (Maliga, 2004). As in bacteria, plastid DNA is organised into DNA/protein complexes termed *plastid nucleoids*, which have to segregate during division. Although little is known at present regarding genome segregation during plastid division, this is an important integral aspect of the plastid division cycle that requires attention.

In plastids, nucleoids are associated with the inner envelope whilst in chloroplasts, nucleoids are associated with the thylakoid membranes (Sato *et al.*, 1993). Little is known about the packaging and organisation of the plastid nucleoids but it has been shown that plastid nucleoids in higher plants contain between 20 and 50 proteins (Jeong *et al.*, 2003). To date, five proteins from the nucleoid complex have been identified including CND41 from tobacco (Nakano *et al.*, 1997; Murakami *et al.*, 2000), PEND from pea (Sato *et al.*, 1993), DPC68 (Cannon *et al.*, 1999), sulphite reductase (SiR) (Sekine *et al.*, 2002) and HU (Kobayashi *et al.*, 2002). The PEND protein is thought to anchor nucleoids to the inner plastid envelope during early stages of development (Sato *et al.*, 1998). More recently, the large coiled-coil protein MFP1 was shown to be localised to the thylakoid membranes with the C-terminal domain exposed to the chloroplast stroma. MFP1 has DNA-binding activity interacting with several regions of the *Arabidopsis* plastid DNA with equal affinity (Jeong *et al.*, 2003). It is possible therefore that during chloroplast division nucleoids are segregated together with the thylakoids during late stages of division.

In *E. coli* both the Min system and the nucleoid itself (nucleoid occlusion) regulate cell division by preventing Z-ring formation at sites other than at midcell (Margolin, 2000). In addition, the Min system has a direct effect on nucleoid segregation (Åkerlund *et al.*, 2002). Min-deficient *E. coli* cells show abnormal nucleoid segregation; however, this is not due to the polar division characteristics *per se* but more probably due to a direct effect by the Min system (Åkerlund *et al.*, 2002). Although it is possible that MinD in plants has an effect on plastid nucleoid segregation, this has to date not been shown.

FtsK is an essential cell division protein in *E. coli* and is recruited to the Z-ring (Begg *et al.*, 1995). The N-terminal domain of FtsK is essential for cell division whilst the C-terminal domain has a presumed role in coupling division with nucleoid segregation (Yu *et al.*, 1998). The FtsK C-terminal domain is similar to the *B. subtilis* SpoIIIE C-terminal domain shown to transport trapped chromosomes through septa (Sharpe and Errington, 1995). However, to date, no FtsK homologue has been identified in cyanobacteria or plants.

4.8   Conclusions and future prospects

Over 30 years ago it was recognised that plastids undergo division inside plant cells. Since then, and particularly during the last 5 years, our understanding of plastid division has increased dramatically and this chapter has summarised our knowledge to date. Through a combination of molecular–genetic and cell biological approaches, research has started to provide answers to fundamental questions surrounding the process of plastid division. Using bacterial cell division as a paradigm, the evolutionary conservation of the plastid division process has become clear. Several key plastid division proteins (FtsZ, MinD, MinE) involved in division site placement have been identified and characterised based on their similarity to bacterial cell division proteins. However, the apparent lack of several crucial bacterial cell division protein homologues in the *Arabidopsis* genome suggests that plants have substituted these for alternative components of eukaryotic origin. The recruitment and integration of components of eukaryotic origin, such as dynamin-related proteins, into the plastid division process is evident from studies on the constriction event and the cloning of *arc5*. Together, these findings suggest that plastid division is achieved through a complex interplay between proteins of both prokaryotic and eukaryotic origin.

Although the basic plastid division framework has now been established, several fundamental questions still remain to be solved. Firstly, how do the different protein components act together during division initiation and to what extent do proteins of eukaryotic origin influence this process? Our knowledge of bacterial cell division will undoubtedly provide clues towards this. Secondly, what is the composition of the different PD rings and how are they coordinated with Z-ring placement? Thirdly, what are the biochemical activities of the different plastid division components and how do these activities affect the division process? All these are questions that can be largely answered with the tools already at hand.

There are also a number of fairly unexplored issues that deserve attention: How do plastids control DNA segregation and thylakoid partitioning during division and are these two processes linked? How do plant cells perceive and regulate total plastid numbers and what controls the plastid expansion process? Finally, how is plastid division integrated into plant cell development?

Although we have just touched the tip of the plastid-division iceberg, the continued efforts towards the isolation of new plastid division components, the dissection of the different protein activities and their coordinated interplay will shed light on several of these exciting questions.

Acknowledgements

I thank Makoto Fujiwara for critical reading of the manuscript. I also thank all members of my laboratory for their hard work and contributing to our knowledge of the subject. Work in my laboratory is supported by grants from The British

Biotechnology and Science Research Council, The Royal Society, The Ann Ambrose Appleby Trust, The John Oldacre Foundation and Higher Education Funding Council for England (HEFCE).

References

Addinall, S.G. and Lutkenhaus, J. (1996) FtsZ-spirals and -arcs determine the shape of the invaginating septa in some mutants of *Escherichia coli*. *Mol Microbiol.*, 22, 231–237.

Adler, H.I., Fisher, W.D., Cohen, A. and Hardigree, A.A. (1967) Miniature *Escherichia coli* cell deficient in DNA. *Proc. Natl. Acad. Sci. U.S.A.*, 57, 321–326.

Åkerlund, T., Gullbrand, B. and Nordstrom, K. (2002) Effect of the Min system on nucleoid segregation in *Escherichia coli*. *Microbiology*, 148, 3213–3222.

Arimura, S. and Tsutsumi, N. (2002) A dynamin-like protein (ADL2b), rather than FtsZ, is involved in *Arabidopsis* mitochondrial division. *Proc. Natl. Acad. Sci. U.S.A.*, 99, 5727–5731.

Baker, M.E., Grundy, W.N. and Elkan, C.P. (1998) Spinach CSP41, an mRNA-binding protein and ribonuclease, is homologous to nucleotide-sugar epimerases and hydroxysteroid dehydrogenases. *Biochem. Biophys. Res. Commun.*, 248, 250–254.

Baumann, P. and Jackson, S.P. (1996) An archaebacterial homologue of the essential eubacterial cell division protein FtsZ. *Proc. Natl. Acad. Sci. U.S.A.*, 93, 6726–6730.

Begg, K.J., Dewar, S.J. and Donachie, W.D. (1995) A new *Escherichia coli* cell division gene, *ftsK. J. Bacteriol.*, 177, 6211–6222.

Bi, E. and Lutkenhaus, J. (1991) FtsZ ring structure associated with division in *Escherichia coli*. *Nature*, 354, 161–164.

Bi, E. and Lutkenhaus, J. (1993) Cell division inhibitors SulA and MinCD prevent formation of the FtsZ ring. *J. Bacteriol.*, 175, 1118–1125.

Bleazard, W., McCaffery, J.M., King, E.J. *et al.* (1999) The dynamin-related GTPase Dnm1 regulates mitochondrial fission in yeast. *Nat. Cell Biol.*, 1, 298–304.

Boasson, R., Laetsch, W.H. and Price, I. (1972) The etioplasts/chloroplast transformation in tobacco: correlation of ultrastructure, replication and chlorophyll synthesis. *Am. J. Bot.*, 59, 217–233.

Bramhill, D. (1997) Bacterial cell division. *Annu Rev Cell Dev Biol.*, 13, 395–424.

Bramhill, D. and Thompson, C.M. (1994) GTP-dependent polymerization of *Escherichia coli* FtsZ protein to form tubules. *Proc. Natl. Acad. Sci. U.S.A.*, 91, 5813–5817.

Bukau, B. and Horwich, A.L. (1998) The Hsp70 and Hsp60 chaperone machines. *Cell*, 92, 351–366.

Cannon, G.C., Ward, L.N., Case, C.I. and Heinhorst, S. (1999) The 68 kDa DNA compacting nucleoid protein from soybean chloroplasts inhibits DNA synthesis *in vitro*. *Plant Mol. Biol.*, 39, 835–845.

Cha, J.H. and Stewart, G.C. (1997) The divIVA minicell locus of *Bacillus subtilis. J. Bacteriol.*, 179, 1671–1683.

Cheetham, M.E. and Caplan, A.J. (1998) Structure, function and evolution of DnaJ: conservation and adaptation of chaperone function. *Cell Stress Chaperones*, 3, 28–36.

Colletti, K.S., Tattersall, E.A., Pyke, K.A., Froelich, J.E., Stokes, K.D. and Osteryoung, K.W. (2000) A homologue of the bacterial cell division site-determining factor MinD mediates placement of the chloroplast division apparatus. *Curr. Biol.*, 10, 507–516.

Cran, D.G. and Possingham, J.V. (1972) Variation of plastid types in spinach. *Protoplasma*, 74, 345–356.

Danino, D. and Hinshaw, J.E. (2001) Dynamin family of mechanoenzymes. *Curr. Opin. Cell Biol.*, 13, 454–460.

de Boer, P., Crossley, R. and Rothfield, L. (1992) The essential bacterial cell-division protein FtsZ is a GTPase. *Nature*, 359, 254–256.

de Boer, P.A., Crossley, R.E., Hand, A.R. and Rothfield, L.I. (1991) The MinD protein is a membrane ATPase required for the correct placement of the *Escherichia coli* division site. *EMBO J.*, 10, 4371–4380.

de Boer, P.A., Crossley, R.E. and Rothfield, L.I. (1988) Isolation and properties of minB, a complex genetic locus involved in correct placement of the division site in *Escherichia coli*. *J. Bacteriol.*, 170, 2106–2112.

de Boer, P.A.J., Crossley, R.E. and Rothfield, L.I. (1989) A division inhibitor and a topological specificity factor coded for by the minicell locus determine proper placement of the division septum in *E. coli. Cell*, 56, 641–649.

de Pereda, J.M., Leynadier, D., Evangelio, J.A., Chacon, P. and Andreu, J.M. (1996) Tubulin secondary structure analysis, limited proteolysis sites, and homology to FtsZ. *Biochemistry*, 35, 14203–14215.

Desai, A. and Mitchison, T.J. (1997) Microtubule polymerization dynamics. *Annu. Rev. Cell Dev. Biol.*, 13, 83–117.

Desai, A. and Mitchison, T.J. (1998) Tubulin and FtsZ structures: functional and therapeutic implications. *Bioessays,* 20, 523–527.

Din, N., Quardokus, E.M., Sackett, M.J. and Brun, Y.V. (1998) Dominant C-terminal deletions of FtsZ that affect its ability to localize in *Caulobacter* and its interaction with FtsA. *Mol. Microbiol.*, 27, 1051–1063.

Dinkins, R., Reddy, M.S., Leng, M. and Collins, G.B. (2001) Overexpression of the *Arabidopsis thaliana MinD1* gene alters chloroplast size and number in transgenic tobacco plants. *Planta*, 214, 180–188.

Douglas, S.E. and Penny, S.L. (1999) The plastid genome of the cryptophyte alga, *Guillardia theta*: complete sequence and conserved synteny groups confirm its common ancestry with red algae. *J. Mol. Evol.*, 48, 236–244.

Duckett, J.G. and Ligorne R. (1993) Plastid-dividing rings in ferns. *Ann. Bot.*, 72, 619–627.

Edwards, D.H. and Errington, J. (1997) The *Bacillus subtilis* DivIVA protein targets to the division septum and controls the site specificity of cell division. *Mol. Microbiol.*, 24, 905–915.

Erickson, H.P. (1995) FtsZ, a prokaryotic homolog of tubulin? *Cell*, 80, 367–370.

Erickson, H.P. (1998) Atomic structures of tubulin and FtsZ. *Trends Cell Biol.*, 8, 133–137.

Erickson, H.P., Taylor, D.W., Taylor, K.A. and Bramhill, D. (1996) Bacterial cell division protein FtsZ assembles into protofilament sheets and mini-rings, structural homologs of tubulin polymers. *Proc. Natl. Acad. Sci. U.S.A.*, 93, 519–523.

Errington, J., Daniel, R.A. and Scheffers, D.J. (2003) Cytokinesis in bacteria. *Microbiol. Mol. Biol. Rev.*, 67, 52–65.

Fu, X., Shih, Y.-L., Zhang, Y. and Rothfield, L. I. (2001) The MinE ring required for proper placement of the division site is a mobile structure that changes its cellular location during the *Escherichia coli* division cycle. *Proc. Natl. Acad. Sci. U.S.A.*, 98, 980–985.

Fujiwara, M. and Yoshida, S. (2001) Chloroplast targeting of chloroplast division FtsZ2 proteins in *Arabidopsis. Biochem. Biophys. Res. Commun.*, 287, 462–467.

Fujiwara, M.T., Nakamura, A., Itoh, R., Shimada, Y., Yoshida, S. and Møller, S.G. (2004) Chloroplast division site placement requires dimerisation of the ARC11/AtMinD1 protein in *Arabidopsis. J. Cell Sci.*, 117, 2399–2410.

Fulgosi, H., Gerdes, L., Westphal, S., Glockmann, C. and Soll, J. (2002) Cell and chloroplast division requires ARTEMIS. *Proc. Natl. Acad. Sci. U.S.A.*, 99, 11501–11506.

Gao, H., Kadirjan-Kalbach, D., Froehlich, J.E. and Osteryoung, K.W. (2003) ARC5, a cytosolic dynamin-like protein from plants, is part of the chloroplast division machinery. *Proc. Natl. Acad. Sci. U.S.A.*, 100, 4328–4333.

Gray, M.W. (1999) Evolution of organellar genomes. *Curr. Opin. Genet. Dev.*, 9, 678–687.

Gu, X. and Verma, D.P. (1996) Phragmoplastin, a dynamin-like protein associated with cell plate formation in plants. *EMBO J.*, 15, 695–704.

Hale, C.A., Meinhardt, H. and de Boer, P.A.J. (2001) Dynamic localization cycle of the cell division regulator MinE in *Escherichia coli*. *EMBO J.*, 20, 1563–1572.

Hale, C.A., Rhee, A.C. and de Boer, P.A. (2000) ZipA-induced bundling of FtsZ polymers mediated by an interaction between C-terminal domains. *J. Bacteriol.*, 182, 5153–5166.

Hashimoto, H. (1986) Double ring structure around the constricting neck of the dividing plastids of *Avena sativa*. *Protoplasma*, 135, 166–172.

Hinshaw, J.E. (2000) Dynamin and its role in membrane fission. *Annu Rev Cell Dev Biol.*, 16, 483–519.

Hinshaw, J.E. and Schmid, S.L. (1995) Dynamin self-assembles into rings suggesting a mechanism for coated vesicle budding. *Nature*, 374, 190–192.

Hirota, Y., Ryter, A. and Jacob, F. (1968) Thermosensitive mutants of *E. coli* affected in the process of DNA synthesis and cell division. *Cold Spring Harbor Symp. Quant. Biol.*, 33, 677–694.

Honda, S.I., Hongladoran-Honda, T., Kwanyuen, P. and Wildman, S.G. (1971) Interpretations on chloroplast reproduction derived correlations between cells and chloroplasts. *Planta*, 97, 1–15.

Hu, Z., Gogol, E.P. and Lutkenhaus, J. (2002) Dynamic assembly of MinD on phospholipid vesicles regulated by ATP and MinE. *Proc. Natl. Acad. Sci. U.S.A.*, 99, 6761–6766.

Hu, Z. and Lutkenhaus, J. (1999) Topological regulation of cell division in *Escherichia coli* involves rapid pole to pole oscillation of the division inhibitor MinC under the control of MinD and MinE. *Mol. Microbiol.*, 34, 82–90.

Hu, Z. and Lutkenhaus, J. (2000) Analysis of MinC reveals two independent domains involved in interaction with MinD and FtsZ. *J. Bacteriol.*, 182, 3965–3971.

Hu, Z. and Lutkenhaus, J. (2001) Topological regulation of cell division in *E. coli*. Spatiotemporal oscillation of MinD requires stimulation of its ATPase by MinE and phospho-lipid. *Mol. Cell*, 7, 1337–1343.

Hu, Z. and Lutkenhaus, J. (2003) A conserved sequence at the C-terminus of MinD is required for binding to the membrane and targeting MinC to the septum. *Mol. Microbiol.*, 47, 345–355.

Hu, Z., Mukherjee, A., Pichoff, S. and Lutkenhaus, J. (1999) The MinC component of the division site selection system in *Escherichia coli* interacts with FtsZ to prevent polymerization. *Proc. Natl. Acad. Sci. U.S.A.*, 96, 14819–14824.

Hu, Z., Saez, C. and Lutkenhaus, J. (2003) Recruitment of MinC, an inhibitor of Z-ring formation, to the membrane in *Escherichia coli*: role of MinD and MinE. *J. Bacteriol.*, 185, 196–203.

Itoh, R., Fujiwara, M., Nagata, N. and Yoshida, S. (2001) A chloroplast protein homologous to the eubacterial topological specificity factor MinE plays a role in chloroplast division. *Plant Physiol.*, 127, 1644–1655.

Jeong, S.Y., Rose, A. and Meier, I. (2003) MFP1 is a thylakoid-associated, nucleoid-binding protein with a coiled-coil structure. *Nucleic Acids Res.*, 31, 5175–5185.

Kanamaru, K., Fujiwara, M., Kim, M. *et al.* (2000) Chloroplast targeting, distribution and transcriptional fluctuation of AtMinD1, a eubacteria-type factor critical for chloroplast division. *Plant Cell Physiol.*, 41, 1119–1128.

Kiessling, J., Kruse, S., Rensing, S.A., Harter, K., Decker, E.L. and Reski, R. (2000) Visualization of a cytoskeleton-like FtsZ network in chloroplasts. *J. Cell Biol.*, 151, 945–950.

Kobayashi, T., Takahara, M., Miyagishima, S.Y. *et al.* (2002) Detection and localization of a chloroplast-encoded HU-like protein that organizes chloroplast nucleoids. *Plant Cell*, 14, 1579–1589.

Koksharova, O.A. and Wolk, C.P. (2002) A novel gene that bears a DnaJ motif influences cyanobacterial cell division. *J. Bacteriol.*, 184, 5524–5528.

Kuroiwa, T., Kuroiwa. H., Sakai, A., Takahashi, H., Toda, K. and Itoh, R. (1998) The division apparatus of plastids and mitochondria. *Int. Rev. Cytol.*, 181, 1–41.

Kuroiwa, H., Mori, T., Takahara, M., Miyagishima, S.Y. and Kuroiwa, T. (2002) Chloroplast division machinery as revealed by immunofluorescence and electron microscopy. *Planta*, 215, 185–190.

Leech, R.M., Thomson, W.W. and Platt-Aloia, K.A. (1981) Observations on the mechanim of chloroplast division in higher plants. *New Phytol.*, 87, 1–9.

Liu, Z., Mukherjee, A. and Lutkenhaus, J. (1999) Recruitment of ZipA to the division site by interaction with FtsZ. *Mol. Microbiol.*, 31, 1853–1861.

Löwe, J. (1998) Crystal structure determination of FtsZ from *Methanococcus jannaschii*. *J. Struct. Biol.*, 124, 235–243.

Löwe, J. and Amos, L.A. (1998) Crystal structure of the bacterial cell-division protein FtsZ. *Nature*, 391, 203–206.

Löwe, J. and Amos, L.A. (1999) Tubulin-like protofilaments in Ca^{2+}-induced FtsZ sheets. *EMBO J.*, 18, 2364–2371.

Lu, C., Reedy, M. and Erickson, H.P. (2000) Straight and curved conformations of FtsZ are regulated by GTP hydrolysis. *J. Bacteriol.*, 182, 164–170.

Lutkenhaus, J. and Addinall, S.G. (1997) Bacterial cell division and the Z ring. *Ann Rev Biochem.*, 66, 93–116.

Lutkenhaus, J. and Sundaramoorthy, M. (2003) MinD and role of the deviant Walker A motif, dimerization and membrane binding in oscillation. *Mol. Microbiol.*, 48, 295–303.

Lutkenhaus, J.F., Wolf-Watz, H. and Donachie, W.D. (1980) Organization of genes in the ftsA-envA region of the *Escherichia coli* genetic map and identification of a new fts locus (ftsZ). *J. Bacteriol.*, 142, 615–620.

Ma, X. and Margolin, W. (1999) Genetic and functional analyses of the conserved C-terminal core domain of *Escherichia coli* FtsZ. *J. Bacteriol.*, 181, 7531–7544.

Maliga, P. (2004) Plastid transformation in higher plants. *Annu. Rev. Plant Physiol. Plant Mol. Biol.*, 55, 289–313.

Maple, J., Chua, N.H. and Møller, S.G. (2002) The topological specificity factor AtMinE1 is essential for correct plastid division site placement in *Arabidopsis*. *Plant J.*, 31, 269–277.

Maple, J., Fujiwara, M.T., Kitahata, N. *et al.* (2004) GIANT CHLOROPLAST 1 is essential for correct plastid division in *Arabidopsis*. *Curr. Biol.*, 14, 776–781.

Margolin, W. (2000) Themes and variations in prokaryotic cell division. *FEMS Microbiol Rev.*, 24, 531–548.

Margolin, W., Wang, R. and Kumar, M. (1996) Isolation of an ftsZ homolog from the archaebacterium *Halobacterium salinarium*: implications for the evolution of FtsZ and tubulin. *J. Bacteriol.*, 178, 1320–1327.

Marrison, J.L., Rutherford, S.M., Robertson, E.J., Lister, C., Dean, C. and Leech, R.M. (1999) The distinctive roles of five different *ARC* genes in the chloroplast division process in *Arabidopsis*. *Plant J.*, 18, 651–662.

Marston, A.L., Thomaides, H.B., Edwards, D.H., Sharpe, M.E. and Errington, J. (1998) Polar localization of the MinD protein of *Bacillus subtilis* and its role in selection of the mid-cell division site. *Genes Dev.*, 12, 3419–3430.

McAndrew, R.S., Froehlich, J.E. Vitha, S. Stokes, K.D. and Osteryoung, K.W. (2001) Colocalization of plastid division proteins in the chloroplast stromal compartment establishes a new functional relationship between FtsZ1 and FtsZ2 in higher plants. *Plant Physiol.*, 127, 1656–1666.

Mita, T., Kanbe, T., Tanaka, K. and Kuroiwa, T. (1986) A ring structure around the dividing plane of the *Cyanidium caldarium* chloroplast. *Protoplasma*, 130, 211–213.

Miyagishima, S., Itoh, R., Toda, K., Kuroiwa, H. and Kuroiwa, T. (1999) Real-time analyses of chloroplast and mitochondrial division and differences in the behavior of their dividing rings during contraction. *Planta*, 207, 343–353.

Miyagishima, S., Itoh, R., Toda, K., Takahashi, H., Kuroiwa, H. and Kuroiwa, T. (1998a) Identification of a triple ring structure involved in plastid division in the primitive red alga *Cyanidioschyzon merolae*. *J. Electron Microsc.*, 47, 269–272.

Miyagishima, S., Itoh, R., Toda, K., Takahashi, H., Kuroiwa, H. and Kuroiwa, T. (1998b) Orderly formation of the double ring structures for plastid and mitochondrial division in the unicellular red alga *Cyanidioschyzon merolae*. *Planta*, 206, 551–560.

Miyagishima, S., Kuroiwa, H. and Kuroiwa, T. (2001a) The timing and manner of disassembly of the apparatuses for chloroplast and mitochondrial division in the red alga *Cyanidioschyzon merolae*. *Planta*, 212, 517–528.

Miyagishima, S., Nishida, K., Mori, T. *et al.* (2003) A plant-specific dynamin-related protein forms a ring at the chloroplast division site. *Plant Cell*, 15, 655–665.

Miyagishima, S., Takahara, M. and Kuroiwa, T. (2001b). Novel filaments 5 nm in diameter constitute the cytosolic ring of the plastid division apparatus. *Plant Cell*, 13, 707–721.

Miyagishima, S., Takahara , M., Mori, T., Kuroiwa, H., Higashiyama, T. and Kuroiwa, T. (2001c) Plastid division is driven by a complex mechanism that involves differential transition of the bacterial and eukaryotic division rings. *Plant Cell*, 13, 2257–2268.

Miyagishima, S., Takahara, M. and Kuroiwa, T. (2001b). Novel filaments 5 nm in diameter constitute the cytosolic ring of the plastid division apparatus. *Plant Cell*, 13, 707–721.

Moehs, C.P., Tian, L., Osteryoung, K.W. and Dellapenna, D. (2001) Analysis of carotenoid biosynthetic gene expression during marigold petal development. *Plant Mol Biol.*, 45, 281–293.

Mori, T., Kuroiwa, H., Takahara, M., Miyagishima, S.Y. and Kuroiwa, T. (2001) Visualization of an FtsZ ring in chloroplasts of *Lilium longiflorum* leaves. *Plant Cell Physiol.*, 42, 555–559.

Mosyak, L., Zhang, Y. Glasfeld. E. *et al.* (2000) The bacterial cell division protein ZipA and its interaction with an FtsZ fragment revealed by X-ray crystallography. *EMBO J.*, 19, 3179–3191.

Mukherjee, A., Dai, K. and Lutkenhaus, J. (1993) *Escherichia coli* cell division protein FtsZ is a guanine nucleotide binding protein. *Proc. Natl. Acad. Sci. U.S.A.*, 90, 1053–1057.

Mukherjee, A. and Lutkenhaus, J. (1994) Guanine nucleotide-dependent assembly of FtsZ into filaments. *J. Bacteriol.*, 176, 2754–2758.

Mukherjee, A. and Lutkenhaus, J. (1999) Analysis of FtsZ assembly by light scattering and determination of the role of divalent metal cations. *J. Bacteriol.*, 181, 823–832.

Murakami, S., Kondo, Y., Nakano, T. and Sato, F. (2000) Protease activity of CND41, a chloroplast nucleoid DNA-binding protein, isolated from cultured tobacco cells. *FEBS Lett.*, 18, 15–18.

Nakano, T., Murakami, S., Shoji, T., Yoshida, S., Yamada, Y. and Sato, F. (1997) A novel protein with DNA binding activity from tobacco chloroplast nucleoids. *Plant Cell*, 9, 1673–1682.

Niemann, H.H., Knetsch, M.L., Scherer, A., Manstein, D.J. and Kull, F.J. (2001) Crystal structure of a dynamin GTPase domain in both nucleotide-free and GDP-bound forms. *EMBO J.*, 20, 5813–5821.

Nogales, E., Downing, K.H., Amos, L.A. and Löwe, J. (1998) Tubulin and FtsZ form a distinct family of GTPases. *Nat. Struct. Biol.*, 5, 451–458.

Oross, J.W. and Possingham, J.V. (1989) Ultrastructural features of the constricted region of dividing chloroplasts. *Protoplasma*, 150, 131–138.

Osteryoung, K.W. and McAndrew, R.S. (2001) The plastid division machine. *Annu. Rev. Plant Physiol. Plant Mol. Biol.*, 52, 315–333.

Osteryoung, K.W., Stokes, K.D., Rutherford, S.M., Percival, A.L. and Lee, W.Y. (1998) Chloroplast division in higher plants requires members of two functionally divergent gene families with homology to bacterial *ftsZ*. *Plant Cell*, 10, 1991–2004.

Osteryoung, K.W. and Vierling, E. (1995) Conserved cell and organelle division. *Nature*, 376, 473–474.

Pichoff, S. and Lutkenhaus, J. (2001) *Escherichia coli* division inhibitor MinCD blocks septation by preventing Z-ring formation. *J. Bacteriol.*, 183, 6630–6635.

Platt-Aloia, K. and Thomson, W.W. (1977) Chloroplast development in young sesame plants. *New Phytol.*, 78, 599–605.

Possingham, J.V. and Lawrence, M.E. (1983) Controls to plastid division. *Int. Rev. Cytol.*, 84, 1–56.

Possingham, J.V. and Saurer, W. (1969) Changes in chloroplast number per cell during leaf development in spinach. *Planta*, 86, 186–194.

Pyke, K.A. (1997) The genetic control of plastid division in higher plants. *Am. J. Bot.*, 84, 1017–1027.

Pyke, K.A. (1999) Plastid division and development. *Plant Cell*, 11, 549–556.

Pyke, K.A. and Leech, R.M. (1991) Rapid image analysis screening procedure for identifying chloroplast number mutants in mesophyll cells of *Arabidopsis thaliana* (L.) Heynh. *Plant Physiol.*, 96, 1193–1195.

Pyke, K.A. and Leech. R.M. (1992) Chloroplast division and expansion is radically altered by nuclear mutations in *Arabidopsis thaliana. Plant Physiol.*, 99, 1005–1008.

Pyke, K.A. and Leech, R.M. (1994) A genetic analysis of chloroplast division and expansion in *Arabidopsis thaliana. Plant Physiol.*, 104, 201–207.

Pyke, K.A., Rutherford, S.M., Robertson, E.J. and Leech, R.M. (1994) *arc6*, a fertile *Arabidopsis* mutant with only two mesophyll cell chloroplasts. *Plant Physiol.*, 106, 1169–1177.

Raskin, D.M. and de Boer, P.A.J. (1997) The MinE ring: an FtsZ-independent cell structure required for selection of the correct division site in *E. coli. Cell*, 91, 685–694.

Raskin, D.M. and de Boer, P.A.J. (1999) Rapid pole-to-pole oscillation of a protein required for directing division to the middle of *Escherichia coli. Proc. Natl. Acad. Sci. U.S.A.*, 96, 4971–4976.

RayChaudhuri, D. and Park, J.T. (1992) *Escherichia coli* cell-division gene ftsZ encodes a novel GTP-binding protein. *Nature*, 359, 251–254.

Reddy, M.S., Dinkins, R. and Collins, G.B. (2002) Overexpression of the *Arabidopsis thaliana* MinE1 bacterial division inhibitor homologue gene alters chloroplast size and morphology in transgenic *Arabidopsis* and tobacco plants. *Planta*, 215, 167–176.

Ridely, S.M. and Leech, R.M. (1970) Division of chloroplasts in an artificial environment. *Nature*, 227, 463–465.

Robertson, E.J., Pyke, K.A. and Leech, R.M (1995) *arc6*, an extreme chloroplast division mutant of *Arabidopsis* also alters proplastid proliferation and morphology in shoot and root apices. *J. Cell Sci.*, 108, 2937–2944.

Robertson, E.J., Rutherford, S.M. and Leech, R.M. (1996) Characterization of chloroplast division using the *Arabidopsis* mutant *arc5. Plant Physiol.*, 112, 149–159.

Rothfield, L., Justice, S. and Garcia-Lara, J. (1999) Bacterial cell division. *Annu. Rev. Genet.*, 33, 423–448.

Rothfield, L.I., Shih, Y.L. and King, G. (2001) Polar explorers: membrane proteins that determine division site placement. *Cell*, 106, 13–16.

Rowland, S.L., Fu, X., Sayed, M.A., Zhang, Y., Cook, W.R. and Rothfield, L.I. (2000) Membrane redistribution of the *Escherichia coli* MinD protein induced by MinE. *J. Bacteriol.*, 182, 613–619.

Rutherford, S.M. (1996) *The Genetic and Physical Analysis of Mutants of Chloroplast Number and Size in Arabidopsis thaliana*, Department of Biology, University of York, York, UK.

Sato, N., Albrieux, C., Joyard, J., Douce, R. and Kuroiwa, T. (1993) Detection and characterization of a plastid envelope DNA-binding protein which may anchor plastid nucleoids. *EMBO J.*, 12, 555–561.

Sato, N., Ohshima, K., Watanabe, A. *et al.* (1998) Molecular characterization of the PEND protein, a novel bZIP protein present in the envelope membrane that is the site of nucleoid replication in developing plastids. *Plant Cell*, 10, 859–872.

Sekine, K., Hase, T. and Sato, N. (2002) Reversible DNA compaction by sulfite reductase regulates transcriptional activity of chloroplast nucleoids. *J. Biol. Chem.*, 277, 24399–24404.

Sharpe, M.E. and Errington, J. (1995) Postseptational chromosome partitioning in bacteria. *Proc. Natl. Acad. Sci.U.S.A.*, 92, 8630–8634.

Shih, Y.L., Le, T. and Rothfield, L. (2003) Division site selection in *Escherichia coli* involves dynamic redistribution of Min proteins within coiled structures that extend between the two cell poles. *Proc. Natl. Acad. Sci. U.S.A.*, 100, 7865–7870.

Stokes, K.D., McAndrew, R.S., Figueroa, R., Vitha, S. and Osteryoung, K.W. (2000) Chloroplast division and morphology are differentially affected by overexpression of *FtsZ1* and *FtsZ2* genes in *Arabidopsis. Plant Physiol.*, 124, 1668–1677.

Strepp, R., Scholz, S. Kruse, S. Speth, V. and Reski, R. (1998) Plant nuclear gene knockout reveals a role in plastid division for the homolog of the bacterial cell division protein FtsZ, an ancestral tubulin. *Proc. Natl. Acad. Sci. U.S.A.*, 95, 4368–4373.

Suefuji, K., Valluzzi, R. and RayChaudhuri, D. (2003) Dynamic assembly of MinD into filament bundles modulated by ATP, phospholipids, and MinE. *Proc. Natl. Acad. Sci. U.S.A.*, 99, 16776–16781.

Sun, Q. and Margolin, W. (1998) FtsZ dynamics during the division cycle of live *Escherichia coli* cells. *J. Bacteriol.*, 180, 2050–2056.

Szeto, T.H., Rowland, S.L., Rothfield, L.I. and King, G.F. (2002) Membrane localization of MinD is mediated by a C-terminal motif that is conserved across eubacteria, archaea, and chloroplasts. *Proc. Natl. Acad. Sci. U.S.A.*, 99, 15693–15698.

Tewinkel, M. and Volkmann, D. (1987) Observations on dividing plastids in the protonema of the moss *Funaria hygrometrica* Sibth. *Planta*, 172, 309–320.

Uehara, T., Matsuzawa, H. and Nishimura, A. (2001) HscA is involved in the dynamics of FtsZ-ring formation in *Escherichia coli* K12. *Genes Cells*, 6, 803–814.

Vitha, S., Froehlich, J.E., Koksharova, O., Pyke, K.A., van Erp, H. and Osteryoung, K.W. (2003) ARC6 is a J-domain plastid division protein and an evolutionary descendant of the cyanobacterial cell division protein Ftn2. *Plant Cell*, 15, 1918–1933.

Vitha, S., McAndrew, R.S. and Osteryoung, K.W. (2001) FtsZ ring formation at the chloroplast division site in plants. *J. Cell Biol.*, 153, 111–119.

Wakasugi, T., Nagai, T., Kapoor, M. *et al.* (1997) Complete nucleotide sequence of the chloroplast genome from the green alga *Chlorella vulgaris*: the existence of genes possibly involved in chloroplast division. *Proc. Natl. Acad. Sci. U.S.A.*, 94, 5967–5972.

Wang, D., Kong, D., Wang, Y., Hu, Y., He, Y. and Sun, J. (2003) Isolation of two plastid division ftsZ genes from *Chlamydomonas reinhardtii* and its evolutionary implication for the role of FtsZ in plastid division. *J. Exp. Bot.*, 54, 1115–1116.

Wang, X., Huang, J., Mukherjee, A., Cao, C. and Lutkenhaus, J. (1997) Analysis of the interaction of FtsZ with itself, GTP, and FtsA. *J. Bacteriol.*, 179, 5551–5559.

Wang, X. and Lutkenhaus, J. (1996) FtsZ ring: the eubacterial division apparatus conserved in archaebacteria. *Mol Microbiol.*, 21, 313–319.

Ward, J.E., Jr. and Lutkenhaus, J. (1985) Overproduction of FtsZ induces minicell formation in *E. coli. Cell*, 42, 941–949.

Yamamoto, K., Pyke, K.A. and Kiss, J.Z. (2002) Reduced gravitropism in inflorescence stems and hypocotyls, but not roots, of *Arabidopsis* mutants with large plastids. *Physiol. Plant*, 114, 627–636.

Yan, K., Pearce, K.H. and Payne, D.J. (2000) A conserved residue at the extreme C-terminus of FtsZ is critical for the FtsA-FtsZ interaction in *Staphylococcus aureus. Biochem. Biophys. Res. Commun.*, 270, 387–392.

Yu, X.C. and Margolin, W. (1997) Ca²⁺-mediated GTP-dependent dynamic assembly of bacterial cell division protein FtsZ into asters and polymer networks in vitro. *EMBO J.*, 16, 5455–5463.

Yu, X.C., Weihe, E.K. and Margolin, W. (1998) Role of the C terminus of FtsK in *Escherichia coli* chromosome segregation. *J. Bacteriol.*, 180, 6424–6428.

Zhao, C.R., de Boer, P.A. and Rothfield, L.I. (1995) Proper placement of the *Escherichia coli* division site requires two functions that are associated with different domains of the MinE protein. *Proc. Natl. Acad. Sci. U.S.A.*, 92, 4313–4317.

5 The protein import pathway into chloroplasts: a single tune or variations on a common theme?

Ute C. Vothknecht and Jürgen Soll

5.1 Introduction

Chloroplasts, like mitochondria, are endosymbiotic organelles. The ancestor of chloroplasts was a once free-living prokaryotic organism, closely related to today's cyanobacteria. Subsequent to being engulfed and internalized by an already mitochondriate host cell, the endosymbiont was turned into an interdependent cell organelle (Mereschkowsky, 1905; Margulis, 1970). In the course of this event, the host cell sustained many of the special features of the endosymbiont inside the new organelle: most importantly, the capacity for oxygenic photosynthesis, but furthermore fatty acid biosynthesis, nitrate reduction, and the biosynthesis of amino acids. It is now believed that the primary endosymbiotic event that created chloroplasts was unique, resulting in a common ancestry of all photosynthetic eukaryotes (Palmer, 2000). Ensuing evolution created a number of different photosynthetic lineages of monophyletic origin. The cyanobacterial ancestor of the chloroplast was a self-contained organism, with its own genome and the machinery to transcribe and translate the encoded information. Gradually much of the genetic information was lost from the new organelle (Martin *et al.*, 2002). Many genes vanished because their gene products were not any longer needed in the cellular environment. Other genes were consecutively transferred to the nucleus of the host cell and were subsequently deleted from the organelle genome. Nevertheless, this gene loss was never completed, leaving the chloroplasts of even the most evolutionary advanced plant with a small circular genome encoding up to 200 proteins and all tRNAs required for organellar translation (Race *et al.*, 1999). The proteome of chloroplasts is, on the other hand, estimated to comprise around 3000 proteins (Leister, 2003). Thus many of the organellar proteins are now encoded by nuclear genes. Indeed, many of the multi-protein complexes inside the chloroplast are patchworks of polypeptides made inside and outside the organelle.

All nuclear-encoded chloroplast proteins are synthesized on cytosolic ribosomes and have to be targeted to and transported into the chloroplast. The targeting process has to be specific, ensuring that only the proteins destined for the chloroplast will enter the organelle. At the same time the mis-targeting of these proteins into other cell compartments has to be avoided. In order to reach the inside of the chloroplast, the proteins have to traverse the two membranes that surround the organelle, the outer and the inner envelope. For this purpose, both membranes contain a proteinaceous import machinery called the Toc (translocon on the outer envelope of chloroplasts) and the Tic (translocon on the inner envelope membrane) complex, respectively. These import machineries must have evolved in concert with the transfer of genes from the organelle to the host nucleus. This review summarizes our current knowledge on the composition and mode of operation of these import complexes. Special consideration is given to the question whether these import complexes are common to all types of plastids at any stage of development, or whether they can be altered in unison with the environmental status of the organelle and the surrounding cell.

5.2   Cytosolic targeting

5.2.1   Targeting by presequence

More than 3000 nuclear genes encode for proteins that reside inside the chloroplast (Martin *et al.*, 2002). The products of these genes are synthesized as cytosolic precursors. Most chloroplast proteins destined for the thylakoid membrane, the thylakoid lumen, the stroma, and the inner envelope membrane have a cleavable N-terminal presequence that is required for targeting to the organelle and across the envelope membranes (Dobberstein, 1977). On the contrary, most of the outer envelope proteins do not posses such a presequence. They are inserted into the membrane from the cytosolic side and the targeting information is contained in the mature part of the protein (Schleiff and Klösgen, 2001).

It is believed that the presequence is the sole requirement for chloroplast targeting. In general, the transit peptides from chloroplast proteins have an overall positive charge and they are enriched in hydroxylated residues. Yet, they display a huge variety in length and primary amino acid sequence. No common secondary structure has been identified for chloroplast presequences either. Instead, they form a random coil in aqueous environments (Emanuelsson and von Heijne, 2001). It has been suggested that interaction with the outer envelope lipids might induce a structural change that allows recognition of the presequence by the translocon (Bruce, 2000).

Photosynthetic eukaryotes have two endosymbiotic organelles, chloroplasts and mitochondria, both of which use cleavable presequences for organelle targeting. Mitochondrial targeting sequences show a much higher conservation than their chloroplast counterparts and are thus much easier to recognize from their amino acid sequence, even though it is not always obvious from *in silico* analysis to which organelle a particular protein is targeted. It has to be assumed that the targeting is highly specific *in vivo* and this notion is supported by several studies (Peeters and Small, 2001). *In vitro*, several proteins have been shown to be mis-targeted into the wrong organelle if only this organelle was offered in the reaction. In most cases, mis-targeting was abolished when both organelles were present in the same reaction. A small number of proteins are targeted to chloroplasts and mitochondria with the aid of an identical presequence. It is little understood how the organism regulates the distribution of these proteins *in vivo*. There is some evidence that dual-targeting presequences have a domain structure that guides the allocation to the different organelles (Silva-Filho *et al.*, 1997; Hedtke *et al.*, 2000; Rudhe *et al.*, 2002). After import into either mitochondria or chloroplast, the presequence has to be spliced off by the respective processing peptidase, MPP (mitochondrial processing peptidase) and SPP (stromal processing peptidase). Thus, it has to be assumed that an identical presequence can be recognized by either of the two proteins.

5.2.2 *Chloroplast import without a presequence*

There is growing suspicion that proteins without presequence might be able to transfer into the chloroplast. As of date there is only one example described in the literature (Miras *et al.*, 2002). During a proteomic approach to identify proteins of the chloroplast inner envelope, Ferro *et al.* (2002) discovered a homolog of quinone oxidoreductase. This finding was somehow unexpected since the deduced protein sequence does not contain a potential chloroplast targeting sequence. Compared to homologs from bacteria, no extra N-terminal extension that could function as a presequence was obvious at all. In a subsequent study, Miras *et al.* (2002) showed immunologically that the protein is localized in the chloroplast envelope. They showed furthermore that the protein is not processed N-terminally after chloroplast import and GFP (green fluorescent protein) fusion proteins lacking the first 59 amino acids could still be transported into the organelle. Instead, import of the protein into chloroplasts seems to depend on intrinsic amino acids. Further studies will have to show whether this protein is just the proverbial exception that proves the rule or whether these studies open up the route to identify many more organelle proteins that are targeted without a presequence.

5.3   The general import pathway

5.3.1 *Toward the chloroplast*

Chloroplast precursor proteins are synthesized posttranslocationally on cytosolic ribosomes. A set of cytosolic components are involved in ascertain that the precursor protein can make its way successfully from the ribosome to the chloroplast. It was shown early on that the import of pLHCP (precursor of light harvesting chlorophyll binding protein) into isolated organelles could be enhanced by the addition of a cytosolic extract (Waegemann *et al.*, 1990). This effect could be pinned down partially to the presence of cytosolic chaperones and the addition of isolated Hsp70 protein would restore some of the import capacity for pLHCP in the absence of cytosol. On the other hand, a requirement for Hsp70 was not observed for small, soluble proteins like the precursor of pFD (ferrodoxin) or pSSU (small subunit of ribulose-1,5-bisphosphate carboxylase/oxygenase) (Pilon *et al.*, 1990, 1992). It was concluded that the function of Hsp70 was primarily in preventing an aggregation of highly hydrophobic membrane proteins. Nevertheless, it is reasonable to assume that chaperones are assisting generally in the import process by preventing the precursor proteins to fold prematurely inside the cytosol. Consequently, it has been shown that many of the targeting signals contain a sequence that allows interaction with Hsp70 proteins (Ivey *et al.*, 2000).

Cytosolic components seem to have a function beyond the prevention of premature folding or aggregation. Many precursor proteins can be phosphorylated by a serine/threonine protein kinase and phosphorylation stimulates the import of these precursors (Waegemann and Soll, 1996). The base for this stimulation lies in the binding of phosphorylated precursor protein to a so-called guidance complex. Radioactive-labeled precursor proteins are synthesized *in vitro* using either reticulocyte lysate or extracts from wheat germ embryos. Since the precursor proteins are normally not further purified, components endogenous to this extracts can have an impact on the import reaction. This possibility is mostly ignored but it was shown that a presequence-binding factor is present in the reticulocyte lysate that enhances mitochondrial protein import (Murakami and Mori, 1990). For chloroplast import, a whole soluble precursor guidance complex could be identified in wheat germ extract (Waegemann *et al.*, 1990; May and Soll, 2000). It consists of Hsp70, 14-3-3 proteins, and other so far unidentified components (May and Soll, 2000). Nonphosphorylated precursor protein will bind to Hsp70 alone, indicating that phosphorylation-dependent binding to the guidance complex occurs via the 14-3-3 protein or one of the unidentified components of the complex. It is not known whether binding to the guidance complex is essential for chloroplast targeting *in vivo*. For several precursor proteins, *in vitro* chloroplast import can be achieved in the absence of the guidance complex, alas with a strongly reduced efficiency (May and Soll, 2000). After the precursor protein has made contact with the import machinery, it is released from the guidance complex. It is not clear whether this is achieved by ATP hydrolysis or dephosphorylation or whether the complex can dissociate spontaneously.

On the contrary, it was also shown that wheat germ lysate contains components that have a negative effect on the import into both chloroplasts and mitochondria (Schleiff *et al.*, 2002a). While these experiments revealed that at least one of the factors is proteinaceous by nature, no such protein has been identified to date.

5.3.2   *The chloroplast translocon*

Once a precursor protein has made contact with the chloroplast surface, a number of subsequent steps are initiated. In general, the import is divided in three distinctive stages: recognition at the chloroplast surface, commitment into the import machinery, and finally the simultaneous translocation across the outer and inner envelope, followed by stromal processing of the targeting sequencing. All three steps are characterized by specific energy requirements.

The process of translocation begins with the recognition of the precursor protein at the outer envelope membrane. This step does not require energy in form of ATP hydrolysis and can be reversed (Perry and Keegstra, 1994; Ma *et al.*, 1996; Kouranov and Schnell, 1997). Nevertheless, it seems to involve a distinct GTP/GDP status of certain components of the import machinery (see below). The recognition is highly specific. This ensures that only the correct proteins can engage the import pathway. While the recognition process is far from understood, it is clear that both the lipid surface itself as well as proteinaceous components of the envelope membrane are involved. However, the role of the envelope lipids in the recognition process remains enigmatic. The chloroplast envelope contains a number of unique lipids, i.e. monogalactosyldiacylglycerol, digalactosyldiacylglycerol, or sulphoquinovosyldicylglycerol. A possible function of the lipids could involve a partitioning of the precursor into the lipid bilayer prior to its interaction with the Toc complex (Bruce, 1998). Thereby, a conformational change would be induced that alters the secondary structure of the precursor in a way to allow its recognition by the Toc complex. This possibility is supported by several *in vitro* and *in vivo* observations. Precursor proteins have been shown to specifically interact with artificial lipid bilayers only if those contain chloroplast-specific galactolipids (van't Hof *et al.*, 1993; van't Hof and de Kruijff, 1995; Pinnaduwage and Bruce, 1996). In the presence of artificial membranes or in hydrophobic solvents that mimic such an environment, presequences adopt an α-helical structure (Chupin *et al.*, 1994; Pinnaduwage and Bruce, 1996; Wienk *et al.*, 2000). Furthermore, analysis of the digalactosyldiacylglycerol-deficient *dgd1* mutant of *Arabidopsis* displayed a decrease in protein translocation into chloroplasts (Chen and Li, 1998). It is noteworthy that the chloroplast envelope is the only plant membrane containing galactolipids that is exposed to the cytosol; the only other galactolipid-containing membranes being the inner chloroplast envelope and the thylakoids (Block *et al.*, 1983a). It is therefore likely that the presence of galactolipids might assist in distinguishing the chloroplast from other potential target membranes inside the cell.

When precursor proteins have been recognized as acceptable candidates for translocation, they can enroll into the actual import machinery (Plate 2). The precursor inserts into the outer envelope via the Toc complex and makes contact with the Tic complex (Waegemann and Soll, 1991; Olsen and Keegstra, 1992; Akita *et al.*, 1997; Kouranov and Schnell, 1997). The formation of this so-called early import intermediate requires ATP as well as GTP and is irreversible (Olsen and Keegstra, 1992; Kessler *et al.*, 1994; Young *et al.*, 1999; Chen *et al.*, 2000). The import process can be arrested at this stage by provision of low amount of ATP (<50 µmol) because further translocation requires higher ATP concentrations (>100 µmol). The precursor protein can then enter the last stage of the import process, the simultaneous translocation through the Toc and Tic complexes (Flügge and Hinz, 1986; Theg *et al.*, 1989; Schnell and Blobel, 1993). Once precursor proteins have reached the stroma the presequence is cleaved off by SSP (Robinson and Ellis, 1984; Richter and Lamppa, 1998).

Protein translocation requires the action of molecular chaperones that keep the precursor protein in an unfolded state throughout the import process (Marshall *et al.*, 1990; Schnell *et al.*, 1994; Kourtz and Ko, 1997; Ivey *et al.*, 2000; May and Soll, 2000). Chaperones involved in the import process include Hsp70 homologs, both in the cytosol and associated with the two envelope membranes. Furthermore, stromal homologs of Hsp60 (Cpn60) and/or Hsp93 (ClpC) allegedly bind the precursor protein upon its entering into the chloroplast and pull it through the membrane (Akita *et al.*, 1997; Nielsen *et al.*, 1997). It is above all the ATP hydrolysis by the stromal chaperones that is responsible for the vast amount of energy that is required in the import process (Theg *et al.*, 1989; Olsen and Keegstra, 1992).

5.3.2.1   Components of the Toc complex

For all proteins engaging the general import pathway, the Toc complex is the entrance gate into the chloroplast. It is here that precursor proteins make their first contact with the envelope membrane and where the transit peptide is recognized prior to translocation.

In the last 10–15 years, the Toc complex has been isolated and its components have been identified (Plate 2). This happened largely with the use of pea chloroplasts as the model system (Waegemann and Soll, 1991; Hirsch *et al.*, 1994; Kessler *et al.*, 1994; Perry and Keegstra, 1994; Schnell *et al.*, 1994; Wu *et al.*, 1994; Sohrt and Soll, 2000). Only lately have these studies been shifted to the analysis of the import apparatus of *Arabidopsis thaliana* owing to our knowledge of the complete genome sequence and the accessibility of this plant to genetic manipulation (The *Arabidopsis* Genomic Initiative, 2000). Thus, if not specifically mentioned, the names of components of Toc and Tic complexes refer to the proteins identified from pea. Homologs from *Arabidopsis* are marked by an "at" prefix.

To our current knowledge, the Toc complex consists of a core comprising three proteins: Toc159, Toc34, and Toc75 (Hirsch *et al.*, 1994; Kessler *et al.*, 1994; Perry and Keegstra, 1994; Schnell *et al.*, 1994; Wu *et al.*, 1994). The core complex has an apparent molecular mass of about 500 kDa (Schleiff *et al.*, 2003b) and seems to consist of one molecule of Toc159 to four molecules each of Toc75 and Toc34. A fourth protein, Toc64, can associate with the Toc core and might be involved specifically in precursor recognition involving the guidance complex (Sohrt and Soll, 2000).

The first protein identified as part of the Toc complex, Toc75, is the single-most abundant protein of the outer envelope of chloroplasts (Joyard *et al.*, 1983). Just like all the components of the chloroplast translocon, it is named for its apparent molecular mass. Toc75 is an integral membrane protein and forms the actual import channel of the Toc complex. This role has been deduced from topological studies predicting that the protein consists primarily of 16 transmembrane β-sheets, with the potential to form a β-barrel structure (Schnell *et al.*, 1994; Hinnah *et al.*, 1997). The alleged role as a protein conductance channel has been corroborated by electrophysiological measurements on heterologously expressed Toc75 that was reconstituted into liposomes (Hinnah *et al.*, 1997). The analysis showed that Toc75 forms a voltage-gated, cation-selective channel. The opening of the conducting pore has been estimated to exceed 20 Å (Hinnah *et al.*, 2002). A pore of this size would be large enough to admit proteins but only if they are in an unfolded state. The role of Toc75 is not only founded by its channel properties. It is further substantiated by cross-linking studies on purified outer envelopes and intact chloroplasts. Toc75 has been shown to interact with precursor protein early in the import process. The interaction increases at later stages of the import when the precursor has been inserted into the import machinery. Cross-linking studies showed that the interaction with the precursor involves both the presequence and the mature part of the protein (Ma *et al.*, 1996). In electrophysiological experiments, heterologously expressed Toc75 was able to distinguish transit peptides from synthetic peptides or mitochondrial transit sequences via a cytosolic precursor binding site (Hinnah *et al.*, 1997, 2002). Thus, Toc75 is able to recognize precursor protein without the assistance of the other Toc components. Toc75 was also found stably associated with both Toc159 and Toc34, even in the absence of precursor protein (Waegemann and Soll, 1991; Seedorf *et al.*, 1995; Kouranov and Schnell, 1997; Nielsen *et al.*, 1997). These data indicate that these three proteins form a stable core of the import apparatus and that the complex is not disassembled in the absence of translocation events.

While the function of Toc75 in outer envelope translocation is quite clear, there is an ongoing debate as for the function of Toc159 and Toc34. Both proteins contain a GTP-binding site and they appear to function as GTP-dependent precursor protein receptors (Hirsch *et al.*, 1994; Kessler *et al.*, 1994; Schnell *et al.*, 1994; Seedorf *et al.*, 1995). Toc159 was originally identified as Toc86 because the protein is very receptive toward proteolytic degradation when chloroplasts are isolated (Waegemann and Soll, 1991; Bölter *et al.*, 1998a; Chen *et al.*, 2000). Many studies have shown that Toc159 directly interacts with precursor proteins. The interaction seems to occur independent from ATP hydrolysis, supporting a role of Toc159 as a primary import receptor (Perry and Keegstra, 1994; Ma *et al.*, 1996; Kouranov and Schnell, 1997). The structure of Toc159 can be separated into three domains. The most N-terminal amino acids (N-domain) are exposed to the cytosol and predominantly acidic. The function of this domain is not known. Isolated chloroplasts in which the N-domain was proteolytically removed have reduced import efficiency. Therefore, this acidic domain could be involved in the binding of positively charged transit peptides (Bölter *et al.*, 1998a). Nevertheless, an atToc159 construct lacking this domain was found to rescue a Δ-Toc159 mutant of *Arabidopsis*, indicating that the function of the N-domain is not fundamental to the viability of the plant (Bauer *et al.*, 2002). Following the N-domain is a functional GTP-binding domain (G-domain) with a motif typical for GTP-binding proteins. Interestingly, outside the GTP-binding motif, Toc159 does not share any sequence homology with other GTP-binding proteins other than Toc34 (Hirsch *et al.*, 1994; Kessler *et al.*, 1994). Thermolysin degradation can remove both the N- and G-domain, leaving a 52-kDa proteolytic fragment. This experiment shows that both domains are protruding into the cytosol (Chen *et al.*, 2000). Removal of the G-domain affects the import efficiency of isolated chloroplasts. Interestingly, it is especially precursor binding and the formation of the early import intermediate that is reduced by the removal of the G-domain while translocation of already bound precursor proteins proceeds unaffected. Toc159 is anchored to the outer envelope by its C-terminal 400 amino acids (M-domain) even though this part does not contain any obvious transmembrane helices (Chen *et al.*, 2000). *In vivo*, Toc159 was found to be resistant to the extraction with salt and alkali, indicating that the M-domain is truly integrated into the outer envelope and not merely associated to it (Hirsch *et al.*, 1994; Kessler *et al.*, 1994). Surprisingly, a fusion protein between only the G-domain and GFP was found predominantly attached to the outer envelope (Bauer *et al.*, 2002). The authors concluded that the G-domain must be able to bind to the outer envelope in the absence of the C-terminus, probably by interaction with other subunits of the import apparatus. atToc159 was also found in a soluble form in the cytosol (Hiltbrunner *et al.*, 2001b). This has led to the notion that Toc159 might act as a precursor receptor well before the chloroplast envelope, doubling as a component of organelle targeting in addition to its function in translocation. In this model, Toc34 acts as a docking site for Toc159, thereby bringing the precursor protein to the transfer channel. Controversial studies place the function of Toc159 after the interaction of the precursor with Toc34, thereby placing the protein at the interface of Toc34 and the import channel (Schleiff *et al.*, 2002b, 2003a).

Toc34 has intriguing similarities to Toc159 in both structure and function. Toc34 contains a GTPase domain with sequence similarity to Toc159 that extends beyond the actual nucleotide-binding site (Kessler *et al.*, 1994). Indeed, their homology places Toc34 and Toc159 in a unique subclass of GTP-binding proteins. The protein is anchored to the outer envelope with an 8-kDa domain close to its C-terminus while the major part of the protein extrudes into the cytosol (Seedorf *et al.*, 1995; Li and Chen, 1996). Like Toc159, Toc34 has been implied in precursor protein recognition. Toc34 interacts with precursor protein independent from energy. This interaction does not require the presence of the other Toc components. Toc34 that was expressed heterologously in a soluble form by omission of the C-terminal membrane anchor was able to bind precursor protein in a highly regulated fashion (Sveshnikova *et al.*, 2000). Precursor binding occurred only in the GTP-bound form of Toc34 and was disrupted by GTP hydrolysis. Phosphorylation of Toc34 leads to a loss of GTP binding and in turn inhibits binding of precursor protein (Jelic *et al.*, 2002). Interaction of Toc34 with precursor protein does not require ATP (Sveshnikova *et al.*, 2000). This indicates that the function of Toc34 precedes even the formation of the early-import intermediate, which is energy-dependent (Kouranov and Schnell, 1997; Young *et al.*, 1999).

Toc64 is the least studied component of the Toc complex. Nevertheless, it is by far not the least intriguing (Sohrt and Soll, 2000). Like the other Toc components, it is quite abundant in outer envelope fractions and it co-purifies with Toc159, Toc75, and Toc34 during sucrose density gradient centrifugation. Furthermore, Toc64 can be co-immunoprecipitated with precursor protein and the other Toc components when import is arrested at the stage of the early-import intermediate (Sohrt and Soll, 2000). Very little is known about the role of Toc64 in the import process. Like the other Toc –components, it is an integral membrane protein. It is anchored to the outer envelope membrane by a N-terminal transmembrane domain and a large portion of its C-terminus is exposed to the cytosol. This portion of the protein contains two different domains. The part closest to the membrane anchor shares significant sequence homology to amidases and aminotransferases. No amidase or aminotransferase activity has yet been shown for Toc64, and so the functionality of this domain remains mysterious. The most C-terminal part of Toc64 contains three tetratricopeptide repeat (TPR) motives. This appears particularly significant since TPR motives have been implied in various protein–protein interactions (Lamb *et al.*, 1995). Components of other protein-targeting systems have been shown to contain TPR motives, including several of the mitochondrial import receptors (Pfanner and Geissler, 2001). By similarity this would place Toc64 as yet another import receptor of the chloroplast outer envelope. Sohrt and Soll (2000) suggested that Toc64 is involved exclusively in the import of proteins whose targeting is dependent on the guidance complex.

5.3.2.2   *Progression at and regulation of the Toc translocon*
In recent years, a clearer view has arisen on the function of the diverse Toc components and the regulation of the translocation. Yet, the same studies also opened up a new debate on the string of events taking place at different stages of the process. There is little debate on the function of Toc75 as the translocon pore of the Toc complex. The specific function of Toc159 and Toc34 on the other hand is less evident. There is general agreement that both subunits directly interact with the precursor protein and with Toc75. Both expose their GTP-binding domains to the cytosol and binding to the precursor is regulated by GTP. They are therefore considered precursor receptors of the Toc complex.

In *Arabidopsis*, Kessler and coworkers found about half the cells content of atToc159 soluble in the cytosol (Hiltbrunner *et al.*, 2001b). Transient overexpression of atToc159–GFP fusion protein in protoplasts also produced a significant amount of GFP fluorescence in the cytosol. Without its GTPase domain, atToc159 remains in its soluble cytosolic form. On the other hand, a construct containing only the GTPase domain fused to GFP can target the protein to the outer envelope. This would imply that GTP binding to the G-domain is required for targeting of Toc159 to the envelope membrane. Other experiments imply that not only GTP binding but also GTP hydrolysis is required for this process. In this view of the import progression on the Toc translocon, the precursor proteins would first interact with soluble Toc159 in the cytosol. Toc34 then acts as a docking site for the precursor-bound Toc159 (Bauer *et al.*, 2002; Smith *et al.*, 2002). Building of a heterodimer between Toc159 and atToc33, the homolog to Toc34, was suggested as an important step in these events (Weibel *et al.*, 2003). It has been shown *in vitro* that atToc33 can form homodimers in a GDP-bound state, facilitating a specific dimerization motif, D1, for this process. An identical motif exists in atToc159, thereby making a potential dimerization between atToc33 and atToc159 feasible. The precursor would then be passed on via Toc34 to the translocon pore Toc75. This model still has to be proven by experimental data.

Other experimental data suggest a different view on the function of the Toc subunits and the regulation of the Toc transfer (Schleiff *et al.*, 2002b). In this model both Toc34 and Toc159 are permanently present in the envelope membrane and remain in close association with each other. Upon arrival at the outer envelope, the precursor protein first makes contact to Toc34. Precursor protein can only interact with Toc34 when it is present in a GTP-bound state (Svesnikova *et al.*, 2000; Schleiff *et al.*, 2003a). Toc34 was shown to have a low endogenous GTPase activity that can be stimulated by the interaction with precursor protein. When Toc34 changes into the GDP-bound form, the affinity to precursor protein is reduced and the precursor is released from Toc34 and passed on to the next Toc receptor protein Toc159. As in the previous model, the formation of a heterodimer between the two receptors is proposed for this step. Toc34 needs to change back to the GTP-bound form before it can bind the next precursor protein. This phase represent an important regulatory point of Toc translocation since phosphorylation of Toc34 by a specific protein kinase will prevent GTP binding and thereby halt renewed precursor recognition (Fulgosi and Soll, 2002).

Toc159 seems to fulfill a dual role in the import process. First, it takes over the precursor protein from Toc34 and it seems to do so in a GTP-dependent manner (Schleiff *et al.*, 2003a). Since Toc159 is the most prominent phosphorylated protein of the outer envelope, a similar regulation as shown for Toc34 could also be controlling Toc159. Likewise, a specific protein kinase was shown to act on the protein (Fulgosi and Soll, 2002). Second, in addition to its receptor function, Toc159 is also part of the actual translocation machinery. GTP hydrolysis by Toc159 is thought to induce a conformational change that assists in shoving precursor protein through the translocation pore. Reconstituted into liposome, Toc159 and Toc75 are sufficient for driven translocation over the lipid bilayer (Schleiff *et al.*, 2003a). This suggests that these two Toc components represent the minimal translocation unit of the Toc complex.

All in all, further studies are required to elucidate the exact mode of operation of the Toc translocon.

5.3.2.3 *Components of the Tic complex*

After precursor proteins have engaged the Toc –complex, the translocation occurs simultaneously across both envelope membranes (Schnell and Blobel, 1993; Alefsen *et al.*, 1994; Kouranov *et al.*, 1998). This requires at least temporary interaction of the Toc and Tic complexes. Nevertheless, the translocation over the inner envelope is a process independent from the Toc translocation. This is stressed by the fact that even though both complexes carry out a principally identical action, i.e. the translocation of a protein across a membrane, Tic and Toc are composed of a complete different set of proteins (Plate 2). To date, seven proteins have been identified as components of the Tic complex: Tic110, Tic62, Tic55, Tic40, Tic32, Tic22, and Tic20 (Kessler and Blobel, 1996; Lübeck *et al.*, 1996; Caliebe *et al.*, 1997; Kouranov *et al.*, 1998; Stahl *et al.*, 1999; Küchler *et al.*, 2002). A Tic core-complex could be isolated from inner envelope vesicles, which is stably associated even in the absence of precursor protein. As deduced from blue –native-polyacrylamide gel electrophoresis (BN-PAGE), the purified Tic complex has a molecular size of about 250 kDa but the exact composition of this complex is still under investigation. Tic110, Tic62, and Tic55 seem to be the major components in this complex. Several smaller proteins are co-purified with the Tic complex on sucrose gradients but can be separated from the core-complex on BN-PAGE (Caliebe *et al.*, 1997; Küchler *et al.*, 2002). Thus, it is still little known about the exact composition of the Tic complex and there are also diverged opinions on the role that the acknowledged components have in the translocation process. Nevertheless, a number of recent studies have brought us to a better understanding about the Tic –translocon, and it has become clear that Tic translocation is regulated in a fashion very different from Toc translocation.

Tic110 was identified early on as part of the Tic complex (Schnell *et al.*, 1994; Wu *et al.*, 1994). It is one of the most abundant proteins in the inner envelope of chloroplasts (Block *et al.*, 1983b). Tic110 was shown to interact with precursor protein and furthermore with all the other alleged components of the Tic complex as well as Toc75 (Kessler and Blobel, 1996; Lübeck *et al.*, 1996; Caliebe *et al.*, 1997; Kouranov *et al.*, 1998; Stahl *et al.*, 1999; Küchler *et al.*, 2002). Despite its early iden- tification, the exact topology of Tic110 is still under debate. Some groups propose Tic110 to be largely exposed into the stroma of the chloroplasts where it is suggested to attract stromal chaperones such as cpn60 and ClpC to the translocation pore. In- teraction with both of these chaperones has been shown experimentally, and it could be mediated by a hydrophobic domain close to the C-terminus of Tic110 (Kessler and Blobel, 1996; Nielsen *et al.*, 1997; Jackson *et al.*, 1998; Inaba *et al.*, 2003). In this model of Tic110 topology and function, two predicted transmembrane helices at its N-terminus would anchor Tic110 into the envelope membrane. A smaller domain would be exposed into the intermembrane space. Because of its cross-linking with Toc75, it has been proposed that this domain promotes the interaction with the Toc complex thereby forming a joint translocation site for the simultaneous transloca- tion of the precursor protein across both membranes (Lübeck *et al.*, 1996). A recent publication by Heins *et al.* (2002) suggests an altogether different topology and role for Tic110. Using heterologously expressed Tic110 reconstituted into liposomes as well as isolated inner envelope vesicles, the authors could show that Tic110 forms a cation-selective channel whose conductivity is sensitive to the presence of transit peptides. They therefore propose Tic110 to be the actual import pore of the Tic complex, a structure formed by β-barrels. Because of its enormous size, it cannot be excluded that Tic110 is responsible for all its alleged functions.

Tic62 is a rather recent addition to the Tic complex. The protein is part of the Tic core-complex isolated via BN-PAGE, where it co-migrates with Tic110 and Tic55 (Küchler *et al.*, 2002). Antisera raised against Tic110 and Tic55 do co-immunoprecipitate Tic62. It is an integral membrane protein that is anchored to the membrane by a putative hydrophobic domain in its N-terminal part. The membrane anchor is preceded by a functional nicotinamide-dinucleotide-binding site (Küchler *et al.*, 2002). Besides, the C-terminal part of Tic62 comprises several highly conserved repetitive sequence modules that allow the protein to associate with ferredoxin–NAD(H) oxidoreductase (FNR) (Plate 2). Binding of Tic62 to FNR in- dicates a role of the protein in redox regulation of the translocation process, a feature that was already suggested for Tic55.

Indeed, Tic55 was identified prior to Tic62 because of its presence in the same BN-PAGE purified complex (Caliebe *et al.*, 1997). Tic55 contains two predicted membrane-spanning domains close to its C-terminal end and the protein extends a large part of its N-terminus into the stroma. A small part of the protein seems to be exposed into the intermembrane space. Tic55 was not only identified as part of the Tic core-complex but also together with precursor protein and several components of both Toc and Tic. Sequence analysis revealed that Tic55 contains a predicted Rieske-type iron–sulfur cluster and a mononuclear-binding site (Caliebe *et al.*, 1997), both of which are facing the stromal site of the inner envelope (Plate 2).

Little is known about the function of Tic40 in the Tic complex. Tic40 does not co-purify with the Tic core-complex but the protein can be cross-linked to Tic110 as well as to precursor protein retained in the import machinery (Stahl *et al.*, 1999). Tic40 is an integral membrane protein that is anchored into the envelope by a membrane-spanning domain close to its N-terminus. The C-terminal part of the protein extrudes into the stroma and comprises a binding site for Hsp70. It was therefore suggested that Tic40 is involved in the association of chaperones with the import machinery. Tertiary structure analysis furthermore identified a potential TRP domain in the C-terminus of the protein (Chou *et al.*, 2003). Tic40 seems to be important but not essential for chloroplast import. *Arabidopsis* deletion mutants of atTic40 are not lethal but they display reduced chloroplast import, which results in slow growth and pale green leaves (Budziszewski *et al.*, 2001; Chou *et al.*, 2003).

Tic32 has been identified only very recently as a component of the Tic complex (Hörmann *et al.*, in press). Like Tic62 and Tic55, Tic32 could to be involved in regulation of the import process, for it contains an NAD(P)-binding site and has homologies to a class of short-chain dehydrogenases.

Tic22 and Tic20 are two small proteins of the inner envelope that can both be cross-linked to precursor protein during the import process. Tic22 is a peripheral component of the inner envelope. Since it was shown to interact with precursor protein before they engage the Tic complex, Tic22 was placed at the intermembrane space between the two envelope membranes. Tic22 could be involved in promoting the contact site between the Toc and Tic complex or it might act as a precursor protein receptor (Kouranov *et al.*, 1998). Tic20, in contrast, has three predicted transmembrane domains and it is found well buried into the inner envelope membrane. It has been suggested as an alternative to Tic110 for the import pore of the translocon (Kouranov and Schnell, 1997; Kouranov *et al.*, 1998). A decrease of the atTic20 content by antisense expression resulted in a defect of import over the inner envelope (Chen *et al.*, 2002). Consequently, the plants appeared pale or white, had a significant reduction in plastidal protein content, and showed abnormal plastidal ultrastructure.

5.3.2.4 *Regulation of Tic import*

There is very little known on the progression of the translocation on the Tic complex or about the regulation of the translocation process. All evidence so far points toward a redox regulation from the stromal site of the envelope membrane. Sensing and conveying the redox status of the photosynthetic chain has long been known as a major regulation circuit for many plastidal processes. Photosynthesis is the major energy-producing process in the chloroplast (and the whole plant, indeed), and so it is important for the cell to monitor its status and regulate the expression and translocation of chloroplast proteins accordingly. Regulation is conveyed via certain elements of the photosynthetic chain that are present in either reduced or oxidized form, depending on the photosynthetic capacity. A major player in this circuit is ferredoxin. It can pass electrons from the photosynthetic machinery to FNR, which in turn activates or inactivates enzymes in a number of biochemical pathways inside the chloroplast. In order to adapt the chloroplast import to the specific requirements of photosynthesis and metabolism, it would make sense to include the import machinery into the regulation circuit and such a regulation was actually shown recently for *in vitro* import into maize chloroplasts (Hirohashi *et al.*, 2001). With at least three of the Tic components containing potential redox-sensing domains, this idea does not seem to be too far fetched. Tic62 contains an FNR-binding site and was shown to associate FNR with the inner envelope (Küchler *et al.*, 2002). Protein import could thereby directly be regulated in correlation to the redox status of the chloroplast (Plate 2). Tic55 with its Rieske-type iron–sulfur center and Tic32 might aid in further fine-tuning this regulation.

5.4 Stromal processes involved in chloroplast protein import

What happens when precursor proteins have passed through the Toc and Tic translocon into the chloroplast? Immediately upon arrival of the N-terminus in the stroma, even before the translocation process is complete, the precursor protein is met by a set of different proteins (Plate 2). The presequence is removed by the stromal processing peptidase (SSP) (Robinson and Ellis, 1984; VanderVere *et al.*, 1995; Richter and Lamppa, 1998). Features determined by the last 10–15 amino acids of the presequence appear to be important for the processing reaction. SSP belongs to a class of metallopeptidases characterized by a His-X-X-Glu-His motif in its catalytic domain. Amino acid changes within this motif result in a loss of processing activity (Richter and Lamppa, 2003). A single SSP seems to be responsible for the processing of the whole range of precursor proteins. Removal of the presequence is essential for chloroplast function. Antisense plants of *Arabidopsis* SSP are seedling lethal (Zhong *et al.*, 2003). In plants with partially impaired SSP activity, nuclear-encoded chloroplast proteins were shown to accumulate in the cytosol. Furthermore, the number of chloroplasts per cell was reduced and organelle development impaired. The function of SSP is independent of the translocation machinery and of other chloroplast proteins, since heterologously expressed SSP can process precursor proteins *in vitro* with the same efficiency as *in vivo*. These experiments also imply that SSP is able to recognize precursor proteins by itself. SSP cleaves the presequence in a single step and the mature protein is released (Richter and Lamppa, 1999). The presequence is subsequently degraded by a second ATP-dependent metalloprotease.

The set of stromal factors involved in protein import furthermore comprises several different chaperones, including homologs of Hsp93 (ClpC), Hsp70, and Hsp60 (Cpn60). Several of these chaperones were shown to interact with the precursor protein instantly upon their entering into the chloroplast and are therefore considered constitutive parts of the import machinery (Marshall *et al.*, 1990; Akita *et al.*, 1997; Kouranov *et al.*, 1998; Jackson-Constan *et al.*, 2001). The chaperones are believed to be involved both in pulling the precursor protein into the chloroplast as well as in the correct folding of the mature protein after processing of the presequence. They would also account for most of the ATP requirement of the later stage of the import process. While binding of chaperones to precursor proteins might play an important role in chloroplast import, it may not be essential. FNR precursor with reduced binding capacity to Hsp70 showed import kinetics very similar to wild-type FNR (Rial *et al.*, 2003).

5.5   The general import pathway: really general?

Protein import into chloroplast has been studied most intensively on the organelles from pea leaves. For a long time, the picture obtained by these studies has been taken for granted for all plastids in all tissues and environmental conditions. First doubts about the existence of such a general import pathway came from the genome sequence of *A. thaliana* (The *Arabidopsis* Genome Initiative, 2000). For many of the known components of the Toc as well as the Tic complex, several homologs were found in the completed genome. This raised the question whether all of these homologs genes where actively transcribed, and if so, whether they encode redundant proteins serving the same activity. Alternatively, they could exist to adapt the import apparatus to changing environmental conditions, developmental stages, or specific requirements of different tissues. Table 5.1 provides a list of homologs of pea Toc and Tic components identified in *A. thaliana*.

The organelle which originally developed from the cyanobacterial endosymbiont was the photosynthetically active chloroplast and in most plastid-harboring eukaryotes, especially green and red algae, this is the only existing form of this organelle. Nevertheless, during the continuous evolution toward vascular plants, this organelle has undergone further differentiation. Meristematic cells contain so-called proplastids. These proplastids represent a reduced form of the organelle. They are surrounded by a double membrane but are depleted of internal membranes, and they can only perform a limited set of functions. Proplastids can differentiate into several forms of plastids with distinct structure and function; these include photosynthetic chloroplasts in all green tissue, amyloplasts for starch storage in roots, and chromoplasts to color flowers and fruits. Because of their common heritage all of these plastid types have the capacity and need for protein import, yet they differ vastly in their specific requirements. It is very likely that it is the need for the import of distinct sets of proteins that is mirrored in the presence of different homologs of translocon components in the genome of *Arabidopsis*.

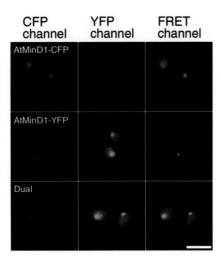

	CFP channel	YFP channel	FRET channel
AtMinD1-CFP			
AtMinD1-YFP			
Dual			

Plate 1 FRET assay for AtMinD1 protein–protein interaction in living chloroplasts. AtMinD1–CFP and/or AtMinD1–YFP were expressed in tobacco leaves by particle bombardment. Fluorescence of CFP and YFP in leaf epidermal chloroplasts was detected by epifluorescence microscopy in the CFP, YFP and FRET channel. In the FRET channel, emission of YFP was detected upon CFP excitation. Single and dual expression of fluorescent protein-tagged AtMinD1 proteins are shown. Bar: 5 μm.

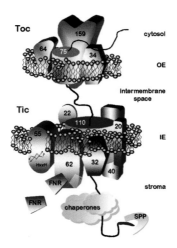

Plate 2 The protein import machinery of chloroplasts consists of two multi-protein complexes that catalyze the translocation across the outer envelope (OE) and the inner envelope (IE) membrane. They are called Toc (translocon on the outer envelope membrane of chloroplasts) and Tic (translocon on the outer envelope membrane of chloroplasts), respectively. The components of this multiple-protein complexes are named by their molecular weight. The Toc translocon consists of a core unit build by Toc159, Toc75, and Toc34. A fourth subunit, Toc64, can associate with the Toc core unit. The Tic complex comprises a total of seven different proteins. The Tic complex namely consists of Tic110, Tic62, Tic55, Tic40, Tic32, Tic22, and Tic20. Translocation occurs simultaneously across both envelope membranes and depends on the presence of a transit peptide, which is removed by the stromal processing peptidase (SPP). Translocation is further assisted by several chaperones.

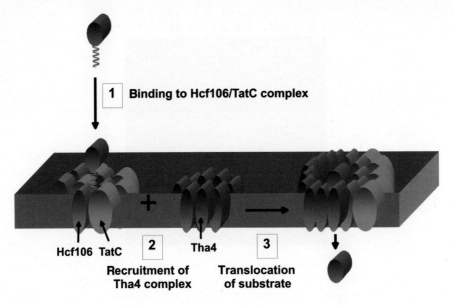

1 **Binding to Hcf106/TatC complex**

Hcf106 TatC **2** Tha4 **3**

Recruitment of **Translocation**
Tha4 complex **of substrate**

Plate 3 Model for the operation of the Tat system. A basic mode of operation has been proposed for the Tat system, based on the known properties of Tat subcomplexes and cross-linking data (see text). In this model, substrates first bind to a large (approx. 500–600 kDa) complex comprising multiple copies of Hcf106 and TatC. These subunits form the binding site for incoming substrate molecules. Binding then triggers the recruitment of a separate, homo-oligomeric Tha4 complex to form the 'supercomplex' capable of translocation. This complex generates a translocation channel (which is completely uncharacterised at present) and the substrate is transported in a folded form. After translocation, the signal peptide is removed by the thylakoidal processing peptidase.

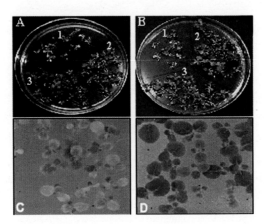

Plate 4 Drought tolerance assays. (A, B) Dehydration/rehydration assay. Three-week-old seedlings were dried for 7 h and rehydrated in MS medium for 48 h. 1. Untransformed; 2,3. T_1 and T_2 chloroplast transgenic lines. (C, D) PEG growth assay. Four-week-old seedlings were grown on MS medium with 6% PEG. (C) Untransformed; (D) T2 chloroplast transgenic line.

Table 5.1 *Arabidopsis* homologs for components of the Toc and Tic complex

Pea protein	*Arabidopsis* homolog	Gene annotation	Comment
Toc75	atToc75-III	At3g46740	P, R
	atToc75-IV	At4g09080	R
	atToc75-I	At1g35860	R
	atToc75-V	At5g19620	P, R
Toc159	atToc159	At4g02510	P, R, M
	atToc132	At2g16640	R
	atToc120	At3g16620	R
	atToc90	At5g20300	R
Toc34	atToc33	At1g02280	P, R, M
	atToc34	At5g05000	P, R, M
Toc64	atToc64-V	At5g09420	P, R
	atToc64-III	At3g17960/70[1]	P, R
	atToc64-I	At1g08980	R
Tic110	atTic110	At1g06940	P, R
Tic62	atTic62	At3g18890	P, R
Tic55	atTic55	At2g24820	P, R
Tic40	atTic40	At5g16620	P, R, M
Tic22	atTic22-IV	At4g33350	P, R
	atTic22-III	At3g23710	R
Tic20	atTic20-I	At1g04940	P, R, M
	atTic20-IV	At4g03320	R

Note: Toc and Tic components are identified by their name and their gene annotation in *Arabidopsis*. The abbreviations in the comment column indicate the extent to which a component has been characterized. P: expression was shown at the protein level; R: expression was shown by the presence of mRNA; M: a mutant has been characterized.

[1] These two loci represent a single gene.

5.5.1   Variation on the Toc complex

A detailed analysis of the *Arabidopsis* genome revealed the presence of at least four homologs to Toc159, called atToc159, atToc132, atToc120, and atToc90 (Bauer *et al.*, 2000; Hiltbrunner, 2001a; Jackson-Constan and Keestra, 2001). atToc159, atToc132, and atToc120 show between 37 and 48% sequence identity to Toc159 from pea and display the three-domain structure characteristic for Toc159. The presence of EST clones and results of RT-PCR indicate that all three genes are transcribed. From the mRNA level, atToc159 is the most abundant of the three homologs in photosynthetic tissue and the only one identified on a protein base. A knockout mutant of atToc159, called *ppi2* (plastid protein import 2), is seedling lethal and has an albino phenotype when grown under heterotrophic conditions (Bauer *et al.*, 2000). atToc159 must therefore be essential for chloroplast development. This is supported by further analyses that showed that the genes for several photosynthetic proteins are downregulated in the ppi2 mutant. In contrast, genes coding for proteins not directly associated with photosynthesis are much less affected and their gene products are still imported into plastids. This led to the suggestion that atToc159 is the specific import receptor for photosynthetic proteins while atToc132 and atToc120 direct the import of other proteins. atToc90 has been identified by homology search of the *Arabidopsis* genome and no expression of this gene is evident so far. From the deduced amino acid sequence, the gene product would be lacking the A-domain but would contain both the G- and the M-domain (Hiltbrunner *et al.*, 2001a).

There are at least four homologs of Toc75 in the genome of *Arabidopsis*. In reference to the chromosome on which they are encoded, they are called atToc75-III, atToc75-I, atToc75-IV, and atToc75-V (Jackson-Constan and Keegstra, 2001, Eckart *et al.*, 2002). atToc75-III is universally expressed in all plant tissues while expression of atToc75-I and atToc75-IV has not been proven yet. Therefore it was assumed that atToc75-III is the homolog to Toc75 in pea and the principle import pore of the Toc complex. This is substantiated by the finding that no homozygot mutant of atToc75-III has been described. If Toc75 were the general import pore, it would be expected that such a mutation would be lethal to the plant. atToc75-V was first identified as its pea homolog Toc75-V (Eckart *et al.*, 2002), but was later shown to be present in *Arabidopsis* envelope membrane as well (Froehlich *et al.*, 2003). This protein is abundant in the outer envelope of pea chloroplast but it does not seem to interact with Toc34 or Toc159. Toc75-V has significant sequence homology to a class of bacterial pore proteins, which are implicated in export processes (Bölter *et al.*, 1998b; Reumann *et al.*, 1999). From the momentary data it cannot be deduced whether Toc75-V is an alternative channel of the chloroplast protein import machinery or whether its function is related to the import or export of other macromolecules.

Two homologs of Toc34, atToc33 and atToc34, could be identified in *Arabidopsis*. These proteins share about 60% sequence homology with Toc34 and subsequent investigation have shown that both of them are functional subunits of the *Arabidopsis* Toc complex (Jarvis *et al.*, 1998; Gutensohn *et al.*, 2000). Nevertheless, the two genes show a very different pattern of expression (Jarvis *et al.*, 1998). The expression level of atToc33 is several orders of magnitude higher than that of atToc34. Furthermore, atToc34 is expressed nearly uniformly in all tissues at all times of plant development, while the expression of atToc33 undergoes a strong induction in photosynthetic tissue during rapid growth. In nonphotosynthetic tissue, atToc33 is downregulated to about the level of atToc34 (Kubis *et al.*, 2003). *In vitro*, atToc34 and atToc33 revealed different affinities for different types of precursor proteins (Jelic *et al.*, 2003) and only atToc33, the assumed homolog to Toc34 of pea, is regulated by phosphorylation. These results were corroborated by the analysis of the *Arabidopsis* mutant *ppi1*, which showed that the disruption of the atToc33 gene locus affected foremost the import of photosynthetic protein into the chloroplasts while the import of housekeeping proteins not related to photosynthesis was not affected (Kubis *et al.*, 2003). In agreement to these results, the *ppi1* mutant displayed defects in chloroplast development but not in root plastids (Jarvis *et al.*, 1998; Yu and Li, 2001). It was therefore concluded that, like the Toc159 homologs, atToc33 and atToc34 are involved in the import of different sets of proteins probably by specific binding to the precursor proteins before import.

The *Arabidopsis* genome contains at least three proteins with homology to Toc64 outside the amidase or TRP domain (Jackson-Constan and Keegstra, 2000). All three genes are represented by ESTs, indicating that the proteins are expressed. atToc64-V is considered to be the homolog of Toc64 of pea because of their sequence similarity. On the other hand, only atToc64-III was identified in a proteomics approach to characterize components of the *Arabidopsis* envelope membrane (Ferro *et al.*, 2003). There is no evidence for the specific function of the Toc64 homologs.

What is the implication of the multiple homologs of the Toc complex? The easiest answer would be that they represent different variations of the Toc complex for different tissues, i.e. plastid forms, or different development stages. This means that dependent on the requirement of the cell, different homologs of the Toc subunits would be expressed and assembled in the plastid envelope. Studies on the expression levels of some of the Toc homologs indicate that the amount of transcript can vary by tissue and stage of development. Nevertheless, expression of one or the other homolog rarely seems to be exclusive. Instead, it appears that multiple isoforms of the Toc subunits exist simultaneously in the same organelle. This would suggest that the composition of the Toc complex is heterogeneous, comprising different isoforms of all subunits at the same time. While such a scenario is easy to imagine for the import receptors, it would be intriguing to see whether this heterogeneity extends to the import pore. Alternatively, distinct Toc complexes with a different subunit composition could exist in the same plastid. These complexes would be responsible for the import of distinct set of proteins and their different quantities could be adapted to reflect the momentary requirement of the organelle.

5.5.2 Variation on the Tic complex

While the genome of *Arabidopsis* seems to contain an abundance on homologs of Toc components, the picture is quite reversed for the Tic complex (Jackson-Constan and Keegstra, 2001; Davila-Aponte *et al.*, 2003). Tic22 and Tic20 are the only components of the Tic complex for which at least two homologs were identified. No information is yet available concerning their functional role. Even extensive database searches and immunological analyses could not reveal more than one isoform of Tic110 (Jackson-Constan and Keegstra, 2001; Davila-Aponte *et al.*, 2003). More importantly, the protein seems to be present in similar amounts in all kinds of tissues. This would be in accordance with the alleged function of Tic110 as the translocation pore of the Tic complex. One could envision that the import pore had to be developed first when the need for back-transfer of proteins into the chloroplast arose. When further into the development of the organelle a need for a more sophisticated and controlled import machinery arose, this was not accomplished by an alteration of the import pore itself. Rather, by the addition of further subunits did nature form a set of different Tic complexes that could be regulated in concert with the varying organelle requirements.

As with Tic110, only single genes are found in *Arabidopsis* coding for Tic40, Tic32, Tic55, and Tic62 (Jackson-Constan and Keegstra, 2001; Küchler *et al.*, 2002). The structures of Tic32, Tic55, and Tic62 suggest a role in redox-regulated translocation on the inner envelope. It is therefore feasible that some Tic components are only found in Tic complexes responsible for translocation of a subset of photosynthetic proteins. They would be absent in Tic complexes involved in the import of other proteins. Such Tic complexes could, on the other hand, contain Tic22 and Tic20, which would explain why different Tic complexes are purified by different groups.

5.6   Conclusion and future prospects

In the last decade, enormous progress has been made to identify the many subunits of the chloroplast protein translocon. For several of the components of the Toc and Tic translocon, first insight into their specific function has been gained. Yet, both the identification of components as well as the elucidation of their function is an ongoing process. The new millennium has seen further challenges in the investigation of protein translocation. On one hand, it has become clear that the translocation process is tightly regulated both on the inside and on the outside of the envelope membrane. An important function of this regulation is to prioritize the import of proteins in direct correlation to the requirement of the organelle and the surrounding cell. Very different modes of regulation are employed at the Toc and the Tic translocon and further investigations are necessary before this process will be completely understood. On the other hand, the idea of a general import pathway that operates in every kind of tissue under all conditions might have to be abandoned in favor of a more complex picture. Translocon complexes of varying composition seem to exist in plastids, depending on the tissue or the developmental state of the plant. Even more they might also occur in one and the same organelle at the same time. It will be one of the future challenges to elucidate the function of this variation in translocon composition.

References

Akita, M., Nielsen. E. and Keegstra, K. (1997) Identification of protein transport complexes in the chloroplastic envelope membranes via chemical cross-linking. *J. Cell Biol.*, 136, 983–994.

Alefsen, H., Waegemann, K. and Keegstra, K. (1994) Analysis of the chloroplast protein import machinery. *J. Plant Physiol.*, 144, 339–345.

Bauer, J., Chen, K., Hiltbunner, A. *et al.* (2000) The major protein import receptor of plastids is essential for chloroplast biogenesis. *Nature*, 403, 203–207.

Bauer, J., Hiltbrunner, A., Weibel, P. *et al.* (2002) Essential role of the G-domain in targeting of the protein import receptor atToc159 to the chloroplast outer membrane. *J. Cell Biol.*, 159, 845–854.

Block, M.A., Dorne, A.J., Joyard, J. and Douce, R. (1983a) Preparation and characterization of membrane fractions enriched in outer and inner envelope membranes from spinach chloroplasts, II: biochemical characterization. *J. Biol. Chem.*, 258, 13281–13286.

Block, M.A., Dorne, A.J., Joyard, J. and Douce, R. (1983b) Preparation and characterization of membrane fractions enriched in outer and inner envelope membranes from spinach chloroplasts, I: electrophoretic and immunochemical analyses. *J. Biol. Chem.*, 258, 13273–13280.

Bölter, B., May, T. and Soll, J. (1998a) A protein import receptor in pea chloroplasts, Toc86, is only a proteolytic fragment of a larger polypeptide. *FEBS Lett.*, 441, 59–62.

Bölter, B., Soll, J., Schulz, A., Hinnah, S. and Wagner, R. (1998b) Origin of a chloroplast protein importer. *Proc. Natl. Acad. Sci. U.S.A.*, 95, 15831–15836

Bruce, B. (1998) The role of lipids in plastid protein transport. *Plant Mol. Biol.*, 38, 223–246.

Bruce, B.D. (2000) Chloroplast transit peptides: structure, function and evolution. *Trends Cell Biol.*, 10, 440–447.

Budziszewski, G.J., Lewis, S.P., Glover, L.W. *et al.* (2001) *Arabidopsis* genes essential for seedling viability: isolation of insertional mutants and molecular cloning. *Genetics*, 159, 1765–1778.

Caliebe, A., Grimm, R., Kaiser, G., Lübeck, J., Soll, J. and Heins, L. (1997) The chloroplastic protein import machinery contains a Rieske-type iron–sulfur cluster and a mononuclear iron-binding protein. *EMBO J.*, 16, 7342–7350.

Chen, K., Chen, X. and Schnell, D.J. (2000) Initial binding of preproteins involving the Toc159 receptor can be bypassed during protein import into chloroplasts. *Plant Physiol.*, 122, 813–822.

Chen, K. and Li, H.M. (1998) A mutant deficient in the plastid lipid DGD is defective in protein import into chloroplasts. *Plant J.*, 16, 33–39.

Chen, X., Smith, M.D., Fitzpatrick, L. and Schnell, D.J. (2002) *In vivo* analysis of the role of atTic20 in protein import into chloroplasts. *Plant Cell*, 14, 641–654.

Chou, M.L., Fitzpatrick, L.M., Tu, S.L. *et al.* (2003) Tic40, a member-anchored co-chaperone homologue in the chloroplast protein translocon. *EMBO J.*, 22, 2970–2980.

Chupin, V., van't Hof, R., and de Kruijff, B. (1994) The transit sequence of a chloroplast precursor protein reorients the lipids in monogalactosyl diglyceride containing bilayers. *FEBS Lett.*, 350, 104–108.

Davila-Aponte, J.A., Inoue, K. and Keegstra, K. (2003) Two chloroplastic protein translocation components, Tic110 and Toc75, are conserved in different plastid types from multiple plant species. *Plant Mol. Biol.*, 51, 175–181.

Dobberstein, B., Blobel, G. and Chua, N.H. (1977) *In vitro* synthesis and processing of a putative precursor for the small subunit of ribulose-1,5-bisphosphate carboxylase in *Chlamydomonas reinhardtii*. *Proc. Natl. Acad. Sci. U.S.A.*, 74, 1028–1085.

Eckart, K., Eichacker, L., Sohrt, K., Schleiff, E., Heins, L. and Soll, J. (2002) A Toc-75-like protein import channel is abundant in chloroplasts. *EMBO Rep.*, 3, 557–562.

Emanuelsson, O. and von Heijne, G. (2001) Prediction of organellar targeting signals. *Biochim. Biophys. Acta*, 1541, 114–119.

Ferro, M., Salvi, D., Brugiere, S. *et al.* (2003) Proteomics of the chloroplast envelope membranes from *Arabidopsis thaliana*. *Mol. Cell Proteomics*, 2, 325–345.

Ferro, M., Salvi, D., Riviere-Rolland, H. *et al.* (2002) Integral membrane proteins of the chloroplast envelope: identification and subcellular localization of new transporters. *Proc. Natl. Acad. Sci. U.S.A.*, 99, 11487–11492.

Flügge, U.I. and Hinz, G. (1986) Energy dependence of protein translocation into chloroplasts. *Eur. J. Biochem.*, 160, 563–567.

Froehlich, J.E., Wilkerson, C.G., Ray, W.K. *et al.* (2003) Proteomic study of the *Arabidopsis thaliana* chloroplastic envelope membrane utilizing alternatives to traditional two-dimensional electrophoresis. *J. Proteome Res.*, 2, 413–425.

Fulgosi, H. and Soll, J. (2002) The chloroplast protein import receptors Toc34 and Toc159 are phosphorylated by distinct protein kinases. *J. Biol. Chem.*, 277, 8934–8940.

Gutensohn, M., Schulz, B., Nicolay, P. and Flügge, U.-I. (2000) Functional analysis of the two *Arabidopsis* homologues of Toc34, a component fo the chloroplast protein import apparatus. *Plant J.*, 23, 771–783.

Hedtke, B., Borner, T. and Weihe, A. (2000) One RNA polymerase serving two genomes. *EMBO Rep.*, 1, 435–440.

Heins, L., Mehrle, S., Hemmler, R. *et al.* (2002) The preprotein conduction channel at the inner envelope membrane of plastids. *EMBO J.*, 21, 2616–2625.

Hiltbrunner, A., Bauer, J., Alvarez-Huerta, M. and Kessler, F (2001a) Protein translocon at the *Arabidopsis* outer chloroplast membrane. *Biochem. Cell Biol.*, 79, 629–635.

Hiltbrunner, A., Bauer, J., Vidi, P.-A. *et al.* (2001b) Targeting of an abundant cytosolic form of the protein import receptor atToc159 to the outer chloroplast membrane. *J. Cell Biol.*, 154, 309–316.

Hinnah, S.C., Hill, K., Wagner, R., Schlicher, T. and Soll, J. (1997) Reconstitution of a chloroplast protein import channel. *EMBO J.*, 16, 7351–7360.

Hinnah, S.C., Wagner, R., Sveshnikova, N., Harrer, R. and Soll, J. (2002) The chloroplast protein import channel Toc75: Pore properties and interaction with transit peptides. *Biophys. J.*, 83, 899–911.

Hirohashi, T., Hase, T. and Nakai, M. (2001) Maize non-photosynthetic ferredoxin precursor is mis-sorted to the intermembrane space of chloroplasts in the presence of light. *Plant Physiol.*, 125, 2154–2163.

Hirsch, S., Muckel, E., Heemeyer, F., von Heijne, G. and Soll, J. (1994) A receptor component of the chloroplast protein translocation machinery. *Science*, 266, 1989–1992.

Hirsch, S. and Soll, J. (1995) Import of a new chloroplast inner envelope protein is greatly stimulated by potassium phosphate. *Plant Mol. Biol.*, 27, 1173–1181.

Hörmann, F., Küchler, M., Sveshnikov, D., Oppermann, U., Yong, L. and Soll, J. (in press) Tic32, an essential component in chloroplast biogenesis. *J. Biol. Chem.*

Inaba, T., Li, M., Alvarez-Huerta, M., Kessler, F. and Schnell, D.J. (2003) atTic110 functions as a scaffold for coordinating the stromal events of protein import into chloroplasts. *J. Biol. Chem.*, 278, 38617–38627.

Ivey, R.A., III, Subramanian, C. and Bruce, B. (2000) Identification of a Hsp70 recognition domain within the rubisco small subunit transit peptide. *Plant Physiol.*, 122, 1289–1299.

Jackson, D.T., Froehlich, J.E. and Keegstra, K. (1998) The hydrophobic domain of Tic110, an inner envelope membrane component of the chloroplastic protein tranloction apparatus, faces the stromal compartment. *J. Biol. Chem.*, 273, 16583–16588.

Jackson-Constan, D. and Keegstra, K. (2001) *Arabidopsis* genes encoding components of the chloroplastic protein import apparatus. *Plant Physiol.*, 125, 1567–1576.

Jarvis, P., Chen, L.J., Li, H., Peto, C.A., Fankhauser, C. and Chory, J. (1998) An *Arabidopsis* mutant defective in the plastid general protein import apparatus. *Science*, 282, 100–103.

Jelic, M., Soll, J. and Schleiff, E. (2003) Two Toc34 homologues with different properties. *Biochemistry*, 42, 5906–5916.

Jelic, M., Sveshnikova, N., Motzkus, M., Hörth, P., Soll, J. and Schleiff, E. (2002) The chloroplast import receptor Toc34 functions as preprotein-regulated GTPase. *Biol. Chem.*, 383, 1875–1883.

Joyard, J., Billecocq, A., Bartlett, S.G., Block, M.A., Chua, N.H. and Douce, R. (1983) Localization of polypeptides to the cytosolic side of the outer envelope membrane of spinach chloroplasts. *J. Biol. Chem.*, 258, 10000–10006.

Kessler, F. and Blobel, G. (1996) Interaction of the protein import and folding machineries in the chloroplast. *Proc. Natl. Acad. Sci. U.S.A.*, 93, 7684–7689.

Kessler, F., Blobel, G., Patel, H.A. and Schnell, D.J. (1994) Identification of two GTP-binding proteins in the chloroplast protein import machineyr. *Science*, 266, 1035–1039.

Kouranov, A., Chen, X., Fuks, B. and Schnell, D.J. (1998) Tic20 and Tic22 are new components of the protein import apparatus at the chloroplast inner envelope membrane. *J. Cell Biol.*, 143, 991–1002.

Kouranov, A. and Schnell, D.J. (1997) Analysis of the interactions of preproteins with the import machinery over the course of protein import into chloroplasts. *J. Cell Biol.*, 139, 1677–1685.

Kourtz, L. and Ko, K. (1997) The early stage of chloroplast protein import involves Com70. *J. Biol. Chem.*, 272, 2808–2813.
Kubis, S., Baldwin, A., Ramesh, P. *et al.* (2003) The *Arabidopsis* ppi1 mutant is specifically defective in the expression, chloroplast import, and accumulation of photosynthetic proteins. *Plant Cell*, 15, 1859–1871.
Küchler, M., Decker, S., Hörmann, F., Soll, J. and Heins, L. (2002) Protein import into chloroplasts involves redox-regulated proteins. *EMBO J.*, 22, 6136–6145.
Lamb, J.R., Tugendreich, S. and Hieter, P. (1995) Tetratrico peptide repeat interactions: to TPR or not to TPR? *Trends Biochem. Sci.*, 20, 257–259.
Leister, D. (2003) Chloroplast research in the genomic age. *Trends Genet.*, 19, 46–47.
Li, H.-M. and Chen, L.-J. (1996) Protein targeting and integration signal for the chloroplastic outer envelope membrane. *Plant Cell*, 8, 2117–2126.
Lübeck, J., Soll, J., Akita, M., Nielsen, E. and Keegstra, K. (1996) Topology of IEP110, a component of the chloroplastic protein import machinery present in the inner envelope membrane. *EMBO J.*, 15, 4230–4238.
Ma, Y., Kouranov, A., LaSala, S.E. and Schnell, D.J. (1996) Two components of the chloroplast protein import apparatus, IAP86 and IAP75, interact with the transit sequence during the recognition and translocation of precursor proteins at the outer envelope. *J. Cell Biol.*, 134, 315–327.
Margulis, L. (1970) *Origin of Eukaryotic Cells*, Yale University Press, New Haven, CT.
Marshall, J.S., DeRocher, A.E., Keegstra, K. and Vierling, E. (1990) Identification of heat shock protein hsp70 homologues in chloroplasts. *Proc. Natl. Acad. Sci. U.S.A.*, 87, 374–378.
Martin, W., Rujan, T., Richly, E. *et al.* (2002) Evolutionary analysis of *Arabidopsis*, cyanobacterial, and chloroplast genomes reveals plastid phylogeny and thousands of cyanobacterial gnes in the nucleus. *Proc. Natl. Acad. Sci. U.S.A.*, 99, 429–441.
May, T. and Soll, J. (2000) 14-3-3 proteins form a guidance complex with chloroplast precursor proteins in plants. *Plant Cell*, 12, 53–64.
Mereschkowsky, C. (1905) Über Natur und Ursprung der Chromatophoren im Pflanzenreiche. *Biol. Centralbl.*, 25, 593–604.
Miras, S., Salvi, D., Ferro, Mm. *et al.* (2002) Non-canonical transit peptide for import into the chloroplast. *J. Biol. Chem.*, 49, 47770–47778.
Murakami, K. and Mori, M. (1990) Purified presequence binding factor (PBF) forms an import-competent complex with a purified mitochondrial precursor protein. *EMBO J.*, 10, 3201–3208.
Nielsen, E., Akita, M., Davila-Aponte, J. and Keegstra, K. (1997) Stable association of chloroplastic precursors with protein translocation complexes that contain proteins from both envelope membranes and a stromal Hsp100 molecular chaperone. *EMBO J.*, 16, 935–946.
Olsen, L.J. and Keegstra, K. (1992) The binding of precursor proteins to chloroplasts requires nucleoside triphosphates in the intermembrane space. *J. Biol. Chem.*, 267, 433–439.
Palmer, J.D. (2000) A single birth of all plastids? *Nature*, 405, 32–33.
Peeters, N. and Small, I. (2001) Dual targeting to mitochondria and chloroplasts. *Biochim. Biophys. Acta*, 1541, 54–63.
Perry, S.E. and Keegstra, K. (1994) Envelope membrane proteins that interact with chloroplastic precursor proteins. *Plant Cell*, 6, 93–105.
Pfanner, N. and Geissler, A. (2001) Versatility of the mitochondrial protein import machinery. *Nat. Rev. Mol. Cell Biol.*, 2, 339–349.
Pilon, M., de Boer, A.D., Knols, S.L. *et al.* (1990) Expression in *Escherichia coli* and purification of a translocation-competent precursor of the chloroplast protein ferredoxin. *J. Biol. Chem.*, 265, 3358–3361.
Pilon, M., de Kruijff, B. and Weisbeek, P.J. (1992) New insights into the import mechanism of the ferredoxin precursor into chloroplast. *J. Biol. Chem.*, 267, 2548–2556.

Pinnaduwage, P. and Bruce, B.D. (1996) *In vitro* interaction between a chloroplast transit peptide and chloroplast outer envelope lipids is sequence-specific and lipid class-dependent. *J. Biol. Chem.*, 271, 32907–32915.

Race, H.L., Herrmann, R.G. and Martin, W. (1999) Why have organelles retained genomes? *Trends Genet.*, 15, 364–370.

Reumann, S., Davila-Aponte, J. and Keegstra, K. (1999) The evolutionary origin of the protein-translocating channel of chloroplastic envelope membranes: identification of a cyanobacterial homolog. *Proc. Natl. Acad. Sci. U.S.A.*, 96, 784–789.

Rial, D., Ottado, J. and Ceccarelli, E.A. (2003) Precursors with altered affinity for Hsp70 in their transit peptides are efficiently imported into chloroplasts. *J. Biol. Chem.*, 278, 46473–46481.

Richter, S. and Lamppa, G.K. (1998) A chloroplast processing enzyme functions as the general stromal processing peptidase. *Proc. Natl. Acad. Sci. U.S.A.*, 95, 7463–7468.

Richter, S. and Lamppa. G.K. (1999) Stromal processing peptidase binds transit peptides and initiates their ATP-dependent turnover in chloroplasts. *J. Cell Biol.*, 147, 33–44.

Richter, S. and Lamppa, G.K. (2003) Structural properties of the chloroplast stromal processing peptidase required for its function in transit peptide removal. *J. Biol. Chem.*, 278, 39497–39502.

Robinson, C. and Ellis, R.J. (1984) Transport of proteins into chloroplasts. Partial purification of a chloroplast protease involved in the processing of important precursor polypeptides. *Eur. J. Biochem.*, 142, 337–342.

Rudhe, C., Clifton, R., Whelan, J. and Glaser, E. (2002) N-terminal domain of the dual-targeted pea glutathion reductase signal peptide controls organellar targeting efficiency. *J. Mol. Biol.*, 324, 577–585.

Schleiff, E., Jelic, M. and Soll, J. (2003a) A GTP-driven motor moves proteins across the outer envelope of chloroplasts. *Proc. Natl. Acad. Sci. U.S.A.*, 100, 4604–4609.

Schleiff, E. and Klösgen, R.B. (2001) Without a little help from "my" friends: direct insertion of proteins into chloroplast membranes? *Biochim. Biophys. Acta*, 1541, 22–33.

Schleiff, E., Motzkus, M. and Soll, J. (2002a) Chloroplast protein import is inhibited by a soluble factor from wheat germ lysate. *Plant Mol. Biol.*, 50, 177–185.

Schleiff, E., Soll, J., Küchler, M., Kühlbrand, W. and Harrer, R. (2003b) Characterization of the translocon of the outer envelope of chloroplasts. *J. Cell Biol.*, 160, 541–551.

Schleiff, E., Soll, J., Sveshinkova, N. *et al.* (2002b) Structural and guanosine triphosphate/diphosphate requirements for transit peptide recognition by the cytosolic domain of the chloroplast outer envelope receptor, Toc34. *Biochemistry*, 41, 1934–1946.

Schnell, D.J. and Blobel, G. (1993) Identification of intermediates in the pathway of protein import into chloroplasts and their localization to envelope contact sites. *J. Cell Biol.*, 120, 103–115.

Schnell, D.J., Kessler, F. and Blobel, G. (1994) Isolation of components of the chloroplast protein import machinery. *Science*, 266, 1007–1012.

Seedorf, M., Waegemann, K. and Soll, J. (1995) A constituent of the chloroplast import complex represents a new type GTP-binding protein. *Plant J.*, 7, 401–411.

Silva-Filho, M.D., Chaumont, R., Seterme, S. and Boutry, M. (1997) Mitochondrial and chloroplast targeting sequences in tandem modify protein import specificity in plant organelles. *Plant Mol. Biol.*, 30, 769–780.

Smith, M.D., Hiltbrunner, A., Kessler, F. and Schnell, D.J. (2002) The targeting of the atToc159 preprotein receptor to the chloroplast outer membrane is mediated by its GTPase domain and is regulated by GTP. *J. Cell Biol.*, 159, 833–843.

Sohrt, K. and Soll, J. (2000) Toc64, a new component of the protein translocon of chloroplasts. *J. Cell Biol.*, 148, 1213–1221.

Stahl, T., Glockmann, C., Soll, J. and Heins, L. (1999) Tic40, a new "old" subunit of the chloroplast protein import translocon. *J. Biol. Chem.*, 274, 37467–37472.

Sveshnikova, N., Soll, J. and Schleiff, E. (2000) Toc34 is a preprotein receptor regulated by GTP and phosphorylation. *Proc. Natl. Acad. Sci. U.S.A.*, 97, 4973–4978.

The *Arabidopsis* Genome Initiative (2000) Analysis of the genome sequence of the flowering plant *Arabidopsis thaliana*. *Nature*, 408, 796–815.

Theg, S.M., Bauerle, C., Olsen, L.J., Selman, B.R. and Keegstra, K. (1989) Internal ATP is the only energy requirement for the translocation of precursor proteins across chloroplastic membranes. *J. Biol. Chem.*, 264, 6730–6736.

VanderVere, P.S., Bennett, T.M., Oblong, J.E. and Lamppa, G.K. (1995) A chloroplast processing enzyme involved in precursor maturation shares a zinc-binding motif with a recently recognized family of metalloendopeptidases. *Proc. Natl. Acad. Sci. U.S.A.*, 92, 7177–7181.

van't Hof, R. and de Kruijff, B. (1995) Transit sequence-dependent binding of the chloroplast precursor protein ferredoxin to lipid vesicles and its implications for membrane stabilty. *FEBS Lett.*, 361, 35–40.

van't Hof, R., van Klompenburg, W., Pilon, M. *et al.* (1993) The transit sequence mediates the specific interactions of the precusor of ferredoxin with chloroplast envelope membrane lipids. *J. Biol. Chem.*, 268, 4037–4042.

Waegemann, K., Paulsen, H. and Soll, J. (1990) Phosphorylation of the transit sequence of chloroplast precursor proteins. *J. Biol. Chem.*, 271, 6545–6554.

Waegemann, K. and Soll, J. (1991) Characterization of the protein import apparatus in isolated outer envelopes of chloroplasts. *Plant J.*, 1, 149–158.

Waegemann, K. and Soll, J. (1996) Phosphorylation of the transit sequence of chloroplast precursor proteins. *J. Biol. Chem.*, 271, 6545–6554.

Weibel, P., Hiltbrunner, A., Brandt, L. and Kessler, F. (2003) Dimerization of Toc-GTPases a the chloroplast protein import machinery. *J. Biol. Chem.*, 278, 37321–37329.

Wienk, H.L., Wechselberger, R.W., Czisch, M. and de Kruijff, B. (2000) Structure, dynamics, and insertion of a chloroplast targeting peptide in mixed micelles. *Biochemistry*, 39, 8219–8227.

Wu, C., Seibert, F.S. and Ko, K. (1994) Identification of chloroplast envelope proteins in close physical proximity to a partially translocated chimeric precursor protein. *J. Biol. Chem.*, 269, 32264–32271.

Young, M.E., Keegstra, K. and Froehlich, J.E. (1999) GTP promotes the formation of early-import intermediates but is not required during the translocation step of protein import into chloroplasts. *Plant Physiol.*, 121, 237–244.

Yu, T.S. and Li, H. (2001) Chloroplast protein translocon components atToc159 and atToc33 are not essential for chloroplast biogenesis in guard cells and root cells. *Plant Physiol.*, 127, 90–96.

Zhong, R., Wan, J., Jin, R. and Lamppa, G. (2003) A pea antisense gene for the chloroplast stromal processing peptidase yields seedling lethals in *Arabidopsis*: survivors show defective GFP import *in vivo*. *Plant J.*, 34, 802–812.

6   Biogenesis of the thylakoid membrane

Colin Robinson and Alexandra Mant

6.1   Introduction

Although heavily involved in photosynthetic light capture, photophosphorylation and carbon dioxide fixation, the chloroplast also carries out an entire array of functions that includes the synthesis of amino acids, chlorophyll and various lipids. The net result is an organelle that has been estimated to contain in the region of 2000 proteins (see Chapter 1 in this volume). This figure is actually derived from studies of the *Arabidopsis* genome, and thus includes proteins that may be specific to other types of plastid, but the bulk of these proteins will certainly be targeted into chloroplasts at some stage. Chloroplast protein import is therefore a major process in plant cell biology (covered in Chapter 5). However, intraorganellar protein *sorting* is equally important because during or after import, these proteins have to be directed to one of a total of six chloroplast sub-compartments (outer and inner envelope membranes, intermembrane space, stroma, thylakoid membrane and thylakoid lumen). In this chapter we consider the processes involved in thylakoid protein biogenesis. This area has attracted interest for many years, partly because some of the thylakoid proteins are so abundant and well-characterised, and partly because the import pathway is intrinsically interesting – these proteins have to traverse both envelope membranes and the soluble stromal phase in order to reach the thylakoid membrane. Many thylakoid proteins are located in the lumenal phase enclosed by this interconnecting membrane, and these have attracted particular attention because their biogenesis requires an additional membrane translocation step. These processes have been studied using a variety of *in vitro* assays, in conjunction with *in vivo* studies on plant mutants, and several of the pathways are now understood in some detail. Here, we review the known pathways for the targeting of proteins into the thylakoid membrane and lumen. However, thylakoid biogenesis involves more than the targeting of individual protein molecules, and we consider the biogenesis of the membrane itself, taking into account current models for the trafficking of lipids to this enormously abundant membrane network.

6.2   Targeting of thylakoid lumen proteins

6.2.1   *The basic two-phase import pathway for lumenal proteins*

The thylakoid lumen contains well-characterised photosynthetic proteins such as plastocyanin, the 33-, 23- and 16-kDa subunits of the photosystem II (PSII) oxygen-evolving complex (OEC33, OEC23 and OEC16) and photosystem I subunit N (PsaN). However, recent proteomic studies using carefully fractionated chloroplasts have revealed the existence of many more proteins (at least 80 were identified, and the lumen potentially contains as many as 200; Peltier *et al.*, 2002; Schubert *et al.*, 2002). These proteins include a surprising number of peptidyl-prolyl *cis–trans* isomerases and proteases, as well as a number of proteins of no known function. All of the known lumenal proteins are encoded in the nucleus, and they are invariably synthesised with bipartite pre-sequences containing two targeting signals in tandem: a 'transit' peptide specifying entry into the chloroplast, followed by a cleavable signal peptide that directs transport across the thylakoid membrane. The only exception to this rule is cytochrome f, which is encoded by chloroplast DNA and synthesised within the chloroplast. Strictly speaking, cytochrome f is a thylakoid membrane protein but the bulk of the protein is located in the lumen, attached to the thylakoid membrane by a C-terminal transmembrane (TM) anchor (Willey *et al.*, 1984). The protein is synthesised with a cleavable signal peptide and discussed in more detail below.

All of the available data suggest that lumenal proteins are initially imported into chloroplasts by the 'standard' route used by stromal proteins. The N-terminal domains of these bipartite pre-sequences appear to be typical transit peptides in terms of length and amino acid composition, and early studies in this field showed that these domains on their own do indeed direct translocation into the stroma (Hageman *et al.*, 1990). Almost invariably, these signals are removed by the stromal processing peptidase (SPP) that removes the signals of imported stromal proteins (Hageman *et al.*, 1986; James *et al.*, 1989). A rare exception is PsaN, which crosses the thylakoid membrane as the full precursor form (Nielsen *et al.*, 1994). Thereafter, the signal peptides direct translocation across the thylakoid membrane, after which they are removed by a thylakoid processing peptidase. This peptidase belongs to the signal peptidase family of serine proteases, and strongly resembles bacterial signal peptidases in terms of cleavage specificity (Halpin *et al*, 1989). This was one of the earliest indications that thylakoid protein transport systems were inherited from cyanobacterial-type progenitors of chloroplasts.

6.2.2 *Lumenal proteins are transported across the thylakoid membrane by two completely different pathways*

Initial studies in this field suggested a common mechanism for the transport of lumenal proteins across the thylakoid membrane, because all of the thylakoid-targeting signals resemble classical 'signal' peptides in structural terms. However, the development of *in vitro* assays for thylakoid protein import led to an unexpected finding: lumenal proteins fall into two distinct groups in terms of import requirements. While some proteins (such as plastocyanin) require ATP and stromal factors, others do not and instead show an absolute reliance on the thylakoidal ΔpH, examples being the OEC23 and OEC16 proteins (Mould and Robinson, 1991; Cline *et al.*, 1992; Klösgen *et al.*, 1992). Further studies have shown that members of these groups compete with each other, but not with members of the other group, for translocation

Tat signal peptides

```
Sp 23K        --NVLNSGVSRRLALTVLIGAAAVGSKVSPADA

Wh 23K        --SDAAVVTSRRAALSLLAGAAAIAVKVSPAAA

Ma 16K        --GDAVAQAGRRAVIGLVATGIVGGALSQAARA

Ara 16Ka      --AQQSEETSRRSVIGLVAAGLAGGSFVQAVLA

Ara 16Kb      --NVSVPESSRRSVIGLVAAGLAGGSFVKAVFA

Bar PSI-N     --VQVAPAKDRRSALLGLAAVFAATAASAGSARA

Cot PSII-T    --RKTEGNNGRREMMFAAAAAAICSVAGVATA

Ara PSII-T    --KEQSSTTMRRDLMFTAAAAAVCSLAKVAMA

Tom PPO       --QDETNSVDRRNVLLGLGGLYGVANAIPLAASA

Ara P29       --SGESLAFHRRDVLKLAGTAVGMELIGNGFINNVGDAKA

Ara Hcf136    --SSSSLSFSRRELLYQSAAVSLSLSSIVGPARA

Ara P16       --TSSSLLWKRRELSLGFMSSLVAIGLVSNDRRRHDANA
```

Sec-type signal peptides

```
Sp 33K        --CVDATKLAGLALATSALIASGANA

Wh 33K        --CADAAKMAGFALATSALLVSGATA

Sp PSI-F      --KLELAKVGANAAAALALSSVLLSSWSVAPDAAMA

Bar PSI-F     --LSASIKTFSAALALSSVLLSSAATSPPPAAA

Bar PC        --ASLGKKAASAAVAMAAGAMLLGGSAMA

Sp PC          --ASLKNVGAAVVATAAAGLLAGNAMA

Ara P17.4     --SLFPLKELGSIACAALCACTLTIASPVIA

Ara DegP      -PFSAVKPFFLLCTSVALSFSLFAASPAVESASA

Ara Avde1     -APLLLKLVGVLACAFLIVPSADA

Sp Tlp40      RSFSVKECAISLALAAALISGVPSLSWERHAEA

Sp CtpA       --YVPSVRLVVGVVLLMSVSVALNQDPSWS

Ara TL11.6    --EEVSKRSLFALVSASLFFVDPALA

Ara TL13.4    --SRFRSKSLSLVFSGALALGLSLSGVGFADA
```

Figure 6.1 Targeting signals for thylakoid lumen proteins. The figure shows the signal peptides of representative lumenal proteins that are targeted by the Tat- or Sec-dependent pathways. The precise start points of the signals are not known, since these signals are preceded by transit

across the thylakoid membrane (Cline *et al.*, 1993). The existence of two entirely separate pathways was finally confirmed by the use of chimeric proteins, where, for example, the pre-sequence of OEC23 was found to redirect mature plastocyanin quantitatively onto the ΔpH-dependent pathway (Henry *et al.*, 1994; Robinson *et al.*, 1994). More detailed studies of lumenal targeting signals revealed that there are subtle but important differences in the signal peptides for these pathways. Signals for the Sec pathway resemble bacterial Sec-type signal peptides in that they comprise three domains: an N-terminal positively charged domain, a hydrophobic core domain and a more polar C-terminal domain ending with the Ala-Xaa-Ala consensus motif recognised by the processing peptidase. The ΔpH-dependent system recognises signals that are startlingly similar in overall structure – they contain the same basic three-domain organisation – but a critical feature is the presence of a twin-arginine motif just before the hydrophobic domain. This motif is essential for targeting by this pathway (Chaddock *et al.*, 1995) and substitution of either arginine (even by lysine) blocks translocation. A selection of lumen-targeting signals is shown in Figure 6.1.

It is now known that roughly equal numbers of lumenal proteins use each of these pathways, and studies in the late 1990s have identified the core components of these translocation systems. Plastocyanin, OEC33 and other proteins follow a Sec-type pathway that minimally involves stromal SecA (Yuan *et al.*, 1994) together with membrane-bound components SecY (Laidler *et al.*, 1995) and SecE (Schuenemann *et al.*, 1999). The Sec pathway has been intensively studied in bacteria, where it is largely responsible for the export of proteins across the plasma membrane (reviewed by Manting and Driessen, 2000). In this export pathway, substrate proteins are synthesised with an N-terminal signal peptide, after which they interact with chaperone molecules that serve to prevent folding of the pre-protein. SecB fulfils this role in *Escherichia coli*, although other proteins presumably carry out this function in some other bacteria where SecB is not present. The pre-protein next interacts with SecA, which hydrolyses ATP and uses the generated energy to push sections of the pre-protein into a membrane-bound channel that comprises SecYEGyajC together with several ancillary proteins of undefined function. The protein is threaded through the membrane in an unfolded state and the signal peptide is cleaved by signal peptide on the *trans* side of the membrane. To date, it appears that the thylakoidal Sec pathway uses a basically similar, but rather slimmed-down, apparatus. SecA plays a vital role, and the core SecYE components have been identified but there is no evidence for *secB* or *secG* genes in the *Arabidopsis* genome. However, there is clear evidence that

---

peptides that direct entry into the chloroplast and the intermediate processing sites are not obvious from the sequence data. Twin-arginine motifs, critical for the functioning of the Tat peptides, and other basic amino acids are shown in boldface. The signals are aligned at either the twin-arginine motifs or the correspondingly located basic residues of Sec-specific signals. Hydrophobic core domains are underlined. All of the signals end with short-chain residues at the −3 and −1 positions, relative to the terminal cleavage site (typically, Ala-Xaa-Ala); this motif is important for recognition by the thylakoidal processing peptidase.

thylakoid Sec substrates are transported in an unfolded state (Hynds *et al.*, 1998), and so the substrate proteins must either be maintained in an unfolded state while in the stroma, or actively unfolded at some stage in the translocation process.

The Sec pathway is also used by chloroplast-encoded pre-cytochrome f. This protein is made with a classical signal peptide and, although the targeting of the authentic pre-protein is difficult to analyse in intact chloroplasts, the later stages of the pathway have been analysed by importing constructs in which a transit peptide is fused in front of pre-cytochrome f. Under these conditions, the thylakoid-targeting of the protein is inhibited by azide (a classical inhibitor of the Sec pathway) but not by proton ionophores that disrupt the ΔpH-dependent pathway (Mould *et al.*, 1997). Further evidence comes from studies on the maize *tha1* mutant (Voelker and Barkan, 1995). The Sec pathway is severely compromised in this mutant and the precursor form of cytochrome f was observed to accumulate in pulse-chase studies.

The other, ΔpH-dependent pathway involves completely different targeting machinery. Voelker and Barkan (1995) isolated a second maize mutant, termed *hcf106*, that is specifically defective in this pathway and, because the *hcf106* gene contained a transposon insertion, this led to the cloning of the first component of this novel pathway. The sequencing of the gene (Settles *et al.*, 1997) produced a major surprise – clear homologues are present in the majority of sequenced genomes from free-living bacteria, and yet there was little at that time to suggest the operation of a second, Sec-independent export pathway in bacteria. It is now known that two pathways do indeed operate in bacteria, just as in thylakoids. The bacterial Sec-independent pathway resembles the thylakoid system in many respects, and likewise recognises substrates bearing twin-arginine signal peptides (reviewed in Robinson and Bolhuis, 2001). Several components of the novel translocation system have been identified in bacteria and plants, and the system has been termed the twin-arginine translocation, or Tat system, in view of the importance of a twin-arginine motif within its substrates.

6.2.3   Unique properties of the Tat system

The Tat system has now been studied in some detail using *in vitro* assays with plant thylakoids and genetic/structural approaches in bacteria, particularly *E. coli* (for a recent review, see Robinson and Bolhuis, 2001). Three primary *tat* genes have been identified; these form an operon in *E. coli* (*tatABC*) and corresponding genes have been identified in plants. The basic structures of the encoded proteins are illustrated in Figure 6.2. In *E. coli*, TatA and TatB are known to be single-span membrane proteins with very short periplasmic N-terminal regions and cytosolic C-terminal domains. The region between the TM span and the soluble domain is predicted to form an amphipathic helix. The two proteins are homologous, particularly in the TM span and amphipathic region, but these proteins nevertheless carry out distinct, essential functions and cannot substitute for one another (Sargent *et al.*, 1998, 1999; Weiner *et al.*, 1998). On the basis of sequence homology and other studies (described below), it is highly likely that the TatA/TatB homologues are Tha4 and Hcf106, respectively

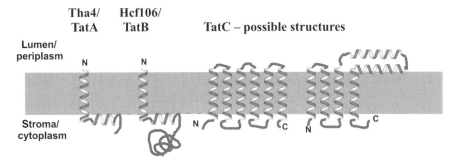

Figure 6.2 Components of the Tat machinery. The diagram shows the basic structures of the Tha4, Hcf106 and TatC components of the Tat system (and the corresponding bacterial counterparts, TatABC). The precise topology of TatC is still uncertain; sequence analysis suggests a six-TM-span model, whereas reporter gene fusions suggest four TM spans. In either case, there is evidence that the N- and C-termini of TatC are located in the chloroplast stroma/bacterial cytoplasm. Tha4 and Hcf106 contain a single TM span and a predicted short amphipathic helical region on the stromal/cytoplasmic surface of the membrane. These subunits are homologous, especially in the TM and amphipathic regions, but carry out very distinct functions.

(Settles *et al.*, 1997; Walker *et al.*, 1999). The third gene in the *E. coli tat* operon is *tatC*, which is also critical for Tat export activity in *E. coli* (Bogsch *et al.*, 1998) and plants (Motohashi *et al.*, 2001). This protein was initially thought to have six TM spans but more recent reporter gene fusions suggest four instead (Gouffi *et al.*, 2002). Both possible topologies are shown in Figure 6.2.

The Tat proteins are unrelated to any other proteins in the database and thus the system is unique in terms of structure. The mechanism of this system is also very different to those of all other known protein transporters, and its most notable attribute is its ability to transport proteins in a folded state. This has been shown biochemically in two different thylakoid studies. In one, Clark and Theg (1997) showed that an internally cross-linked bovine pancreatic trypsin inhibitor construct could be transported by the Tat pathway using an attached signal peptide, when unfolding of the protein could not possibly occur. In the other study, Hynds *et al.* (1998) showed that dihydrofolate reductase could be transported across the thylakoid membrane, together with a bound folate analogue in the active site. This represents strong evidence that the protein must have remained largely, if not completely, folded during the translocation process.

In bacteria, it is clear that the Tat pathway is again involved in the transport of folded proteins. The primary Tat substrates in *E. coli* are periplasmic redox proteins that bind one of a range of cofactors such as FeS, molybdopterin or NiFe centres (Berks, 1996; Santini *et al.*, 1998; Sargent *et al.*, 1998). Importantly, these cofactors are inserted enzymatically in the cytosol, necessitating the subsequent export of the protein in a folded form. While indirect, these findings represent very compelling evidence that the Tat system transports either fully folded proteins, or

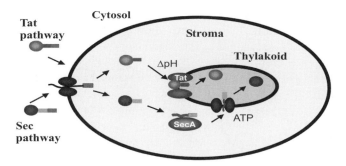

Figure 6.3 Pathways for the targeting of thylakoid lumen proteins. Nuclear-encoded thylakoid lumen proteins are synthesised in the cytosol on free ribosomes and imported across the envelope membranes by the standard import pathway. They are synthesised with bipartite pre-sequences containing two signals in tandem: a transit peptide, which directs import into the organelle, and which is usually removed in the stroma; and a thylakoid signal peptide. As shown in Figure 6.1, lumenal proteins contain either a Sec- or Tat-specific signal peptide. Proteins on the Sec pathway interact with SecA in the stroma, which hydrolyses ATP and effectively pushes the pre-protein through a membrane-bound translocon in an unfolded state. The translocon is known to contain SecY and SecE but is otherwise poorly characterised. It is also unclear whether the substrate protein remains unfolded in the stromal phase, perhaps with the aid of (unidentified) chaperone molecules, or whether it is actively unfolded at the membrane. Substrates on the Tat pathway, on the other hand, have been shown to refold in the stroma and are not believed to interact with any dedicated Tat components until they reach the membrane. Here, they are transported by a membrane-bound Tat system that contains Tha4, Hcf106 and TatC. The Tat system is able to translocate substrates in a fully folded state, although it is not clear whether this method is always employed. The thylakoidal ΔpH is essential for translocation, at least in isolated thylakoids and intact isolated chloroplasts.

at least large folded domains, and it should be noted that some of the transported proteins are very large – in excess of 100 kDa. While the Tat system is now well known for its role in the export of redox proteins in bacteria, it should be pointed out that many Tat substrates do not bind any form of cofactor. This applies particularly to chloroplasts, where the vast majority of Tat substrates do not bind cofactors. It therefore appears likely that the Tat system serves to transport two categories of proteins: those that have to fold prior to translocation, for example to bind cofactors, and those that simply fold too quickly or tightly for the Sec system to handle. The contrasting characteristics of the Sec- and Tat-dependent thylakoid targeting pathways are illustrated in Figure 6.3.

6.2.4 Tat structure and mechanism

Studies on the thylakoid and bacterial Tat systems have begun to unravel some of the more detailed aspects of the Tat system structure and operation. Tagging of the TatC subunit has enabled purification of a approximately 500–600-kDa TatABC complex from *E. coli* (Bolhuis *et al.*, 2001), confirming that these subunits do form a large complex and raising the possibility that these are indeed the only important Tat subunits. TatB and TatC were found to be present in a 1:1 ratio within this complex, and appear to form a structural and functional unit. It was noted, however, that the vast majority of the TatA did not co-purify with the TatBC subunits and this subunit appears to form a separate homo-oligomeric complex of about 460 kDa (Porcelli *et al.*, 2002).

Studies on the thylakoid Tat complex suggest a broadly similar organisation. The complex has not been purified but blue-native gels show the Hcf106 and TatC subunits to co-migrate in a large complex (Cline and Mori, 2001), consistent with the pairing of the TatB–TatC subunits in bacteria, although the TatA homologue Tha4 was not detected in this complex, but instead ran as a separate band. *In vitro* import assays have begun to point to a likely rationale for these findings. Bound Tat substrate was found to cross-link to Hcf106 and TatC, strongly suggesting that these subunits form the initial binding site for precursors (Cline and Mori, 2001). Interestingly, Tha4 was found only to cross-link to the Hcf106–TatC subunits during ongoing protein transport, and in the presence of a ΔpH (Mori and Cline, 2002). Taken together, these results suggest a model in which substrates first bind to an Hcf106/TatC complex (or TatBC complex in bacteria), triggering the recruitment of a separate Tha4 (equivalent to bacterial TatA) complex to generate the active translocation complex. This model is illustrated in Plate 3, but it must be stressed that further work is required to fill the numerous gaps in our knowledge, and at this point the actual translocation mechanism is very poorly understood. We do not know, for example, which subunits actually form the translocation channel.

6.3 The targeting of thylakoid membrane proteins

The thylakoid membrane contains some of the most abundant membrane proteins found in nature and the membrane itself accounts for well over 95% of total membrane in the chloroplast. Proteins are thus targeted into this membrane on a very large scale and, as with lumenal proteins, these processes have been intensively studied in an effort to understand the targeting pathways involved. Many of these studies have relied on *in vitro* reconstitution assays, and thylakoids do appear to be particularly amenable to this form of analysis. These studies have been backed up by analysis of plant mutants in some cases.

6.3.1 The signal recognition particle dependent pathway

The first assays for the insertion *in vitro* of a membrane protein into isolated thylakoids were developed in 1986 (Cline, 1986) using the major light-harvesting chlorophyll-binding protein, Lhcb1, as substrate (surprisingly, this was the only substrate analysed during the following decade). It was shown that insertion required the presence of stromal protein factor(s) and nucleoside triphosphates, and subsequent studies have identified the key players in the insertion mechanism. The two central requirements in the stromal phase of the insertion pathway are for signal recognition particle (SRP; Li *et al.*, 1995) and FtsY (Kogata *et al.*, 1999). In bacteria, these factors play a vital role in the biogenesis of plasma membrane proteins (reviewed by Dalbey and Kuhn, 2000) and the operation of a related pathway in chloroplasts is another example of *conservative sorting* – the use of translocation pathways and apparatus that have been inherited from the prokaryotic ancestor of the organelle. Studies in *E. coli* have shown that SRP probably binds co-translationally to the nascent chains of membrane proteins, almost certainly binding to the particularly hydrophobic regions that are unique to membrane proteins; the signal peptides of exported periplasmic proteins appear not to be sufficiently hydrophobic to interact stably with SRP. A broadly similar situation appears to apply to the targeting of Lhcb1. This protein is synthesised only with a transit peptide specifying chloroplast import, and does not contain a cleavable thylakoid-targeting signal (Lamppa, 1988; Viitanen *et al.*, 1988). The information specifying thylakoid-targeting must therefore reside in the mature protein, and it appears highly likely that SRP recognises one of the three TM spans in the mature Lhcb1 protein.

In bacteria, SRP next interacts with its partner protein, FtsY, at the membrane surface and insertion then involves the hydrolysis of GTP by both SRP and FtsY, together with the participation of membrane-bound translocation machinery, usually the Sec translocon and a component termed YidC (reviewed by Dalbey and Kuhn, 2000). Thylakoid-insertion of Lhcb1 likewise requires FtsY, GTP hydrolysis and a homologue of YidC known as Albino3 (Alb3; Moore *et al.*, 2000), which is described in more detail below.

There are, however, significant differences between the bacterial and thylakoid SRP pathways. First, Lhcb1 insertion occurs post-translationally, since the protein is released into the stroma following import into the chloroplast. Secondly, the structure of the SRP molecules is rather different. In bacteria, SRP comprises a 4.5 S RNA molecule together with a 54-kDa subunit that is often termed Ffh. Chloroplastic SRP contains no RNA and contains instead the 54-kDa subunit (SRP54) together with a novel subunit, SRP43 (Schuenemann *et al.*, 1998). These subunits co-purify in a 1:1 ratio when expressed in *E. coli*, and because the SRP is active in thylakoid-insertion assays (Groves *et al.*, 2001) this strongly suggests that the SRP contains only these subunits. The role of SRP43 has not been fully resolved but it has been shown to bind to a short sequence within Lhcb1, known as the L18 peptide (DeLille *et al.*, 2000; Tu *et al.*, 2000). Possibly, this type of chloroplastic SRP (cpSRP) is adapted to the obligatorily post-translational insertion pathway and it was suggested that stable binding to Lhcb1 may involve the simultaneous interactions of SRP43 with the L18 sequence and SRP54 with the hydrophobic targeting signal. Another important point is that SRP is used only for the insertion of a minority of thylakoid membrane proteins (see below) and this may be a mechanism that ensures interaction only with correct substrates that contain the L18 sequence.

The stromal events in the cpSRP pathway appear only to involve interaction of Lhcb1 with cpSRP and FtsY (Tu *et al.*, 1999) but the mode of action between these factors is poorly understood at present (the bacterial SRP pathway is likewise vague at this point). There have been suggestions that FtsY may act as a release factor once insertion into the membrane is underway, and *E. coli* FtsY does have a propensity to bind to membrane surfaces (reviewed by Seluanov and Bibi, 1997) but the precise mode of interaction between SRP and FtsY remains to be elucidated. It should also be emphasised that a completely different mechanism has been proposed for the biogenesis of LHC proteins. In this model, the LHC proteins insert stably into the chloroplast envelope; the addition of chlorophyll at this point is key to the insertion mechanism, and no stromal intermediates are involved (see later).

Once at the membrane, other factors come into play. In chloroplasts, a multi-spanning membrane protein, Alb3 plays a vital role in the cpSRP insertion pathway. *Arabidopsis* mutants lacking this factor are unable to develop thylakoids and exhibit a seedling-lethal phenotype (Sundberg *et al.*, 1997) and *in vitro* insertion assays have shown that pre-incubation of thylakoids with antibodies to Alb3 blocks the insertion of Lhcb1 (Moore *et al.*, 2000). Once again, this is consistent with the endosymbiotic origin of the chloroplast and its SRP pathway, because homologues of Alb3 play similarly crucial roles in bacteria and mitochondria. Alb3 belongs to a group of proteins known as the Oxa1 family, following elegant studies on the role of Oxa1 in the mitochondrial inner membrane (reviewed by Kuhn *et al.*, 2003). Oxa1 is essential for the efficient insertion of proteins from the matrix side of the membrane, and this includes several mitochondrially encoded proteins as well as nuclear-encoded proteins that are first imported into the matrix before inserting into the inner membrane. A related protein, YidC, plays a similarly important role in *E. coli*, where it is required for the insertion, or at least the *efficient* insertion, of a very wide range of plasma membrane proteins (Kuhn *et al.*, 2003). There are, however, differences between these systems. YidC appears to act in two distinct capacities in *E. coli*: on its own, for some membrane proteins, and in conjunction with the membrane-bound Sec apparatus for others (especially those on the SRP pathway). Yeast and mammalian mitochondria, on the other hand, do not contain Sec components and it is believed that Oxa1 operates on its own as a homo-oligomer.

The situation in chloroplasts has yet to be resolved. While the thylakoid membrane does contain Sec machinery, there is no evidence as yet that it is involved in the insertion of typical membrane proteins, including those that are inserted by the cpSRP pathway. Indeed, antibodies to SecY block the translocation of lumenal Sec substrates but do not affect the insertion of Lhcb1 (Mori *et al.*, 1999), suggesting that the SecYE translocon is not involved in the cpSRP pathway and raising the possibility that Alb3 acts on its own, like Oxa1. Further studies are certainly required in this area, however, since some of these experiments are rather indirect.

6.3.2   *Most thylakoid membrane proteins are inserted by an SRP-independent, possibly spontaneous pathway*

The SRP pathway is the dominant mechanism for the insertion of membrane proteins in bacteria, and the discovery of the cpSRP pathway initially suggested a similarly 'global' role in thylakoid biogenesis. A wide-ranging role was furthermore suggested by studies which showed cross-linking of a chloroplast-encoded protein (D1) to SRP54, during the apparently co-translational insertion of this protein into the thylakoids (Nilsson *et al.*, 1999). Surprisingly, this is not the case, and it is now known that the vast majority of thylakoid proteins do not use this pathway but instead use one of at least two alternative pathways. The first, and minor, pathway involves the use of cleavable signal peptides. CFoII, PsbW, PsbX and PsbY are all synthesised with bipartite pre-sequences that very much resemble those of imported thylakoid lumen proteins. It was initially anticipated that these proteins would use the Sec-dependent pathway (none of these proteins contain twin-arginine signal peptides) but *in vitro* insertion assays have shown that these proteins do not require SecA, SRP or any form of nucleoside triphosphate for efficient insertion (Michl *et al.*, 1994; Lorkovic *et al.*, 1995; Kim *et al.*, 1998, Thompson *et al.*, 1999). Significantly, protease-treatment of thylakoids destroys the Tat and Sec translocons and completely blocks the insertion of Lhcb1, but has no effect on the insertion of these precursor proteins (Robinson *et al.*, 1996). On the basis of these results it was suggested that these proteins may insert spontaneously into the thylakoid membrane bilayer, without any input from proteinaceous translocation machinery. Rather than interacting with protein targeting apparatus (the usual role of signal peptides), the signal peptides of these proteins may instead serve as additional hydrophobic regions that help the remainder of the protein to insert into the bilayer. Inhibition of the thylakoid processing peptidase blocks maturation of these proteins (Thompson *et al.*, 1998) and leads to the accumulation of loop intermediates, suggesting that the signal peptides may enable these proteins to insert more efficiently as 'helical hairpins'.

While CFoII, PsbW and PsbX insert as simple hairpins, each containing a single TM span and signal peptide, PsbY is a truly remarkable protein. The primary translation product encodes two separate, related mature proteins, each of which is preceded by a cleavable signal peptide (Mant and Robinson, 1998). This is the only known polyprotein in higher plant chloroplasts. After import into the chloroplast, the N-terminal transit signal is removed and the remainder of the protein inserts into the thylakoid in a double-loop conformation, after which the protein is cleaved twice by the thylakoidal processing peptidase and once by an unknown protease on the stromal face of the membrane (Thompson *et al.*, 1999).

The suggestions of a spontaneous insertion mechanism for these proteins gained further credence when it was shown that the Alb3 protein is not involved in their biogenesis (Woolhead *et al.*, 2001). Incubation of thylakoids with anti-Alb3 antibodies blocked insertion of Lhcb1, as found previously, and two further members of the light-harvesting chlorophyll-binding protein (LHC) family (Lhcb4 and Lhcb5), strongly suggesting that they too use the cpSRP pathway. However, these antibodies did not inhibit the insertion of the signal-peptide containing PsbX, PsbY and PsbW. Most importantly of all, tests on a range of other thylakoid membrane proteins have confirmed that they do not utilise the cpSRP pathway or indeed any of the known targeting pathways. Most thylakoid membrane proteins are synthesised with only a single cleavable targeting signal that specifies entry into the chloroplast, and studies by Kim *et al.* (1999) initially showed that several such proteins insert into thylakoids by a Sec/SRP/Tat/NTP-independent pathway. Woolhead *et al.* (2001) showed that Alb3 is not required for the insertion of several of this type of protein, including members of the LHC family such as PsbS, and Mant *et al.* (2001) showed that none of the known translocation apparatus is required for the insertion of the two-TM-span PsaK protein into thylakoids.

More recently, these *in vitro* studies have received important backing from work on a *Chlamydomonas reinhardtii* mutant that specifically lacks one of the two Alb3-related proteins present in this unicellular organism. Bellafiore *et al.* (2002) found that the light-harvesting systems of both photosystems were massively depleted in this mutant, in agreement with the importance of Alb3 for Lhcb1/4/5, but that the vast majority of thylakoid membrane proteins were present in normal amounts. The conclusion is that SRP, FtsY and Alb3 are required only for a small group of abundant LHC proteins. These studies represent a major surprise. Firstly, homologues of SRP, FtsY and Alb3 play enormously important roles in bacteria and the Oxa1 protein is similarly important in mitochondria, yet these proteins are used for a select few substrates in thylakoids. Secondly, the vast majority of thylakoid membrane proteins do not need any known form of translocation apparatus or any form of free energy for their insertion, and this raises the strong possibility that they insert directly into the thylakoid membrane using a so-called 'spontaneous' mechanism. This would be unprecedented among mainstream protein insertion pathways, and the immediate question is how such a mechanism could maintain pathway specificity. The only obvious answer is to invoke the highly unusual nature of thylakoid membrane lipids. Whereas most membranes are composed primarily of phospholipids, galactolipids account for over 80% of thylakoid lipid and these lipids are furthermore highly unsaturated (Block *et al.*, 1983; Douce and Joyard, 1990). It is certainly possible that membrane proteins could require less 'assistance' to insert into this type of bilayer (for review, see Bruce, 1998) and furthermore possible that thylakoid membrane proteins could have evolved to be compatible with the types of lipids they encounter during insertion. Against this model, it is known that the inner envelope membrane is very similar to the thylakoid membrane in terms of lipid composition, and it would therefore be difficult for thylakoid proteins to avoid inserting into the inner face of the envelope after import into the chloroplast. One cannot rule out the possibility that these SRP-independent proteins are transported from the envelope to the thylakoid via vesicles, as suggested for LHC proteins (see below). However, there is no evidence for this form of transport and it has been clearly shown that (a) these proteins do insert very efficiently into isolated thylakoids and (b) transient stromal forms can be observed during chloroplast import assays.

Further studies are certainly required to resolve this issue, for example using liposomes composed of thylakoid-type lipids. Whatever the eventual explanation, the thylakoid clearly differs from other prokaryotic/organellar membranes in terms of membrane protein biogenesis, since novel protein insertion machinery must operate if a lipid-directed spontaneous insertion mechanism does not after all apply. It is also curious that the SRP/Alb3 pathway should be used exclusively by proteins that bind chlorophyll. Might this be a mechanism to coordinate the targeting of this group of proteins with cofactor insertion? This would imply a physical linkage of the targeting apparatus and chlorophyll-insertion machinery, but there is no evidence for this arrangement as yet. Figure 6.4 shows the basic features of the three known pathways for thylakoid membrane protein insertion, although it should be emphasised that chloroplast-encoded proteins are not well understood in this respect, and have been omitted from this scheme.

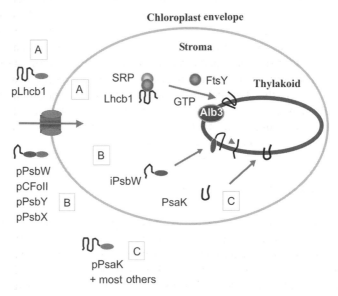

Figure 6.4   Pathways for the targeting of nuclear-encoded thylakoid membrane proteins. Pathway A represents the SRP-dependent pathway, used by light-harvesting chlorophyll-binding proteins such as Lhcb1. These proteins are synthesised with transit peptides that specify import into the chloroplast. After import, the transit peptides are removed and the substrate interacts with cpSRP, which comprises SRP54 and SRP43 subunits, and its partner protein, FtsY. Both of these factors hydrolyse GTP during their mode of action. These factors direct the targeting of Lhcb1 to the membrane and insertion occurs in a poorly understood process that depends heavily on membrane-bound Alb3. Pathways B and C represent distinct pathways that require none of the known protein targeting apparatus, and which may therefore involve the spontaneous insertion of the substrates in the thylakoid membrane. Pathway B is used by a small number of thylakoid membrane proteins (to date, CFoII, PsbX, PsbW and PsbY) that are synthesised with cleavable signal-type peptides. These sequences somehow aid the insertion of the proteins in loop conformations (possibly by simply providing additional, transient hydrophobic regions). After insertion, cleavage by thylakoidal processing peptidase yields the mature protein. Pathway C is used by the majority of thylakoid membrane proteins; PsaK is shown as an example. Here, the protein is synthesised with a transit peptide that is remobed in the stroma; the protein then inserts into the membrane as the mature-size form with no apparent input from any of the known targeting factors in the stroma or membrane, and without the necessity of any form of free energy.

6.4 Biogenesis of the thylakoid membrane

The above discussion has focused on the targeting and insertion or translocation of individual nuclear- and chloroplast-encoded proteins. However, it is important to understand the biogenesis of the thylakoid membrane in broader terms because this membrane network accounts for the bulk of chloroplast lipid and its biogenesis is poorly understood. What do we know about the thylakoid membranes in their entirety: how are they generated? Are they synthesised *de novo* by assembling lipids, pigments and proteins in the chloroplast stroma? Or, are they derived from pre-existing membranes? At present, most of the evidence points to the thylakoid lipid bilayer being constructed from pre-existing membranes, namely, the chloroplast envelope. However, this model is likely to be over-simplified, and recent research findings encourage us to reappraise the mainstream view that nuclear-encoded proteins are always threaded through the chloroplast envelope and targeted to the thylakoids via a soluble stromal intermediate stage. Here, we survey recent advances in our understanding of thylakoid development and construction, and outline the many questions that remain unanswered. We also consider the possibility, raised in some studies, that the targeting of thylakoid proteins may be linked to the trafficking of lipids or the insertion of pigments in the envelopes.

6.5 Biosynthesis of chloroplast lipids

Thylakoid glycerolipids are assembled in the chloroplast envelope from fatty acids, glycerol and a polar head group (reviewed by Douce and Joyard, 1994). By far the greatest proportion of thylakoid lipids is contributed by the neutral galactolipids, mono- and digalactosyldiacylglycerol (MGDG and DGDG), around 55 and 30%, respectively (Block *et al.*, 1983, Douce and Joyard, 1990). The negatively charged lipids, sulphoquinovosyldiacylglycerol (SL) and phosphatidylglycerol (PG), account for another 14%, while phosphatidylinositol (PI) accounts for about 1% of the total. The lipid content of the thylakoid membrane is basically indistinguishable from that of the inner chloroplast envelope, which is the point of embarkation for lipids on their way to the thylakoids (Douce and Joyard, 1990).

Many of the key enzymes involved in chloroplast lipid biosynthesis have been cloned, such as the major MGDG synthase (MGD1; Shimojima *et al.*, 1997), which catalyses the synthesis of MGDG from diacylglycerol and UDP-galactose (Miège *et al.*, 1999), and the major DGD synthase (DGD1; Dörmann *et al.*, 1999), which catalyses the formation of DGDG from two molecules of MGDG. To complicate matters, chloroplast lipids can obtain their diacylglycerol moiety from two different sources: either the chloroplast (known as the prokaryotic, or plastid route) or the endoplasmic reticulum (eukaryotic, or ER route) (reviewed recently by Dörmann and Benning, 2002). Not only that, the fatty acids of ER-derived diacylglycerol are first made in the plastid, and so journey out to the ER before eventually returning to the chloroplast in the form of a transport metabolite whose identity has not been firmly established (Jorasch and Heinz, 1999). The extent to which plants utilise plastid-derived diacylglycerol varies phylogenetically, with some species relying entirely upon the ER to supply diacylglycerol (Mongrand *et al.*, 1998). Plants are also able to cope when the plastid pathway is disrupted, as in the case of the *act1* mutant of *Arabidopsis* that lacks the plastid acyltransferase (Kunst *et al.*, 1988). These mutants are not stunted and simply use the ER pathway to supply diacylglycerol (Dörmann *et al.*, 1999).

The topology of chloroplast lipid biosynthesis means that there must be a vast movement of precursors, intermediates and finished lipids during the lifetime of the organelle. Recently, Xu *et al.* (2003) identified a permease-like protein, TGD1, in the outer chloroplast envelope, which is involved in import of lipid precursors from the ER to the chloroplast, probably as part of a larger transporter complex. The mechanisms by which lipids are transported between the ER and the plastid have long been sought (Jorasch and Heinz, 1999), and the discovery of TGD1 represents a stride forward in understanding this process. The *Arabidopsis* mutant *tgd1-1* was identified by means of a novel, high-throughput chromatographic screen, based on the identification of 'unnatural' oligogalactolipids that are produced when the ER to plastid lipid transfer pathway is disrupted; screening methods such as this can be expected to yield more candidates for the interorganelle lipid trafficking process.

The major site of MGDG synthesis is in the inner envelope (Block *et al.*, 1983; Miège *et al.*, 1999), whereas the enzyme DGD1 is located in the outer envelope (Froehlich *et al.*, 2001). There may be no need to invoke a mechanism to transfer MGDG to the outer envelope for its use in DGDG synthesis, because *Arabidopsis* harbours two further MGDG synthases (MGD2 and MGD3) in the outer envelope (Awai *et al.*, 2001). It is tempting to speculate that these enzymes catalyse the synthesis of MGDG precursors, which are then converted to DGDG by DGD1. The question is – how are these DGDG molecules transported from the outer to the inner envelope membranes? One possibility is that DGD1 itself mediates the process, because the enzyme is composed of two domains, of which only the second, C-terminal one bears homology to plant and bacterial glycosyltransferases (Dörmann *et al.*, 1999). The N-terminal domain may function in the transfer of lipid to the inner membrane (Dörmann and Benning, 2002). The same authors also suggest that envelope-located ABC transporters could carry out lipid transfer, given that the ABC transporter MsbA, a 'lipid flippase', has been implicated in the movement of lipids from the plasma membrane to the outer membrane in *E. coli* (Chang and Roth, 2001; Doerrler *et al.*, 2001; Peelman *et al.*, 2003).

6.6 Thylakoid biogenesis during chloroplast development

One of the most popular systems for studying the development of chloroplasts and the thylakoid network has been monocotyledonous grasses, such as barley (*Hordeum vulgare*) and wheat (*Triticum aestivum*) (reviewed by Mullet, 1988). This is because the meristem at the leaf base generates a gradient of cells, with the youngest at the meristem and the oldest at the leaf tip. The youngest cells are almost colourless, and contain undifferentiated proplastids, while the deeply green, oldest cells contain fully mature chloroplasts. Microscopic analyses have shown that proplastids contain very little in the way of an internal membrane system, but that illumination prompts the accumulation of arrays of membrane sacs, which gradually coalesce to form stacked granal membranes and the elongated, interconnecting stromal lamellae (von Wettstein, 1959). In darkness, however, proplastids develop into etioplasts, containing few internal membranes, but a prominent crystalline structure, called the prolamellar body (Gunning, 1965), an example of which is shown in Figure 6.5. The prolamellar body consists of the enzyme protochlorophyllide: NADPH-oxidoreductase complexed with lipids, particularly MGDG, and protochlorophyllide (for recent reviews, see Aronsson et al., 2003; Staehelin, 2003). The conical molecular shape of MGDG, with its relatively small head group and two unsaturated fatty acids means that it favours non-bilayer structures, such as hexagonal and cubic phases (Bruce, 1998), and is probably a major factor in determining the distinctive, ordered form of the prolamellar body (Bruce, 1998). Once etioplasts are illuminated, the prolamellar body disperses and is converted into thylakoids (Henningsen and Boynton, 1970). Our usual view of chloroplasts is in cross-section, which can make it difficult to picture the three-dimensional structure of the thylakoid network. In a mature chloroplast, it is currently believed that the entire thylakoid

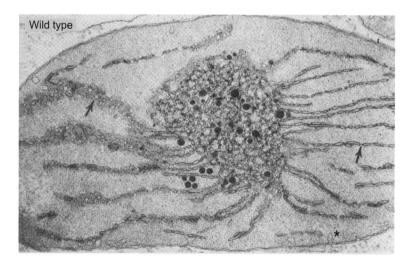

Figure 6.5   Electron micrograph of a plastid from a dark-grown barley seedling, illuminated for 1 min with white light. The prolamellar body has already started to disperse into primary lamellar layers, which contain perforations (marked by arrows). An example of a vesicle budding from the inner envelope membrane is marked with an asterisk. Scale: ×31,000. Reproduced from Henningsen et al. (1993), with kind permission of D. von Wettstein and The Royal Danish Academy of Sciences and Letters, Copenhagen.

network consists of one membrane, intricately folded to enclose one, continuous lumenal compartment. The way in which this membrane is folded is the subject of some discussion (Mustárdy and Garab, 2003; Staehelin, 2003) and there are competing models describing thylakoid architecture (Paolillo, 1970; Arvidsson and Sundby, 1999).

A key observation made during the early ultrastructural studies of chloroplast development was that the arrays of membrane sacs appear to be formed by budding of vesicles from the inner envelope membrane (Mühlethaler and Frey-Wyssling, 1959; von Wettstein, 1959). Over the years, vesicles have been observed in a variety of plastids (Douce and Joyard, 1979; Carde *et al.*, 1982; Westphal *et al.*, 2003), while Hugueney *et al.* (1995) observed apparent vesicle budding and fusion in ultrathin sections of bell pepper (*Capsicum annuum*) plastids undergoing the chloroplast to chromoplast developmental transition. This ripening process involves the complete breakdown of the thylakoid network and the synthesis of membranes and fibrils containing carotenoids and a new complement of proteins (Camara and Brangeon, 1981). Another drastic reorganisation of plastid internal structure occurs when prolamellar bodies formed in leaf proplastids at night are transformed into thylakoids when illuminated the following day; the transformation is accompanied by budding of vesicles from the inner envelope membrane (Henningsen *et al.*, 1993; von Wettstein *et al.*, 1995; von Wettstein, 2001). All these reports of vesicles in plastids have essentially been 'snapshots' of normal developmental processes, but it is also possible to see vesicles in mutants where normal thylakoid development has been disrupted, e.g. in a number of the *xantha* and *albina* barley mutants (Henningsen *et al.*, 1993). Figure 6.6 shows an etioplast from the *xantha-g*[45] mutant, where vesicles are clearly visible in the region of the inner envelope membrane.

6.6.1 *Proposed mechanisms for moving lipid to the thylakoids*

The ultrastructural studies discussed above led to the suggestion that lipids synthesised in the chloroplast envelope are transported to the thylakoids by the budding of vesicles from the inner envelope and their subsequent fusion to form thylakoid lamellae, in a manner reminiscent of cytosolic vesicle transport systems (Douce and Joyard, 1979; Carde *et al.*, 1982; Wellburn, 1982). The cargo of the vesicles need not necessarily be limited to thylakoid-destined lipids, but might conceivably contain certain pigments, membrane-inserted proteins or soluble proteins enclosed within the vesicle interior (Morré *et al.*, 1991a, b, Joyard *et al.*, 1998, von Wettstein, 2001, Vothknecht and Westhoff, 2001). Before discussing the evidence supporting the vesicular hypothesis, it should be pointed out that other mechanisms for the transfer of lipid to the thylakoids have been proposed. It could be envisaged that specialised lipid transfer proteins remove lipids from the envelope and promote their incorporation into the thylakoids, as similar systems exist in the cytosol (Rogers and Bankaitis, 2000). Evidence supporting this theory is scant, however, although a thylakoid lumen protein of unknown function has been reported to bear homology to cyanobacterial proteins involved in glycolipid localisation (Kieselbach

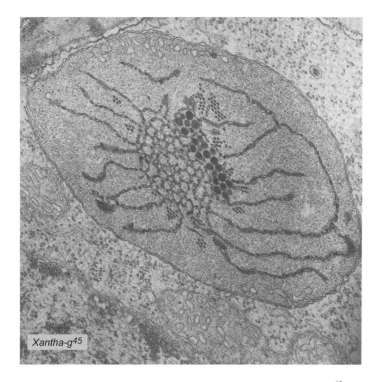

Xantha-g⁴⁵

Figure 6.6 Electron micrograph of a plastid from a dark-grown barley *xantha-g*⁴⁵ mutant plant, which is defective in pigment accumulation. The crystalline prolamellar body shown here is more loosely arranged than that of a typical wild-type etioplast. The leaf was fed δ-aminolevulinic acid (precursor of protochlorophyllide) in darkness for 24 h, during which time bundles of tubules in hexagonal array (shown in transverse section) were formed. Note the prominent vesicles budding from the inner envelope membrane. Scale: ×34,000. Reproduced from Henningsen *et al*. (1993), with kind permission of D. von Wettstein and The Royal Danish Academy of Sciences and Letters, Copenhagen.

et al., 1998). The same studies that observed vesicles budding from the inner envelope also reported that nascent internal membranes were sometimes attached to the inner envelope during the earliest stages of chloroplast development (Mühlethaler and Frey-Wyssling, 1959; von Wettstein, 1959), which would imply a lateral transfer of lipids (and possibly other components) to the thylakoids. It is often stated that such connections between the inner envelope and thylakoids are not observed in mature chloroplasts, but we have observed occasional connections between a thylakoid lamella and the inner envelope in *Arabidopsis* seedling chloroplasts where the granal stacks are already well established (A. Schulz and A. Mant, unpublished observations, 2000). Studies utilising radioactively labelled lipid precursors concluded that the primary mechanism of lipid transport from the envelope to the thylakoids is lateral transfer (Rawyler *et al.*, 1992, 1995), because DGDG is exported less readily than MGDG, newly synthesised and exported lipids display the same TM asymmetry as bulk thylakoid lipids, and the rapid export of lipids takes place with identical efficiencies in the light and dark, implying no requirement for ATP. There is, however, no a priori reason to exclude the existence of more than one mechanism for lipid transfer in chloroplasts – vesicle transport and lateral transfer may take place in parallel, or at different plastid developmental stages, or in response to differing environmental cues. The mechanisms for transferring lipid to the thylakoids may differ between young, greening chloroplasts, and mature chloroplasts carrying out maintenance, adaptation and repair of the thylakoids.

6.6.2   Chloroplast vesicle transport: clues from the cytoplasm

One of the first experimental studies to address the phenomenon of vesiculation in chloroplasts discovered that cold treatment of leaf tissue results in the accumulation of quite substantial vesicles in the space between the inner envelope and the thylakoids, with many of them appearing as invaginations of the inner envelope membrane (Morré et al., 1991b). Observations were similar in pea (Pisum sativum), soybean (Glycine max), spinach (Spinacia oleracea), tobacco (Nicotiana tabacum) and even in expanding beech (Fagus silvatica) leaves that had experienced a cold night in late spring. These findings were interesting, because cold treatment characteristically induces the accumulation of vesicles in the endomembrane system of animal cells (Tartakoff, 1986). Vesicles continue to bud from the ER, but their fusion with the Golgi apparatus is inhibited at temperatures between 10 and 20°C. Raising the temperature reverses the inhibition, and this was also true for the plant tissue; the vesicles disappeared, although individual vesicles could occasionally be seen both before and after cold treatment. More recent work has confirmed the observations in Arabidopsis thaliana (Kroll et al., 2001; Westphal et al., 2001b).

The issue of whether vesicles derived from the inner envelope actually constitute transport intermediates, capable of fusion with a target membrane was pursued in a cell-free system (Morré et al., 1991a). A donor and target membrane system was set up in vitro, where radioactively labelled lipids were transported from isolated envelopes to thylakoids immobilised on nitrocellulose strips. Transfer was temperature dependent, and stimulated by one or more stromal proteins, and addition of ATP. Extending these experiments, Räntfors et al. (2000) found that release of galactolipids from isolated envelopes was dependent on stromal protein(s) and stimulated by hydrolysable nucleoside triphosphates and acyl-CoA. A vesicular mechanism for transporting lipid from the envelopes seems a reasonable interpretation of these data, although the authors suggest some of the transfer may take place via release of lipid monomers (Räntfors et al., 2000).

Use of inhibitor compounds has been central to the study of cytoplasmic membrane trafficking in eukaryotic cells, each inhibitor affecting one or more protein-mediated steps in the budding, transport or fusion of vesicles. An isolated chloroplast system has been established in pea, where it was possible to test a wide range of inhibitors to see whether they could affect accumulation or dissipation of vesicles and thus provide a clue to the identity of any proteins involved in vesiculation (Westphal *et al.*, 2001b). Vesicle accumulation could be induced over a period of 20 min by incubating the chloroplasts at 4°C (one might suppose that vesiculation in an isolated chloroplast is limited by the supply of lipid precursors and the surface tension of the envelope membranes). The dissipation of accumulated vesicles upon raising the temperature to 22°C was markedly inhibited by microcystin LR and to a lesser extent by ophiobolin A and W7. These three compounds inhibit protein phosphatase 1, calmodulin and antagonise Ca^{2+} binding, respectively (reviewed by Holmes *et al.*, 2002; Burgoyne and Clague, 2003): protein phosphatase 1 is involved in lipid bilayer fusion during yeast vacuole formation (Peters *et al.*, 1999), while calmodulin and Ca^{2+} form part of the complex mediating homotypic membrane fusion (Burgoyne and Clague, 2003). Therefore, it was proposed that the inhibition of vesicle dissipation represents an inhibition of vesicle fusion, mediated by homologues of protein phosphatase 1 and Ca^{2+}-calmodulin. Various non-hydrolysable GTP analogues, on the other hand, inhibited the accumulation of vesicles in chloroplasts incubated at 4°C, which was interpreted as an inhibition of vesicle budding from the inner envelope membrane. A number of GTP-binding proteins are involved in cytoplasmic membrane trafficking, including the protein dynamin (reviewed by Sever, 2002). The *Arabidopsis* genome encodes a large number of dynamin-like proteins (Dombrowski and Raikhel, 1995; Kang *et al.*, 1998; Park *et al.*, 1998; Mikami *et al.*, 2000; Jin *et al.*, 2001; Kang *et al.*, 2001; Arimura and Tsutsumi, 2002; Miyagishima *et al.*, 2003; Gao *et al.*, 2003; Kang *et al.*, 2003), of which two, ADL2A and ARC5, have been localised to the chloroplast (Kang *et al.*, 1998; Kim *et al.*, 2001; Gao *et al.*, 2003), the latter on the cytosolic face of the outer envelope membrane (Gao *et al.*, 2003).

'Trapping' accumulated vesicles by microcystin LR or other inhibitors might provide a means to separate the vesicles from other chloroplast membranes so that the lipid composition can be analysed, and putative associated proteins isolated – although the yield from such a purification procedure may well be tiny.

6.6.3 *Potential protein players*

To date, three proteins have been identified as candidates for involvement in chloroplast membrane trafficking.

6.6.3.1 *Plastid fusion and/or translocation factor in chromoplasts*

The first protein was identified using an *in vitro* assay to measure the mixing of contents from two populations of *Capsicum annuum* chromoplast vesicles (Hugueney *et al.*, 1995). One population of vesicles contained the pigment antheraxanthin, and the second contained the enzyme capsanthin synthase, which could convert antheraxanthin to the red pigment capsanthin. The most likely explanation for the observed conversion of antheraxanthin to capsanthin upon mixing the two populations was fusion between the vesicles. Fusion was strongly stimulated by a 72-kDa stromal protein, named plastid fusion and/or translocation factor (Pftf). Using anti-bodies against Pftf, the authors identified a cDNA clone from a *Capsicum* library, which encoded a protein with significant homology to bacterial FtsH, and eukaryotic *N*-ethylmaleimide-sensitive factor (NSF). FtsH is a membrane-anchored metallo-protease, performing quality control in chloroplasts, and a putative function in the assembly of thylakoid protein complexes (reviewed by Adam and Clarke, 2002), while NSF is a soluble protein required for vesicle fusion (reviewed by May *et al.*, 2001). The biological roles of FtsH and NSF seemed to support the idea that Pftf mediated vesicle fusion in the *in vitro* assay, or carried out some membrane translo-cation function that enabled mixing of the contents of the two vesicle populations. Interestingly, Summer and Cline (1999) demonstrated that Pftf is an integral mem-brane protein in chromoplasts, and considered its close homology to FtsH to be the more important indicator of likely function. The apparent discrepancy between the isolation of Pftf from chromoplast stroma, and its localisation as an integral membrane protein has not yet been reconciled, and at the time of writing, there are no new data on the function of Pftf.

6.6.3.2 Dynamin-like proteins

Dynamin and dynamin-like proteins form a superfamily of high-molecular mass GTP-binding proteins present in algae, fungi, animals and plants; a related open reading frame has also been reported in *E. coli* (reviewed by Van der Bliek, 1999a). They are involved in many different cellular processes, but typically take part in shaping or moving membranes. Dynamin I, first discovered in rat (*Rattus norvegicus*) brain, has been intensively studied, and shown to play a central role in receptor-mediated endocytosis (Sever, 2002). It self-polymerises to form a spiral around the neck of invaginated plasma membrane, and changes conformation during a GTP binding and hydrolysis cycle, resulting in scission of the membrane and release of the vesicle into the cytosol. Whether dynamin causes scission mechanically, or indirectly by regulating other proteins, has been the subject of a lively debate (Van der Bliek, 1999b). Association of dynamin I with the plasma membrane is achieved via a pleckstrin homology domain that binds to phosphatidylinositol 4,5-bisphosphate with high affinity (reviewed by Lee and Lemmon, 2001). Members of the dynamin family show close homology in their N-terminal domains, which include the GTP-binding motifs, but are more diverse at the C-termini, where there are domains for protein-protein interactions (Van der Bliek, 1999a).

ADL2A is a 90-kDa dynamin-like protein, synthesised with an N-terminal pre-sequence resembling an envelope transit peptide (Kang *et al.*, 1998). While the full-length protein itself has not been imported into chloroplasts *in vitro*, a fusion protein between ADL2A and green fluorescent protein (GFP) localised to chloro-plasts when expressed in soybean suspension cells, and the first 35 amino acids of the protein were able to target a GFP fusion protein into the chloroplasts of tobacco protoplasts. An antibody raised against a small peptide epitope in a non-conserved region of ADL2A recognised a protein with an apparent molecular mass of 100 kDa in the envelope membranes of chloroplasts (Kim *et al.*, 2001). ADL2A was expressed in *E. coli*, and characterised biochemically: like other dynamins, it can form spiral coiled-coil structures and rings, suggesting that it too self-polymerises, and it binds phosphatidylinositol 4-phosphate specifically, despite lacking an obvious pleckstrin homology domain that mediates lipid binding in other dynamins. Interestingly, phosphatidylinositol 4-kinase activity is confined to the outer envelope of spinach chloroplasts (Siegenthaler *et al.*, 1997; Bovet *et al.*, 2001), suggesting that ADL2A may interact with the outer envelope membrane, from the cytosolic or intermembrane space sides. The existence of a dynamin-like protein in chloroplasts with similar biochemical properties to dynamin I encourages consideration of whether ADL2A is involved in budding off vesicles from the inner envelope membrane, in a manner analogous to endocytosis.

Such speculation is dangerous, however, for no functional information is yet available for ADL2A. It is important to analyse the phenotypes of mutant plants either lacking ADL2A entirely (knockout lines), or with reduced expression of the gene (antisense or RNA interference lines). If the protein is essential for chloroplast development, a knockout line may not be viable, and so creating plants with reduced expression of the protein is an attractive strategy. Given the large number of conserved dynamin-like genes in the *Arabidopsis* genome, any antisense or RNA interference approach must attempt to avoid down-regulating the expression of other members of the family simultaneously by post-transcriptional gene silencing (Vaucheret and Fagard, 2001), or should at least monitor the expression of other family members.

ADL2A is very closely related to ADL2B, a protein that has been localised to *Arabidopsis* mitochondria (Arimura and Tsutsumi, 2002). By means of the localisation of GFP fusion proteins, and the expression of dominant negative mutants of ADL2B, it was proposed that ADL2B is involved in mitochondrial division. This function would fit nicely with the known role of the closely related yeast (*Saccharomyces cerevisiae*) protein Dnm1p, which mediates mitochondrial fission (reviewed by Shaw and Nunnari, 2002). Given the close homology between ADL2B and ADL2A, it must therefore be considered whether ADL2A might be involved in plastid division, rather than the generation of thylakoid membranes. Plastid division has received a great deal of attention (see Chapter 4); many of the proteins involved have been identified aided by the analysis of an extensive collection of chloroplast division mutants (Pyke and Leech, 1992; Reski, 1998), or by exploitation of the model system, *Cyanidioschyzon merolae* (Kuroiwa, 1998). Division involves the tightly regulated assembly and disassembly of a number of concentric protein rings on the plastid surface, in the envelope, and within the stroma. Plastids have retained part of the cyanobacterial division machinery, including FtsZ and Min proteins (reviewed by Osteryoung, 2001), but until recently it was considered that host-derived dynamin-like proteins were absent from the division machinery (Erickson, 2000). The discovery that the *Arabidopsis ARC5* chloroplast division mutant lacks a cytosolic dynamin-like protein, which in the wild type forms a ring at the chloroplast constriction point (Gao *et al.*, 2003; see also Miyagishima *et al.*, 2003, for similar observations in *C. merolae*), invites consideration of the possibility that ADL2A is a

member of one of the inner plastid division rings. However, yet another yeast mitochondrial dynamin, Mgm1p, which is located in the inner mitochondrial space and
associated with the inner membrane, functions in maintenance of mitochondrial
morphology and shaping the inner membrane (Shaw and Nunnari, 2002). Thus,
mitochondrial dynamins perform a range of complex (and not fully determined)
functions, and so a role for ADL2A in thylakoid biogenesis cannot be excluded.

ADL1A, another member of the *Arabidopsis* dynamin family, was also implicated
in the biogenesis of thylakoid membranes (Park *et al.*, 1998). Immunolocalisation
experiments suggested that the protein is mainly associated with the thylakoid membranes, and dominant negative mutants expressing deletion constructs of the *ADL1A*
gene possessed yellow leaves and chloroplasts with very few thylakoids. However,
the antibody was raised against the highly conserved N-terminal, GTPase domain,
which recognises more than one dynamin-like protein (Kang *et al.*, 2001). A specific antibody against a non-conserved peptide epitope detected ADL1A at the cell
plate in dividing cells, and in punctate, subcellular structures, but not in chloroplasts (Kang *et al.*, 2001) – in agreement with the location of a soybean homologue,
phragmoplastin (Gu and Verma, 1996). Analysis of knockout lines lacking ADL1A
and other closely related ADL1 proteins has shown that the protein is essential for
cytokinesis and polar cell expansion, but does not have any detectable influence
on chloroplast development (Kang *et al.*, 2001, 2003). The yellow leaf phenotype
of the *ADL1A* dominant negative mutants might be explained by down-regulated
expression of other dynamin-like genes along with *ADL1A*, such as *ADL2A*. In the
realm of speculation, one could imagine the involvement of a dynamin-like protein
in trafficking lipid from the ER to the plastid; disruption of this pathway, particularly
in plants reliant on the ER pathway, might stunt the development of chloroplasts. At
present, though, the balance of evidence is against a direct involvement of ADL1A
in thylakoid biogenesis.

6.6.3.3   Vesicle-inducing protein in plastids, and cyanobacteria

VIPP, the vesicle-inducing protein in plastids, was first identified as a 37-kDa protein of unknown function in pea (Li *et al.*, 1994); a function in the transfer of
galactolipids from the envelope to the thylakoids was proposed, based partly on the
protein's location in the envelopes and thylakoids (Li *et al.*, 1994). Some years later,
Kroll *et al.* (2001) investigated an *Arabidopsis* T-DNA insertion mutant unable to
grow photoautotrophically on soil, but that could be encouraged to grow on agar
media supplemented with sucrose. The plants were of normal size, but pale green
and unable to carry out normal photosynthetic electron transport, as a result of a
severe reduction in the content of photosystems I and II and the cytochrome b_6/f
complex. The thylakoids were reduced and irregularly shaped, and cold treatment
of leaf tissue failed to elicit vesicle accumulation in the space between the inner
envelope and the thylakoids. The T-DNA insertion turned out to lie just upstream
of the gene encoding VIPP (*VIPP*), causing a drastic reduction in VIPP expression. The mutant phenotype could, to a large extent, be rescued by transformation
with a wild-type copy of *VIPP*, including restoration of vesiculation. As in pea,

VIPP could be detected in the inner envelope and the thylakoid membrane (Kroll *et al.*, 2001), despite being a hydrophilic protein. It is synthesised with an apparent N-terminal envelope transit peptide, and the mature protein is about 32 kDa in size. The phenotypic observations led the authors to suggest that VIPP is involved in vesicular transport between the inner envelope membrane and the thylakoids, although the protein shows no homology to any proteins participating in cytoplasmic membrane trafficking (Kroll *et al.*, 2001).

Intriguingly, VIPP is also found in the plasma membrane of cyanobacteria, including *Synechocystis* sp. PCC6803 (Westphal *et al.*, 2001a). Disruption of the gene leads to cells containing a few, sparsely distributed thylakoids instead of the usual, ordered arrays of membranes, and an inability to carry out light-stimulated oxygen evolution (Westphal *et al.*, 2001a). The discovery of VIPP in cyanobacteria raises the question of whether the protein is directly involved in the production and/or transport of membrane vesicles, because no vesicle transport system has been described in cyanobacteria (Westphal *et al.*, 2003).

6.6.4 Do vesicles carry a protein cargo?

A central tenet of the general import model is that nuclear-encoded thylakoid proteins proceed through a soluble stromal phase after being threaded through the chloroplast envelope membranes. It is likely that most, if not all, newly imported proteins associate with one or more chaperones in the stroma, and those proteins on the Sec or SRP-dependent pathways will also interact with soluble translocation factors, such as cpSecA or cpSRP. Evidence for the stromal intermediate phase comes from a wide range of studies, including experiments where thylakoid translocation is inhibited *in organello*, resulting in the accumulation of precursor protein in the stroma, which is then shown to be a productive transport intermediate once inhibition is lifted (Cline *et al.*, 1993; Creighton *et al.*, 1995). The fact that *in vitro* translated precursor proteins can insert into, or be translocated across, isolated thylakoid membranes also demonstrates that soluble proteins can be substrates for the thylakoid translocation machinery.

That proteins might travel to the thylakoids in vesicles, either inserted in the vesicle membrane or as soluble bulk cargo, is a rather controversial idea. Nevertheless, a number of interesting observations demand consideration of whether this takes place under certain circumstances, or generally, *in vivo*.

Using the model photosynthetic organism *Chlamydomonas reinhardtii*, Hoober and co-workers have shown that accumulation of LHCII in chloroplasts takes place only in the presence of chlorophyll synthesis, and that interaction with chlorophyll is necessary to stabilise newly imported LHCII and prevent its retrograde movement to the cytosol (reviewed in Hoober and Eggink, 1999, 2001; summarised in Figure 6.7). Furthermore, kinetic and ultrastructural investigations suggest that LHCII apoprotein inserts and assembles in the envelope, rather than the thylakoid membrane, and that nascent thylakoids bud from the chloroplast envelope. The authors suggest that this is the mainstream mechanism of thylakoid biogenesis in greening plastids,

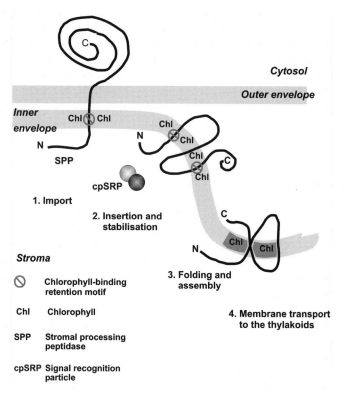

Figure 6.7 An alternative model of the insertion and assembly of LHCII. This is based on experimental work on *Chlamydomonas reinhardtii*. The figure is adapted from the model presented by Hoober and Eggink (1999). The precursor protein is imported through the chloroplast envelope, so that the envelope transit peptide and processing site is accessible to SPP. A retention motif lodges the first TM span in the inner envelope membrane, and binding of chlorophyll molecules restricts further movement. The remainder of the protein is imported through the outer membrane and inserted into the inner membrane, possibly with the assistance of cpSRP. The third TM span contains another chlorophyll-binding retention motif. Rearrangement of the TM spans and folding of the protein, concomitant with further chlorophyll binding, leads to a stably inserted LHC II molecule. Membrane is subsequently transferred to the thylakoid network, possibly by means of vesicle budding and fusion.

and that *in vivo* cpSRP interactions could take place at the chloroplast envelope (Hoober and Eggink, 1999). *Arabidopsis* plants deficient in functional cpSRP54 display a virescent phenotype: that is, the seedlings are very pale, but gradually become green, achieving almost wild-type content of antenna proteins and pigments (Pilgrim *et al.*, 1998). The phenotype strongly implies a mechanism compensating for the lack of cpSRP54, and it was suggested that light-harvesting proteins were being transported to the thylakoids in vesicles (Amin *et al.*, 1999).

Careful purification and separation of cyanobacterial thylakoids from plasma membrane led to another surprising discovery, that some core subunits of photosystems I and II, complexed with pigments, could be detected in the plasma membrane of *Synechocystis* sp. PCC 6803 cells (Zak *et al.*, 2001). The authors propose that photosystem subcomplexes are assembled in the plasma membrane, but that subsequent steps take place in the thylakoid membranes. Vesicle transport, or lateral transfer through connecting membranes, is suggested to translocate the subcomplexes to the thylakoids.

6.7 Concluding remarks

A great deal is now known about the biogenesis of the thylakoid, and many of the studies into protein targeting pathways have revealed the targeting mechanisms in considerable detail. We also understand the evolution of the pathways in most cases – the Sec, Tat and SRP pathways have clearly been inherited from the chloroplast's prokaryotic ancestor, with only minor modifications in some cases. The 'spontaneous' insertion pathway for the majority of thylakoid membrane proteins is a notable exception in this context; the precise mechanism is not yet understood and it remains to be determined whether an analogous mechanism operates in photosynthetic prokaryotes.

Nevertheless, many aspects of thylakoid biogenesis remain vague, perhaps surprisingly given the sheer abundance of this membrane. We know quite a lot about the synthesis of thylakoid-type lipids but remarkably little about their transport to the thylakoid. Here, the significance of chloroplast vesicles is an area that needs considerable attention, and it must not be forgotten that the proposed vesicle transport pathways may also deliver proteins to the thylakoid. Finally, it is worth emphasising that the means of separating envelope and thylakoid proteins is still poorly understood. These membranes carry out very different functions and it will be most important to understand how, and where, the efficient separation of these targeting pathways diverges.

References

Adam, Z. and Clarke, A.K. (2002) Cutting edge of chloroplast proteolysis. *Trends Plant Sci.*, 7, 451–456.

Amin, P., Sy, D.A.C., Pilgrim, M.L., Parry, D.H., Nussaume, L. and Hoffman, N.E. (1999) *Arabidopsis* mutants lacking the 43- and 54-kilodalton subunits of the chloroplast signal recognition particle have distinct phenotypes. *Plant Physiol.*, 121, 61–70.

Arimura, S. and Tsutsumi, N. (2002) A dynamin-like protein (ADL2b), rather than FtsZ, is involved in *Arabidopsis* mitochondrial division. *Proc. Natl. Acad. Sci. U.S.A.*, 99, 5727–5731.

Aronsson, H., Sundqvist, C. and Dahlin, C. (2003) POR – import and membrane association of a key element in chloroplast development. *Physiol. Plant*, 118, 1–9.

Arvidsson, P.-O. and Sundby, C. (1999) A model for the topology of the chloroplast thylakoid membrane. *Aust. J. Plant Physiol.*, 26, 687–694.

Awai, K., Maréchal, E., Block, M.A. *et al.* (2001) Two types of MGDG synthase genes, found widely in both 16:3 and 18:3 plants, differentially mediate galactolipid syntheses in photosynthetic and nonphotosynthetic tissues in *Arabidopsis thaliana*. *Proc. Natl. Acad. Sci. U.S.A.*, 98, 10960–10965.

Bellafiore, S., Ferris, P., Naver, H., Gohre, V. and Rochaix, J.D. (2002) Loss of Albino3 leads to the specific depletion of the light-harvesting system. *Plant Cell*, 14, 2303–2314.

Berks, B. (1996) A common export pathway for proteins binding complex redox cofactors? *Mol. Microbiol.*, 22, 393–404.

Block, M.A., Dorne, A.J., Joyard, J. and Douce, R. (1983) Preparation and characterization of membrane fractions enriched in outer and inner envelope membranes from spinach chloroplasts, II: biochemical characterization. *J. Biol. Chem.*, 258, 13281–13286.

Bogsch, E.G., Sargent, F, Stanley, N.R., Berks, B.C., Robinson, C. and Palmer, T. (1998) An essential component of a novel bacterial protein export system with homologues in plastids and mitochondria. *J. Biol. Chem.*, 273, 18003–18006.

Bolhuis, A., Mathers, J.E., Thomas, J.D., Barrett, C. and Robinson, C. (2001) TatB and TatC form a structural and functional unit of the twin-arginine translocase of *Escherichia coli*. *J. Biol. Chem.*, 276, 20213–20219.

Bovet, L., Muller, M.-O. and Siegenthaler, P.-A. (2001) Three distinct lipid kinase activities are present in spinach chloroplast envelope membranes: phosphatidylinositol phosphorylation is sensitive to wortmannin and not dependent on chloroplast ATP. *Biochem. Biophys. Res. Commun.*, 289, 269–275.

Bruce, B.D. (1998) The role of lipids in plastid protein transport. *Plant Mol. Biol.*, 38, 223–246.

Burgoyne, R.D. and Clague, M.J. (2003) Calcium and calmodulin in membrane fusion. *Biochim. Biophys. Acta*, 1641, 137–143.

Camara, B. and Brangeon, J. (1981) Carotenoid metabolism during chloroplast to chromoplast transformation in *Capsicum annuum* fruit. *Planta*, 151, 359–364.

Carde, J.P., Joyard, J. and Douce, R. (1982) Electron microscope studies of envelope membranes from spinach plastids. *Biol. Cell*, 44, 315–324.

Chaddock, A.M., Mant, A., Karnauchov, I. *et al.* (1995) A new type of signal peptide: central role of a twin-arginine motif in transfer signals for the ΔpH-dependent thylakoidal protein translocase. *EMBO J.*, 14, 2715–2722.

Chang, G. and Roth, C.B. (2001) Structure of MsbA from *E. coli*: a homolog of the multidrug resistance ATP binding cassette (ABC) transporters. *Science*, 293, 1793–1800.

Clark, S.A. and Theg, S.M. (1997) A folded protein can be transported across the chloroplast envelope and thylakoid membranes. *Mol. Biol. Cell*, 8, 923–934.

Cline, K. (1986) Import of proteins into chloroplasts. Membrane integration of a thylakoid precursor protein reconstituted in chloroplast lysates. *J. Biol. Chem.*, 261, 14804–14810.

Cline, K., Ettinger, W.F. and Theg, S.M. (1992) Protein-specific energy requirements for protein transport across or into thylakoid membranes. Two lumenal proteins are transported in the absence of ATP. *J. Biol. Chem.* 267, 2688–2696.

Cline, K., Henry, R., Li, C. and Yuan, J. (1993) Multiple pathways for protein transport into or across the thylakoid membrane. *EMBO J.*, 12, 4105–4114.

Cline, K. and Mori, H. (2001) Thylakoid ΔpH-dependent precursor proteins bind to a cpTatC-Hcf106 complex before Tha4-dependent transport. *J. Cell Biol.*, 154. 719–729.

Creighton, A.M., Hulford, A., Mant, A., Robinson, D. and Robinson, C. (1995) A monomeric, tightly folded stromal intermediate on the ΔpH-dependent thylakoidal protein transport pathway. *J. Biol. Chem.*, 270, 1663–1669.

Dalbey, R.E. and Kuhn, A. (2000) Evolutionarily related insertion pathways of bacterial, mitochondrial, and thylakoid membrane proteins. *Annu Rev Cell Dev Biol.*, 16, 51–87.

DeLille, J., Peterson, E.C., Johnson, T., Moore, M., Kight, A. and Henry, R. (2000) A novel precursor recognition element facilitates posttranslational binding to the signal recognition particle in chloroplasts. *Proc. Natl. Acad. Sci. U.S.A.*, 97, 1926–1931.

Doerrler, W.T., Reedy, M.C. and Raetz, C.R.H. (2001) An *Escherichia coli* mutant defective in lipid export. *J. Biol. Chem.*, 276, 11461–11464.

Dombrowski, J.E. and Raikhel, N.V. (1995) Isolation of a cDNA encoding a novel GTP-binding protein of *Arabidopsis thaliana*. *Plant Mol. Biol.*, 28, 1121–1126.

Dörmann, P., Balbo, I. and Benning, C. (1999) *Arabidopsis* galactolipid biosynthesis and lipid trafficking mediated by DGD1. *Science*, 284, 2181–2184.

Dörmann, P. and Benning, C. (2002) Galactolipids rule in seed plants. *Trends Plant Sci.*, 7, 112–118.

Douce, R. and Joyard, J. (1979) Structure and function of the plastid envelope. *Adv. Bot. Res.*, 7, 1–116.

Douce, R. and Joyard, J. (1990) Biochemistry and function of the plastid envelope. *Annu. Rev. Cell Biol.*, 6, 173–216.

Douce, R. and Joyard, J. (1994) Biosynthesis of thylakoid membrane lipids. In *Advances in Photosynthesis, Volume 4. Oxygenic Photosynthesis: The Light Reactions* (eds D.R. Ort and C.F. Yocum), Kluwer Academic Publishers, Dordrecht, pp. 69–101.

Erickson, H.P. (2000) Dynamin and FtsZ: missing links in mitochondrial and bacterial division. *J. Cell Biol.*, 148, 1103–1105.

Froehlich, J.E., Benning. C. and Dörmann, P. (2001) The digalactosyldiacylglycerol (DGDG) synthase DGD1 is inserted into the outer envelope membrane of chloroplasts in a manner independent of the general import pathway and does not depend on direct interaction with monogalactosyldiacylglycerol synthase for DGDG biosynthesis. *J. Biol. Chem.*, 276, 31806–31812.

Gao, H.B., Kadirjan-Kalbach, D., Froehlich, J.E. and Osteryoung, K.W. (2003) ARC5, a cytosolic dynamin-like protein from plants, is part of the chloroplast division machinery. *Proc. Natl. Acad. Sci. U.S.A.*, 100, 4328–4333.

Gouffi, K., Santini, C.L. and Wu, L.F. (2002) Topology determination and functional analysis of the *Escherichia coli* TatC protein. *FEBS Lett.*, 525, 65–70.

Groves, M.R., Mant, A., Kuhn, A. *et al.* (2001) Functional characterisation of recombinant chloroplast signal recognition particle. *J. Biol. Chem.*, 276, 27778–27786.

Gu, X. and Verma, D.P.S. (1996) Phragmoplastin, a dynamin-like protein associated with cell plate formation in plants. *EMBO J.*, 15, 695–704.

Gunning, B.E.S. (1965) The greening process in plastids, 1: The structure of the prolamellar body. *Protoplasma*, 60, 111–130.

Hageman, J., Baecke, C., Ebskamp, M., Pilon, R., Smeekens, S. and Weisbeek, P. (1990) Protein import and sorting inside the chloroplast are independent processes. *Plant Cell*, 2, 479–494.

Hageman, J., Robinson, C., Smeekens, S. and Weisbeek, P. (1986) A thylakoid processing peptidase is required for maturation of the lumen protein plastocyanin. *Nature*, 324, 567–569.

Halpin, C., Elderfield, P.D., James. H.E., Zimmermann, R., Dunbar, B. and Robinson, C. (1989) The reaction specificities of the thylakoidal processing peptidase and *Escherichia coli* leader peptidase are identical. *EMBO J.*, 8, 3917–3921.

Henningsen, K.W. and Boynton, J.E. (1970) Macromolecular physiology of plastids, VIII: pigment and membrane formation in plastids of barley greening under low light intensity. *J. Cell Biol.*, 44, 290–304.

Henningsen, K.W., Boynton, J.E. and von Wettstein, D. (1993) Mutants at *xantha* and *albina* loci in relation to chloroplast biogenesis in barley (*Hordeum vulgare* L.) R. *Dan. Acad. Sci. Lett. Biol. Skrifter*, 42, 1–349.

Henry, R., Kapazoglou, A., McCaffery, M. and Cline, K. (1994) Differences between lumen targeting domains of chloroplast transit peptides determine pathway specificity for thylakoid transport. *J. Biol. Chem.*, 269, 10189–10192.

Holmes, C.F.B, Maynes, J.T., Perreault, K.R., Dawson, J.F. and James M.N.G. (2002) Molecular enzymology underlying regulation of protein phosphatase-1 by natural toxins. *Curr. Med. Chem.*, 9, 1981–1989.

Hoober, J.K. and Eggink, L.L. (1999) Assembly of light-harvesting complex II and biogenesis of thylakoid membranes in chloroplasts. *Photosynth. Res.*, 61, 197–215.

Hoober, J.K. and Eggink, L.L. (2001) A potential role of chlorophylls *b* and *c* in assembly of light-harvesting complexes. *FEBS Lett.*, 489, 1–3.

Hugueney, P., Bouvier, F., Badillo, A., D'Harlingue, A., Kuntz, M. and Camara, B. (1995) Identification of a plastid protein involved in vesicle fusion and/or membrane protein translocation. *Proc. Natl. Acad. Sci. U.S.A.*, 92, 5630–5634.

Hynds, P.J., Robinson, D. and Robinson, C. (1998) The Sec-independent twin-arginine translocation system can transport both tightly folded and malfolded proteins across the thylakoid membrane. *J. Biol. Chem.*, 273, 34868–34874.

James, H.E., Bartling, D., Musgrove, J.E., Kirwin, P.M., Herrmann, R.G. and Robinson, C. (1989) Transport of proteins into chloroplasts. Import and maturation of precursors to the 33 kDa, 23 kDa and 16 kDa proteins of the photosynthetic oxygen-evolving complex. *J. Biol. Chem.*, 264, 19573–19576.

Jin, J.B., Bae, H., Kim, S.J. *et al.* (2003) The *Arabidopsis* dynamin-like proteins ADL1C and ADL1E play a critical role in mitochondrial morphogenesis. *Plant Cell*, 15, 2357–2369.

Jin, J.B., Kim, Y.A., Kim, S.J. *et al.* (2001) A new dynamin-like protein, ADL6, is involved in trafficking from the *trans*-Golgi network to the central vacuole in *Arabidopsis*. *Plant Cell*, 13, 1511–1525.

Jorasch, P. and Heinz, E. (1999) The enzymes for galactolipid biosynthesis are nearly all cloned: so what next? *Trends Plant Sci.*, 4, 469–471.

Joyard, J., Teyssier, E., Miège, C. *et al.* (1998) The biochemical machinery of plastid envelope membranes. *Plant Physiol.*, 118, 715–723.

Kang, B.-H., Busse, J.S. and Bednarek, S.Y. (2003) Members of the *Arabidopsis* dynamin-like gene family, ADL1 are essential for plant cytokinesis and polarized cell growth. *Plant Cell*, 15, 899–913.

Kang, B.-H., Busse, J.S., Dickey, C., Rancour, D.M. and Bednarek, S.Y. (2001) The *Arabidopsis* cell plate-associated dynamin-like protein, ADL1Ap, is required for multiple stages of plant growth and development. *Plant Physiol.*, 126, 47–68.

Kang, S.G., Jin, J.B., Piao, H.L. *et al.* (1998) Molecular cloning of an *Arabidopsis* cDNA encoding a dynamin-like protein that is localized to plastids. *Plant Mol. Biol.*, 38, 437–447.

Kieselbach, T., Mant, A., Robinson, C. and Schröder, W.P. (1998) Characterisation of an *Arabidopsis* cDNA encoding a thylakoid lumen protein related to a novel 'pentapeptide repeat' family of proteins. *FEBS Lett.*, 428, 241–244.

Kim, S.J., Jansson, S., Hoffman, N.E., Robinson, C. and Mant, A. (1999) Distinct 'assisted' and 'spontaneous' mechanisms for the insertion of polytopic chlorophyll-binding proteins into the thylakoid membrane. *J. Biol. Chem.*, 274, 4715–4721.

Kim, S.J., Robinson, C. and Mant, A. (1998) Sec/SRP-independent insertion of two thylakoid membrane proteins bearing cleavable signal peptides. *FEBS Lett.*, 424, 105–108.

Kim, Y.-W., Park, D.-S., Park, S.-C., Kim, S.H., Cheong, G.-W. and Hwang, I. (2001) *Arabidopsis* dynamin-Like 2 that binds specifically to phosphatidylinositol 4-phosphate assembles into a high-molecular weight complex *in vivo* and *in vitro*. *Plant Physiol.*, 127, 1243–1255.

Klösgen, R.B., Brock, I.W., Herrmann, R.G. and Robinson, C. (1992) Proton gradient-driven import of the 16 kDa oxygen-evolving complex protein as the full precursor protein by isolated thylakoids. *Plant Mol. Biol.*, 18, 1031–1034.

Kogata, N., Nishio, K., Hirohashi, T., Kikuchi, S. and Nakai, M. (1999) Involvement of a chloroplast homologue of the signal recognition particle receptor protein, FtsY, in protein targeting to thylakoids. *FEBS Lett.*, 329, 329–333.

Kroll, D., Meierhoff, K., Bechtold, N. *et al.* (2001) *VIPP1*, a nuclear gene of *Arabidopsis thaliana* essential for thylakoid membrane formation. *Proc. Natl. Acad. Sci. U.S.A.*, 98, 4238–4242.

Kuhn, A., Stuart, R., Henry, R. and Dalbey, R.E. (2003) The Alb3/Oxa1/YidC protein family: membrane-localized chaperones facilitating membrane protein insertion? *Trends Cell Biol.*, 13, 510–516.

Kunst, L., Browse, J. and Somerville, C.R. (1988) Altered regulation of lipid biosynthesis in a mutant of *Arabidopsis* deficient in chloroplast glycerol-3-phosphate acyltransferase activity. *Proc. Natl. Acad. Sci. U.S.A.*, 85, 4143–4147.

Kuroiwa, T. (1998) The primitive red algae *Cyanidium caldarium* and *Cyanidioschyzon merolae* as model system for investigating the dividing apparatus of mitochondria and plastids. *Bioessays*, 20, 344–354.

Laidler, V., Chaddock, A.M., Knott, T.G., Walker, D. and Robinson, C. (1995) A SecY homolog in *Arabidopsis thaliana*. Sequence of a full-length cDNA clone and import of the precursor protein into chloroplasts. *J. Biol. Chem.*, 270, 17664–17667.

Lamppa, G.K. (1988) The chlorophyll *a/b*-binding protein inserts into the thylakoids independent of its cognate transit peptide. *J. Biol. Chem.*, 263, 14996–14999.

Lee, A. and Lemmon, M.A. (2001) Analysis of phosphoinositide binding by pleckstrin homology domain from dynamin. *Methods Enzymol.*, 329, 457–468.

Li, H.M., Kaneko, Y. and Keegstra, K. (1994) Molecular-cloning of a chloroplastic protein associated with both the envelope and thylakoid membranes. *Plant Mol. Biol.*, 25, 619–632.

Li, X., Henry, R., Yuan, J., Cline, K. and Hoffman, N.E. (1995) A chloroplast homologue of the signal recognition particle subunit SRP54 is involved in the post-translational integration of a protein into thylakoid membranes. *Proc. Natl. Acad. Sci. U.S.A.*, 92, 3789–3793.

Lorkovic, Z.J., Schröder, W.P., Pakrasi, H.B., Irrgang, K.-D., Herrmann, R.G. and Oelmüller, R. (1995) Molecular characterisation of PSII-W, the only nuclear-encoded component of the photosystem II reaction centre. *Proc. Natl. Acad. Sci. U.S.A.*, 92, 8930–8934.

Mant, A. and Robinson, C. (1998) An *Arabidopsis* cDNA encodes an apparent polyprotein of two non-identical thylakoid membrane proteins that are associated with photosystem II and homologous to algal *ycf32* open reading frames. *FEBS Lett.*, 423, 183–188.

Mant, A., Woolhead, C.A. Moore, M., Henry, R. and Robinson, C. (2001) Insertion of PsaK into the thylakoid membrane in a 'horse-shoe' conformation occurs in the absence of signal recognition particle, nucleoside triphosphates or functional Albino3. *J. Biol. Chem.*, 276, 36200–36206.

Manting, E.H. and Driessen, A.J. (2000) *Escherichia coli* translocase: the unravelling of a molecular machine. *Mol. Microbiol.*, 37, 226–238.

May, A.P., Whiteheart, S.W. and Weis, W.I. (2001) Unraveling the mechanism of the vesicle transport ATPase NSF, the *N*-ethylmaleimide-sensitive factor. *J. Biol. Chem.*, 276, 21991–21994.

Michl, D., Robinson, C., Shackleton, J.B., Herrmann, R.G. and Klösgen, R.B. (1994) Targeting of proteins to the thylakoids by bipartite presequences: CF0II is imported by a novel, third pathway. *EMBO J.*, 13, 1310–1317.

Miège, C., Maréchal, E., Shimojima, M. *et al.* (1999) Biochemical and topological properties of type A MGDG synthase, a spinach chloroplast envelope enzyme catalyzing the synthesis of both prokaryotic and eukaryotic MGDG. *Eur. J. Biochem.*, 265, 990–1001.

Mikami, K., Iuchi, S., Yamaguchi-Shinozaki, K. and Shinozaki, K. (2000) A novel *Arabidopsis thaliana* dynamin-like protein containing the pleckstrin homology domain. *J. Exp. Bot.*, 51, 317–318.

Miyagishima, S., Nishida, K., Mori, T. *et al.* (2003) A plant-specific dynamin-related protein forms a ring at the chloroplast division site. *Plant Cell*, 15, 655–665.

Mongrand, S., Besoule, J.-J., Cabantous, F. and Cassagne, C. (1998) The C16:3/C18:3 fatty acid balance in photosynthetic tissues from 468 plant species. *Phytochemistry*, 49, 1049–1064.

Moore, M., Harrison, M.S., Peterson, E.C. and Henry, R. (2000) Chloroplast Oxa1p homolog albino3 is required for post-translational integration of the light harvesting chlorophyll-binding protein into thylakoid membranes. *J. Biol. Chem.*, 275, 1529–1532.

Mori, H. and Cline, K. (2002) A twin arginine signal peptide and the pH gradient trigger reversible assembly of the thylakoid Delta pH/Tat translocase. *J. Cell Biol.*, 157, 205–210.

Mori, H., Summer, E.J., Ma, X. and Cline, K. (1999) Component specificity for the thylakoidal Sec and delta pH-dependent protein transport pathways. *J. Cell Biol.*, 146, 45–55.

Morré, D.J., Morré, J.T., Morré, S.R., Sundqvist, C. and Sandelius, A.S. (1991a) Chloroplast biogenesis: cell-free transfer of monogalactosylglycerides to thylakoids. *Biochim. Biophys. Acta*, 1070, 437–445.

Morré, D.J., Selldén, G., Sundqvist, C. and Sandelius, A.S. (1991b) Stromal low-temperature compartment derived from the inner membrane of the chloroplast envelope. *Plant Physiol.*, 97, 1558–1564.

Motohashi, R., Nagata, N., Ito, T. *et al.* (2001) An essential role of a TatC homologue of a Delta pH-dependent protein transporter in thylakoid membrane formation during chloroplast development in *Arabidopsis thaliana*. *Proc. Natl. Acad. Sci. U.S.A.*, 98, 10499–10504.

Mould, R.M., Knight, J.S., Bogsch, E. and Gray, J.C. (1997) Azide-sensitive thylakoid membrane insertion of chimeric cytochrome f constructs imported by isolated pea chloroplasts. *Plant J.*, 11, 1051–1058.

Mould, R.M. and Robinson, C. (1991) A proton gradient is required for the transport of two lumenal oxygen-evolving proteins across the thylakoid membrane. *J. Biol. Chem.*, 266, 12189–ss12193.

Mould, R.M., Shackleton, J.B. and Robinson, C. (1991) Requirements for the efficient import of two lumenal oxygen-evolving complex proteins into isolated thylakoids. *J. Biol. Chem.*, 266, 17286–17289.

Mühlethaler, K. and Frey-Wyssling, A. (1959) Entwicklung und struktur der proplastiden. *J. Biophys. Biochem. Cytol.*, 6, 507–512.

Mullet, J. (1988) Chloroplast development and gene expression. *Annu. Rev. Plant Physiol. Plant Mol. Biol.*, 39, 475–502.

Mustárdy, L. and Garab, G. (2003) Granum revisited. A three-dimensional model – where things fall into place. *Trends Plant Sci.*, 8, 117–122.

Nielsen, V.S., Mant, A., Knoetzel, J., Møller, B.L. and Robinson, C. (1994) Import of barley photosystem I subunit N into the thylakoid lumen is mediated by a bipartite presequence lacking an intermediate processing site; role of the delta pH in translocation across the thylakoid membrane. *J. Biol. Chem.*, 269, 3762–3766.

Nilsson, R., Brunner, J., Hoffman, N.E. and van Wijk, K.J. (1999) Interactions of ribosome nascent chain complexes of the chloroplast-encoded D1 thylakoid membrane protein with cpSRP54. *EMBO J.*, 18, 733–742.

Osteryoung, K.W. (2001) Organelle fission in eukaryotes. *Curr. Opin. Microbiol.*, 4, 639–646.

Paolillo, D.J. (1970) The three-dimensional arrangement of inter-granal lamellae in chloroplasts. *J. Cell Sci.*, 6, 243–255.

Park, J.M., Cho, J.H., Kang, S.G. *et al.* (1998) A dynamin-like protein in *Arabidopsis thaliana* is involved in biogenesis of thylakoid membranes. *EMBO J.*, 17, 859–867.

Peelman, F., Labeur, C., Vanloo, B. *et al.* (2003) Characterization of the ABCA transporter subfamily: identification of prokaryotic and eukaryotic members, phylogeny and topology. *J. Mol. Biol.*, 325, 259–274.

Peltier, J.B., Emanuelsson, O., Kalume, D.E. *et al.* (2002) Central functions of the lumenal and peripheral thylakoid proteome of *Arabidopsis* determined by experimentation and genome-wide prediction. *Plant Cell*, 14, 211–236.

Peters, C., Andrews, P.D., Stark, M.J.R. *et al.* (1999) Control of the terminal step of intracellular membrane fusion by protein phosphatase I. *Science*, 285, 1084–1087.

Pilgrim, M.P., van Wijk, K.-J., Parry, D.H., Sy, D.A.C. and Hoffman, N.E. (1998) Expression of a dominant negative form of cpSRP54 inhibits chloroplast biogenesis in *Arabidopsis*. *Plant J.*, 13, 177–186.

Porcelli, I., de Leeuw, E., Wallis, R. *et al.*(2002) Characterization and membrane assembly of the TatA component of the *Escherichia coli* twin-arginine protein transport system. *Biochemistry*, 41, 13690–13697.

Pyke, K.A. and Leech, R.M. (1992) Chloroplast division and expansion is radically altered by nuclear mutations in *Arabidopsis thaliana*. *Plant Physiol.*, 99, 1005–1008.

Räntfors, M., Evertsson, I., Kjellberg, J.M. and Sandelius, A.S. (2000) Intraplastidial lipid trafficking: regulation of galactolipid release from isolated chloroplast envelope. *Physiol. Plant*, 110, 262– 270.

Rawyler, A., Meylan, M. and Siegenthaler, P.-A. (1992) Galactolipid export from envelope to thylakoid membranes in intact chloroplasts. I. Characterization and involvement in thylakoid lipid asymmetry. *Biochim. Biophys. Acta*, 1104, 331–341.

Rawyler, A., Meylan-Bettex, M. and Siegenthaler, P.-A. (1995) (Galacto) lipid export from envelope to thylakoid membranes in intact chloroplasts. II. A general process with a key role for the envelope in the establishment of lipid asymmetry in thylakoid membranes. *Biochim. Biophys. Acta*, 1233, 123–133.

Reski, R. (1998) *Physcomitrella* and *Arabidopsis*: the David and Goliath of reverse genetics. *Trends Plant Sci.*, 3, 209–210.

Robinson, C. and Bolhuis, A. (2001) Protein targeting by the twin-arginine translocation pathway. *Nat. Rev. Mol. Cell. Biol.*, 2, 350–355.

Robinson, C., Cai, D., Hulford, A. *et al.* (1994) The presequence of a chimeric construct dictates which of two mechanisms are utilized for translocation across the thylakoid membrane: evidence for the existence of two distinct translocation systems. *EMBO J.*, 13, 279–285.

Robinson, D., Karnauchov, I., Herrmann, R.G., Klösgen, R.B. and Robinson, C. (1996) Protease- sensitive thylakoidal import machinery for the Sec-, ΔpH- and signal recognition particle- dependent protein targeting pathways, but not for CF_0II integration. *Plant J.*, 10, 149–155.

Rogers, D.P. and Bankaitis, V.A. (2000) Phospholipid transfer proteins and physiological functions. *Int. Rev. Cytol.*, 197, 35–81

Samuelson, J.C., Chen, M., Jiang, F. *et al.* (2000) YidC mediates membrane protein insertion in bacteria. *Nature*, 406, 637–641.

Santini, C.L., Ize, B., Chanal, A., Müller, M. Giordano, G. and Wu, L.F. (1998) A novel Sec-independent periplasmic protein translocation pathway in *Escherichia coli*. *EMBO J.*, 17, 101–112.

Sargent, F., Bogsch, E.G., Stanley, N.R. *et al.* (1998) Overlapping functions of components of a bacterial Sec-independent export pathway. *EMBO J.*, 17, 3640–3650.

Sargent, F., Stanley, N.R., Berks, B.C. and Palmer, T. (1999) Sec-independent protein translocation in *Escherichia coli*. A distinct and pivotal role for the TatB protein. *J. Biol. Chem.*, 274, 36073–36082.

Schubert, M., Petersson, U.A., Haas, B.J., Funk, C., Schroder, W.P. and Kieselbach, T. (2002) Proteome map of the chloroplast lumen of *Arabidopsis thaliana*. *J. Biol. Chem.*, 277, 8354–8365.

Schuenemann, D., Amin, P., Hartmann, E. and Hoffman, N.E. (1999) Chloroplast SecY is complexed to SecE and involved in the translocation of the 33-kDa, but not the 23-kDa subunit of the oxygen- evolving complex. *J. Biol. Chem.*, 274, 12177–12182.

Schuenemann, D., Gupta, S., Persello-Cartieaux, F. *et al.* (1998) A novel signal recognition particle targets light-harvesting proteins to the thylakoid membrane. *Proc. Natl. Acad. Sci. U.S.A.*, 95, 10312–10316.

Seluanov, A. and Bibi, E. (1997) FtsY, the prokaryotic signal recognition particle receptor homo-logue, is essential for the biogenesis of membrane proteins. *J. Biol. Chem.*, 272, 2053–2055.

Settles, M.A., Yonetani, A., Baron, A., Bush, D.R., Cline, K. and Martienssen, R. (1997) Sec-independent protein translocation by the maize Hcf106 protein. *Science*, 278, 1467–1470.

Sever, S. (2002) Dynamin and endocytosis. *Curr. Opin. Cell Biol.*, 14, 463–467.

Shaw, J.M. and Nunnari, J. (2002) Mitochondrial dynamics and division in budding yeast. *Trends Cell Biol.*, 12, 178–184.

Shimojima, M., Ohta, H., Iwamatsu, A., Masuda, T., Shioi, Y. and K.-I. Takamiya. (1997) Cloning of the gene for monogalactosyldiacylglycerol synthase and its evolutionary origin. *Proc. Natl. Acad. Sci. U.S.A.*, 94, 333–337.

Siegenthaler, P.-A., Müller, M.-O. and Bovet, L. (1997) Evidence for lipid kinase activities in spinach chloroplast envelope membranes. *FEBS Lett.*, 416, 57–60.

Staehelin, L.A. (2003) Chloroplast structure: from chlorophyll granules to supra-molecular architecture of thylakoid membranes. *Photosynth. Res.*, 76, 185–196.

Summer, E.J. and Cline, K. (1999) Red bell pepper chromoplasts exhibit *in vitro* import competency and membrane targeting of passenger proteins from the thylakoidal Sec and ΔpH pathways but not the chloroplast signal recognition pathway. *Plant Physiol.*, 119, 575–584.

Sundberg, E., Slagter, J.G., Fridborg, I., Cleary, S.P., Robinson, C. and Coupland, G. (1997) *ALBINO3*, an *Arabidopsis* nuclear gene essential for chloroplast differentiation, encodes a chloroplast protein that shows homology to proteins present in bacterial membranes and yeast mitochondria. *Plant Cell*, 9, 717–730.

Tartakoff, A.M. (1986) Temperature and energy dependence of secretory protein transport in the exocrine pancreas. *EMBO J.*, 5, 1477–1482.

Thompson, S.J., Kim, S.J. and Robinson, C. (1998) Sec-independent insertion of thylakoid membrane proteins: analysis of insertion forces and identification of a loop intermediate. *J. Biol. Chem.*, 273, 18979–18983.

Thompson, S.J., Robinson, C and Mant, A. (1999) Dual signal peptides mediate the Sec/SRP-independent insertion of a thylakoid membrane polyprotein, PsbY. *J. Biol. Chem.*, 274, 4059–4066.

Tu, C.J., Peterson, E.C., Henry, R. and Hoffman, N.E. (2000) The L18 domain of light-harvesting chlorophyll proteins binds to chloroplast signal recognition particle 43. *J. Biol. Chem.*, 275, 13187–13190.

Tu, C.J., Schuenemann, D. and Hoffman, N.E. (1999) Chloroplast FtsY, chloroplast signal recognition particle, and GTP are required to reconstitute the soluble phase of light-harvesting chlorophyll protein transport into thylakoid membranes. *J. Biol.Chem.*, 274, 27219–27224.

Van der Bliek, A.M. (1999a) Functional diversity in the dynamin family. *Trends Cell Biol.*, 9, 96–102.

Van der Bliek, A.M. (1999b) Is dynamin a regular motor or a master regulator? *Trends Cell Biol.*, 9, 253 –254.

Vaucheret, H. and Fagard, M. (2001) Transcriptional gene silencing in plants: targets, inducers and regulators. *Trends Genet.*, 17, 29–35.

Viitanen, P.V., Doran, E.R. and Dunsmuir, P. (1988) What is the role of the transit peptide in thylakoid integration of the light-harvesting chlorophyll *a/b* protein? *J. Biol. Chem.*, 263, 15000– 15007

Voelker, R. and Barkan, A. (1995) Two nuclear mutations disrupt distinct pathways for targeting proteins to the chloroplast thylakoid. *EMBO J.*, 14, 3905–3914.

von Wettstein, D. (1959) The effect of genetic factors on the submicroscopic structures of the chloroplast. *J. Ultrastruc. Res.*, 3, 235–237.

von Wettstein, D. (2001) Discovery of a protein required for photosynthetic membrane assembly. *Proc. Natl. Acad. Sci. U.S.A.*, 98, 3633–3635.

von Wettstein, D., Gough, S. and Kannangara, C.G. (1995) Chlorophyll biosynthesis. *Plant Cell*, 7, 1039–1057.

Vothknecht, U.C. and Westhoff, P. (2001) Biogenesis and origin of thylakoid membranes. *Biochim. Biophys. Acta*, 1541, 91–101.

Walker, M.B., Roy, L.M., Coleman, E., Voelker, R. and Barkan, A. (1999) The maize tha4 gene functions in sec-independent protein transport in chloroplasts and is related to hcf106, tatA, and tatB. *J. Cell. Biol.*, 147, 267–276.

Weiner, J.H., Bilous, P.T., Shaw, G.M. *et al.* (1998) A novel and ubiquitous system for membrane targeting and secretion of cofactor-containing proteins. *Cell*, 93, 93–101.

Wellburn, A.R. (1982) Bioenergetic and ultrastructural changes associated with chloroplast development. *Int. Rev. Cytol.*, 80, 133–191.

Westphal, S., Heins, L., Soll, J. and Vothknecht, U.C. (2001a) *Vipp1* deletion mutant of *Synechocystis*: A connection between bacterial phage shock and thylakoid biogenesis? *Proc. Natl. Acad. Sci. U.S.A.*, 98, 4243–4248.

Westphal, S., Soll, J. and Vothknecht, U.C. (2001b) A vesicle transport system inside chloroplasts. *FEBS Lett.*, 506, 257–261.

Westphal., S., Soll, J. and Vothknecht, U.C. (2003) Evolution of chloroplast vesicle transport. *Plant Cell Physiol.*, 44, 217–222.

Willey, D.L., Auffret, A.D. and Gray, J.C. (1984) Structure and topology of cytochrome f in pea chloroplast membranes. *Cell*, 36, 555–562.

Woolhead, C.A., Thompson, S., Moore, M. *et al.* (2001) Distinct Alb3-dependent and -independent pathways for thylakoid membrane protein insertion. *J. Biol. Chem* 276, 40841–40846.

Xu, C., Fan, J., Riekhof, W., Froehlich, J.E. and Benning, C. (2003) A permease-like protein involved in ER to thylakoid lipid transfer in *Arabidopsis*. *EMBO J.*, 22, 2370–2379.

Yuan, J., Henry, R., McCaffery, M. and Cline, K. (1994) SecA homolog in protein transport within chloroplasts: evidence for endosymbiont-derived sorting. *Science*, 266, 796–798.

Zak, E., Norling, B., Maitra, R., Huang, F., Andersson, B. and Pakrasi, H. (2001) The initial steps of biogenesis of cyanobacterial photosystems occur in plasma membranes. *Proc. Natl. Acad. Sci. U.S.A.*, 98, 13443–13448.

7   The chloroplast proteolytic machinery

Zach Adam

7.1   Introduction

Intracellular proteolytic processes are inherent to the function of any biological system. Chloroplasts are not an exception to this rule. Proteolysis is involved in the biogenesis and maintenance of chloroplasts, and hence is considered a vital factor that influences photosynthesis and other functions of the chloroplast, under both optimal and adverse growth conditions. As the great majority of chloroplast proteins are targeted to the organelle posttranslationally, their signal peptides are removed after import by specific peptidases. This proteolytic cleavage of a single peptide bond is essential for the maturation of the precursor form of the protein. Some plastid-encoded proteins also need to be processed in order to become active. Once chloroplast proteins are found in their final location, the level of different proteins is adjusted during development or in response to different environmental conditions. This adjustment involves either increased transcription and/or translation rates, resulting in an increase in the level of a specific protein, or proteolytic degradation that leads to a decrease in its level. For instance, under low-light conditions, plants increase the size of their photosynthetic antenna in order to maximize the amount of light energy that is absorbed. When such plants are exposed to high light, they adjust the size of their main antenna complex, the light-harvesting complex (LHC) of photosystem II (PSII), by proteolytic degradation of specific chlorophyll a/b binding proteins. A converse transition is accompanied by degradation of the early light-inducible protein (ELIP).

Functional proteins are damaged during their activity and become nonfunctional, and are thus degraded in a process that can be described as "protein quality control." Chloroplast proteins, residing in a highly oxidative environment, are very prone to such damage. In fact, oxidative damage is the primary reason for photoinhibition– inhibition of photosynthesis under increasing light intensity. Rapid degradation of oxidatively damaged proteins, such as the D1 protein of PSII reaction center, is a central component in their repair mechanism. Another manifestation of quality control is the removal of unassembled proteins that lack a protein partner or a prosthetic group. Whether the missing component is due to a mutation, inhibition of synthesis, or nutrient deficiency, the consequence is proteolytic degradation of the existing partner. Leaf senescence and fruit ripening involve breakdown of chloroplasts or transition of chloroplasts to other types of plastids. In both cases, massive degradation of the existing protein repertoire, primarily of proteins involved in photosynthesis, occurs. "Timing proteins," usually involved in controlling gene expression, are also expected to be short-lived, although their identity in chloroplasts is not well established yet.

This chapter intends to review these processes in chloroplasts, to describe the known and characterized chloroplast proteases and peptidases, and those that are predicted to reside in the organelle, to describe possible roles of chloroplast proteases in developmental processes and maintenance, and raise hypotheses to additional functions of these enzymes. Previous reviews, discussing some of these issues, can be found in Adam (1996, 2000), Andersson and Aro (1997), Clarke (1999), and Adam and Clarke (2002).

7.2 Proteolytic processes in chloroplasts

7.2.1 Processing of precursor proteins

The chloroplast genome encodes only a small fraction of its proteome. In *Arabidopsis*, 87 proteins are encoded in the chloroplast genome (Sato *et al.*, 1999), whereas 2000–3000 of the nuclear gene products are predicted to target to chloroplasts (The *Arabidopsis* Genome Initiative, 2000). All nuclear-encoded chloroplast-targeted proteins, with the exception of those targeted to the outer envelope membrane, are synthesized as precursors in the cytosol, and imported posttranslationally into the organelle, where they are sorted to their suborganellar final location. During this process, their N-terminal targeting sequence is removed by a single proteolytic cleavage. Most proteins are processed only once in the stroma, but those targeted to the thylakoid lumen are processed first in the stroma, and then once again in the lumen (cf. Dalbey and Robinson, 1999; Jarvis and Soll, 2002). A few chloroplast-encoded proteins such as cytochrome f (cyt f) and the D1 protein of PSII reaction center are also synthesized with N- or C-terminal extensions, respectively. Although the function of these extensions is not known, they need to be removed before these proteins become functional (Trost *et al.*, 1997).

7.2.2 Degradation of oxidatively damaged proteins

Not all the light energy absorbed by the photosynthetic antenna is dissipated by the photochemical process, especially under high-light conditions. As a consequence, reactive oxygen species are generated. Although scavenging mechanisms exist in the organelle to neutralize the harmful effect of these, oxidative damage to chloroplast proteins is a common phenomenon. In fact, oxidative damage is considered the main cause for photoinhibition (Barber and Andersson, 1992). The primary site of photoinhibition is PSII, but PSI is also damaged and inhibited (Tjus *et al.*, 1998). Photoinhibition in PSII has been well characterized and reviewed (cf. Andersson and Aro, 2001). The most sensitive component of this complex is the D1 protein of the reaction center, whose oxidation is believed to lead to a conformational change and loss of function. A repair cycle of PSII was evolved that involves increased transcription and translation of the *psbA* gene encoding the D1 protein (Melis, 1999). However, a prerequisite for incorporation of the newly synthesized D1 copy into the inhibited complex is the proteolytic removal of the damaged copy. The operation of the repair cycle under photoinhibitory conditions is manifested by the rapid turnover rate of the D1 protein, a phenomenon that led to the discovery and characterization of photoinhibition (Kyle *et al.*, 1984; Mattoo *et al.*, 1984; Barber and Andersson, 1992; Prasil *et al.*, 1992).

7.2.3   Adjustment of antenna size

Plants have evolved mechanisms to minimize the potential damage of high-light intensity. Short-term mechanisms include leaf and chloroplast movement that reduce the amount of energy absorbed by the photosynthetic antenna (Wada *et al.*, 2003), as well as state transition – the shuttle of LHCII antenna subunits between PSII and PSI, to balance the energy input between the two photosystems (Allen, 1992). A longer term adaptation mechanism involves reducing the antenna size upon increase in light intensity by transfer from shade to full light or from a cloudy to a sunny period. In this case, the reduction in the antenna size involves proteolytic degradation of specific antenna subunits (Lindahl *et al.*, 1995; Yang *et al.*, 1998).

7.2.4   Degradation of partially assembled proteins

Inherently stable chloroplast proteins can be destabilized under certain conditions. When one subunit of a protein complex is missing, the other subunits will be degraded rapidly. This was shown for both soluble as well as membrane-bound complexes (cf. Schmidt and Mishkind, 1983; Leto *et al.*, 1985), and it could result from a mutation or inhibition of protein synthesis in either the chloroplast or the cytosol. Prosthetic groups also have a stabilizing effect on their cognate proteins. In the absence of chlorophyll for instance, chlorophyll-binding proteins are proteolytically degraded (cf. Apel and Kloppstech, 1980; Bennett, 1981; Mullet *et al.*, 1990; Kim *et al.*, 1994). The lack of as little as a single copper ion also leads to the degradation of plastocyanin (Merchant and Bogorad, 1986; Li and Merchant, 1995). Thus, it appears that in the absence of major or minor components of a single protein or a protein complex, structural changes occur that lead to protein susceptibility to proteolysis.

7.2.5   Senescence and transition from chloroplasts to other types of plastids

Plant cell plastids, including chloroplasts, can change from one type of plastid to the other. For instance, chloroplasts develop from either proplastids or etioplasts. Chromoplasts develop from either proplastids or chloroplasts. These changes, especially those involving transformation of chloroplasts, are accompanied by major changes in the respective proteome. The ripening of fruits, or senescence of leaves, is characterized by transformation of chloroplasts into chromoplasts, a process that involves massive degradation of the protein components of the photosynthetic apparatus, although the details of these proteolytic processes are not known (Matile, 2001).

7.2.6 Timing proteins

A common theme in the control of gene expression is the appearance of certain regulators at a certain point in time. Their appearance is achieved by regulated transcription and translation of their genes. However, to limit their activity, these proteins need to be modified or disappear. The half-life of many such regulators is inherently very short, ensuring that once their expression is ceased, their level drops down immediately. Such proteins are often designated "timing proteins" (Gottesman, 1996). They are widespread among different biological systems, and thus, are expected to function in chloroplasts as well (Adam, 2000). This is a class of potential chloroplast substrates for proteolysis whose identity still needs to be revealed.

7.3 Identified and characterized chloroplast proteases and peptidases

Although examples for processing or degradation of specific chloroplast proteins have been accumulating for more than 20 years (cf. Adam, 1996), the identity of the components of the chloroplast proteolytic machinery has started to unravel only in the past 10 years or so. Given the prokaryotic origin of chloroplasts, it is not surprising that chloroplast proteases and peptidases are all homologs of bacterial ones. However, whereas most bacterial proteases and peptidases are encoded by single genes, many of the chloroplast ones are encoded by multigene families, a phenomenon whose functional significance is not clear yet. The following is a description of the biochemical properties of the chloroplast enzymes and their bacterial orthologs.

7.3.1 Processing peptidases

Nuclear-encoded chloroplast proteins are synthesized with a transit peptide that targets them posttranslationally to the chloroplast. This sequence is essential for targeting the precursor protein to the organelle, but it is processed during or shortly after the import of the protein. This processing step is carried out by an enzyme known as the chloroplast stromal processing peptidase (SPP). It is a ~140-kDa metallopeptidase with a His-X-X-Glu-His zinc-binding motif at its catalytic site. It performs a single endoproteolytic cleavage step that releases the mature form of a wide range of chloroplast precursor proteins (Vandervere et al., 1995; Richter and Lamppa, 1998, 1999, 2003; Zhong et al., 2003). The bound transit peptide is then further cleaved once by SPP to release a subfragment that is degraded to completion by a protease with characteristics of a metalloprotease that requires ATP (Richter and Lamppa, 1999). A very recent report demonstrated that a recombinant zinc-metalloprotease, which could be targeted to both mitochondria and chloroplasts, was able to degrade both mitochondrial and chloroplast targeting sequences (Bhushan *et al.*, 2003). Although this reaction did not require ATP, this protease is a strong candidate for the protease that dispose of cleaved transit peptides after import of precursor proteins into the organelle. SPP is encoded by a single gene in *Arabidopsis*. Inhibiting its expression by antisense constructs resulted in a high proportion of lethal seedlings, suggesting that it functions as the general stromal peptidase (Zhong *et al.*, 2003).

Proteins that are targeted to the thylakoid lumen are synthesized with a bipartite transit peptide. Its N-terminal region mediates translocation into the stroma, where it is cleaved by SPP to yield an intermediate form of the protein. This form is further translocated by the thylakoid translocation machinery, followed by removal of the thylakoid transfer domain by an enzyme designated the thylakoid processing peptidase (TPP) (Chaal *et al.*, 1998). TPP is homologous to bacterial Type I leader peptidases that uses a Ser-Lys catalytic dyad.

Some chloroplast-encoded proteins such as the D1 protein of PSII reaction center or cyt *f* are synthesized with either C- or N-terminal extensions, respectively. The function of these extensions is not known, but they need to be removed before the protein can assemble into the respective complex. The C-terminal extension of the D1 protein is processed by the CtpA peptidase that is located in the lumen (Shestakov *et al.*, 1994; Inagaki *et al.*, 1996; Oelmuller *et al.*, 1996; Trost *et al.*, 1997; Yamamoto *et al.*, 2001). Similar to TPP, it uses a Ser-Lys catalytic dyad. It appears that the activity of CtpA is limited to this process only, and other lumenal proteins are insensitive to it. Which enzyme is responsible for the maturation of cyt *f* is not known.

7.3.2 Clp protease

The Clp protease is an ATP-dependent serine protease complex, composed of proteolytic subunits and cognate ATPases. It is characterized best in *Escherichia coli*, where the core of the complex is composed of two heptameric rings of the 21-kDa proteolytic subunit ClpP. In this barrel-like structure, the active sites, composed of the catalytic triad Ser-His-Asp, are found within a central pore of $\sim$51 Å in diameter, whose entrance is only $\sim$10 Å wide (Wang *et al.*, 1997). This implies that the catalytic chamber is inaccessible to most globular proteins, unless they are unfolded prior to degradation. Indeed, *in vitro* assays demonstrated that the proteolytic subcomplex can degrade short peptides, but is incapable of degrading proteins. These can be degraded only when the regulatory subunit is present. The pore of the catalytic subcomplex is capped from both sides by hexameric rings of ATPases, either ClpA or ClpX. These differ from each other by their size, 83 and 46 kDa, and the number of ATP-binding domains, 2 or 1, respectively. These ATPases are responsible for substrate recognition and binding, and confer specificity to the complex.

Once substrates are bound, they are unfolded and fed into the catalytic subcomplex in a process that requires the hydrolysis of ATP (cf. Gottesman, 1996; Porankiewicz *et al.*, 1999).

Subunits of the chloroplast Clp protease are found mostly soluble in the stroma (Shanklin *et al.*, 1995; Halperin and Adam, 1996; Ostersetzer *et al.*, 1996), but a small fraction is also associated with thylakoids (Peltier *et al.*, 2001). ClpP is the only protease encoded in the chloroplast genome, but in addition to this copy, designated ClpP1, five more homologs are encoded in the nucleus (Clarke, 1999; Adam *et al.*, 2001). Four of these, ClpP3, ClpP4, ClpP5, and ClpP6, are targeted to the chloroplast, whereas ClpP2 is found in mitochondria (Halperin *et al.*, 2001b; Zheng *et al.*, 2002; Peltier *et al.*, 2004). The *Arabidopsis* genome contains four additional homologs of ClpP, designated ClpR, but they do not contain a perfect catalytic triad (Adam *et al.*, 2001). Thus, they are not expected to function as proteolytic subunits, and their function is still unknown. Nevertheless, they are found together with ClpP subunits in a 350-kDa complex (Peltier *et al.*, 2001).

The regulatory subunit of the chloroplast Clp protease is encoded by two highly similar homologs of ClpA, designated ClpC1 and ClpC2, each having two ATP-binding domains. Although they could not be found associated with complexes that were isolated from chloroplasts (Peltier *et al.*, 2001, 2004), immunoprecipitation of ClpC from stroma, supplemented with ATP-γS, could precipitate ClpP together with it (Halperin *et al.*, 2001a). This observation suggested that the association between the proteolytic and the regulatory subcomplexes is dependent on the binding of ATP. Another plant homolog of ClpA is ClpD. Its transcript is upregulated by different stress conditions (see Adam and Clarke, 2002), but similar to ClpB, the bacterial homolog of ClpA, there is no indication for its association with ClpP, and it might fulfill a chaperone function only. Plants have homologs of ClpX as well, but these are targeted to mitochondria where they might associate with ClpP2 (Halperin *et al.*, 2001b). However, there is no experimental support for this suggestion yet.

7.3.3 FtsH protease

FtsH is a membrane-bound metalloprotease complex. Most available information on this protease comes from the *E. coli* enzyme and its three yeast mitochondrial orthologs, known as AAA proteases (cf. Langer, 2000). In *E. coli*, FtsH is the only ATP-dependent protease that is essential for survival. Unlike Clp protease, FtsH proteolytic and ATPase domains are found on the same 71-kDa polypeptide. The N-terminus contains one or two transmembrane α-helices that anchor the protein to the respective membrane: the cytoplasmic one in *E. coli* and the inner membrane in yeast mitochondria. The hydrophobic domain is followed by the ATPase domain, and the zinc-binding motif H-E-X-X-H, which serves as the active site of the enzyme, is found toward the C-terminus of the protein. FtsHs form hexamers, but although the crystal structure of the ATPase domain was determined (Krzywda *et al.*, 2002), the overall structure, and the relative arrangement of the proteolytic domain with respect to the ATPase one, is not known. Nevertheless, it is assumed that like other

ATP-dependent proteases, the active sites of FtsH are self-compartmentalized, and access to them is regulated by the ATPase domain of the protein.

Chloroplast FtsH is found in the thylakoid membrane with its functional domains exposed to the stroma (Lindahl *et al.*, 1996). Plant nuclear genomes contain multiple FtsH genes, at least 12 in *Arabidopsis*. (Four additional genes, encoding homologous proteins with impaired zinc-binding domains, are also found in the *Arabidopsis* genome (Sokolenko *et al.*, 2002). These could potentially function as chaperones, but not as proteases.) Out of the 12 FtsH genes, the products of 3 are targeted to mitochondria, whereas the other 9 are capable of entering the chloroplast, as revealed by transient expression assay of GFP (green fluorescent protein) fusions (Sakamoto *et al.*, 2003). However, separation of thylakoid membrane proteins by 2D-PAGE, followed by mass spectrometry analysis, revealed that only four isozymes were accumulated in *Arabidopsis* leaves grown under optimal conditions, FtsH1, FtsH2, FtsH5, and FtsH8. Out of these, FtsH2 is by far the most abundant species (Sinvany-Villalobo *et al.*, in press). It is possible that the other isozymes are expressed under different environmental conditions, but this has not been demonstrated yet. Similar to bacterial and mitochondrial FtsHs, *Arabidopsis* FtsHs also form hexamers. These are composed of at least FtsH2 and FtsH5 (Sakamoto *et al.*, 2003; Yu *et al.*, 2004), and possibly other FtsHs as well. However, the stoichiometry of the different subunits in the complex is not known. The apparent differential abundance of the different FtsHs suggests that homomeric complexes, probably of FtsH2, may also exist, but it is not known whether these have different functions from the heteromeric complexes or not.

7.3.4 DegP

Bacterial DegP (or HtrA) is a serine protease complex peripherally attached to the periplasmic side of the plasma membrane. It is characterized best in *E. coli*, where it is essential for survival at elevated temperatures (Strauch *et al.*, 1989; Lipinska *et al.*, 1990; Skorko-Glonek *et al.*, 1995; Pallen and Wren, 1997; Sassoon *et al.*, 1999). Determination of its three-dimensional structure revealed that it forms a hexamer made of two staggered trimers (Clausen *et al.*, 2002; Krojer *et al.*, 2002). The 48-kDa monomer is composed of two domains. The N-terminal one is the proteolytic domain, where the typical catalytic triad of serine proteases, Ser-Asp-His, is found. The C-terminus of the protein contains two PDZ domains, implicated in protein–protein interaction (Fanning and Anderson, 1996; Ponting, 1997), and substrate recognition and binding in the context of proteases (Levchenko *et al.*, 1997). DegP has two homologs in *E. coli*, DegQ and DegS, but these are less characterized (Kolmar *et al.*, 1996; Waller and Sauer, 1996). Another homolog of the DegP protease, designated HtrA2, is found in mitochondria, where it is involved in apoptosis. This enzyme forms a trimer whose three-dimensional structure was also determined (Gray *et al.*, 2000; Li *et al.*, 2002; Ramesh *et al.*, 2002).

Although the proteolytic activity of bacterial DegP is independent of ATP, it has a chaperone activity as well. Interestingly, at normal growth temperature DegP is active as a chaperone, whereas the proteolytic activity dominates at elevated temperatures (Spiess *et al.*, 1999). This temperature-dependent switch between the two different activities can be now explained by the structure of the protein. At normal growth temperature, the active site is blocked by segments of the protein itself, and only upon a thermal-induced conformational change it becomes accessible to substrates (Clausen *et al.*, 2002; Krojer *et al.*, 2002).

The first plant homolog of DegP, designated DegP1, was found peripherally attached to the lumenal side of the thylakoid membrane (Itzhaki *et al.*, 1998). Unlike the bacterial DegP that has two PDZ domains in tandem, DegP1 contains only one such domain. Similar to the bacterial enzyme, it forms hexamers and its activity is stimulated by high temperature (Chassin *et al.*, 2002). Modeling DegP1 structure based on the structures of *E. coli* DegP and the mitochondrial HtrA2 revealed that it fits better the mitochondrial one. As shown in Figure 7.1, access to the catalytic cleft is prevented by one loop of the proteolytic domain and the entire PDZ domain. Another two homologs, designated DegP5 and DegP8, were predicted to reside also

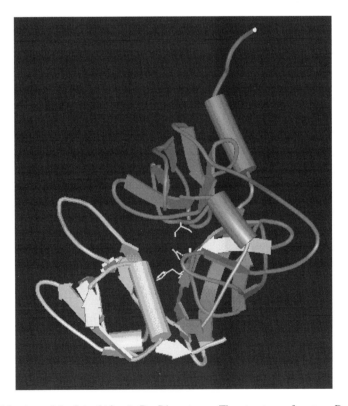

Figure 7.1 A model of *Arabidopsis* DegP1 protease. The structure of mature DegP1 was modeled based on the three-dimensional structure of the mitochondrial HtrA2 protease. The proteolytic domain is in dark gray, with the N-terminus of the protein at the top. The side chains of the catalytic triad Ser-Asp-His, from top to bottom, are indicated. The PDZ domain is in light gray (N. Adir and Z. Adam, unpublished data, 2003).

in the lumen, and proteomic analyses confirmed this prediction (Adam *et al.*, 2001; Peltier *et al.*, 2002; Schubert *et al.*, 2002). DegP5 is very similar to DegP1, but does not contain a PDZ domain, whereas DegP8 is somewhat less similar, but does have a similar domain, and hence, its mature size is almost identical to that of DegP1, ~35 kDa. Chloroplasts contain at least one more homolog, DegP2, peripherally attached to the stromal side of the thylakoid membrane (Haussuhl *et al.*, 2001). Other plant homologs are predicted to reside in mitochondria, and maybe other cellular sites (Adam *et al.*, 2001).

7.4   Predicted chloroplast proteases and peptidases

In addition to the above proteases and peptidases, chloroplasts should contain other less identified enzymes. One such protease, whose existence in chloroplasts was already demonstrated, is SppA. This is a light-induced serine protease that is peripherally attached to the thylakoid membrane on its stromal side (Lensch *et al.*, 2001). The existence of other proteases and peptidases in the chloroplast can be predicted from their N-terminal sequences by bioinformatic tools, but experimental evidence supporting these predictions is still lacking. Among these, one can find Lon protease, a homolog of the bacterial ATP-dependent serine protease. *Arabidopsis* contains four homologous genes, the products of two of them are expected to locate in mitochondria and the other two in chloroplasts (Adam *et al.*, 2001). The presence of a mitochondrial Lon was already demonstrated (Sarria *et al.*, 1998), and the presence of the chloroplast homolog is supported by immunoblot analysis (O. Ostersetzer and Z. Adam, unpublished data, 2001).

As for peptidases, two copies of a processing metallopeptidase are expected to be associated with chloroplast membranes, and so is one copy of a serine leader peptidase (Sokolenko *et al.*, 2002). Several aminopeptidases, mostly methionine aminopeptidases, are predicted to reside in the chloroplast stroma as well (Sokolenko *et al.*, 2002), but there is no further experimental support for these predictions.

7.5   Roles of identified proteases in development and maintenance

7.5.1   ClpCP

The first circumstantial evidence for the importance of chloroplast proteases for viability came from the analysis of the plastid genome of the nonphotosynthetic parasitic plant *Epifagus virginiana*. This genome has lost all its photosynthetic genes, but one of the few remaining genes was *clpP* (Depamphilis and Palmer, 1990; Wolfe *et al.*, 1992), suggesting that it plays an essential role in the function of plastids. This suggestion was corroborated when the *clpP* gene was found to be essential for the growth of *Chlamydomonas* (Huang *et al.*, 1994). This is also true for chloroplasts of higher plants. Complete segregation was impossible in attempts to disrupt the chloroplast *clpP* gene (Shikanai *et al.*, 2001). In this study, it was also demonstrated that etioplast development in the dark was inhibited, suggesting that ClpP was essential for nonphotosynthetic plastid functions as well. Complete removal of the chloroplast-encoded ClpP was recently achieved using the CRE/lox recombination system in tobacco (Kuroda and Maliga, 2003). This resulted in seedlings with white cotyledons that could not develop further even when grown on sucrose, suggesting a fundamental role for ClpP in shoot development.

Repression of the expression of nuclear-encoded ClpP also has negative effects on plant development. As shown in Figure 7.2, expression of an antisense construct of *clpP4* in *Arabidopsis* resulted in severe growth inhibition and lack of chloroplast development, especially in the mid-rib region of the leaf (B. Zheng and A.K. Clarke, personal communication, 2003). The regulatory subunit of the Clp complex,

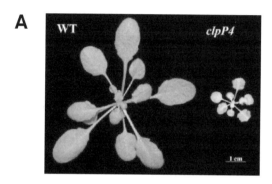

Figure 7.2 *Arabidopsis clpP4* antisense line. The full-length cDNA of *clpP4* was expressed in *Arabidopsis* plants in antisense orientation, and transgenic plants were screened for repression of the ClpP4 protein using a specific antibody (Zheng *et al.*, 2002). (A) Four-week-old wild-type (WT) and *clpP4* antisense lines. (B) Three-month-old *clpP4* antisense line. Courtesy of B. Zheng and A.K. Clarke.

Figure 7.3 *Arabidopsis clpC1* T-DNA insertion lines. Two independent T-DNA insertion lines in the *clpC1* gene, with insertions in the fourth and seventh exons, were obtained from the Slak and Syngenta collections, respectively. Six-week-old homozygous lines and wild type are presented. Courtesy of L.E. Sjogren and A.K. Clarke.

ClpC, is also essential for proper development. T-DNA insertion lines in the *clpC1* gene, one of the two *clpC* genes in *Arabidopsis*, show a severe chlorotic phenotype (Figure 7.3; L.E. Sjogren and A.K. Clarke, personal communication, 2003). These lines contain ~30% of ClpC, compared with wild-type plants, that can be attributed to the expression of the *clpC2* gene, which is not impaired.

Only little is known about ClpCP substrates at the biochemical level. Since the chloroplast-encoded ClpP could not be deleted from the chloroplast genome of *Chlamydomonas reinhardtii*, Majeran *et al.* (2000, 2001) attenuated the level of its expression by modifying its start codon. This manipulation resulted in a strain that was impaired in degradation of subunits of the cyt b_6/f complex, PSII subunits, and Rubisco under nutrient starvation and light stress. Other possible substrates of ClpCP are mistargeted proteins in the stroma. Intentional targeting of the lumenal OE33 protein to the stroma led to its degradation, both *in vitro* and *in vivo*, by an enzyme with characteristics of ClpCP, or the Clp complex that was immunoprecipitated from the stroma (Halperin and Adam, 1996; Halperin *et al.*, 2001a; Levy *et al.*, 2004).

7.5.2   FtsH

Analysis of variegated mutants in *Arabidopsis* revealed that mutations in the *ftsH2* and *ftsH5* genes were responsible for the *var2* and *var1* phenotypes, respectively (Chen *et al.*, 2000; Takechi *et al.*, 2000; Sakamoto *et al.*, 2002; Sakamoto, 2003). As can be seen in Figure 7.4, whereas the cotyledons of the FtsH2 mutant look like wild type, the first true leaves are yellow. Subsequent leaves are variegated, with decreasing ratio of yellow to green sectors, but yellow sectors can be seen even in mature plants. This phenotype suggests that FtSH is involved in early stages of chloroplast development, primarily in the development of thylakoids. However, the

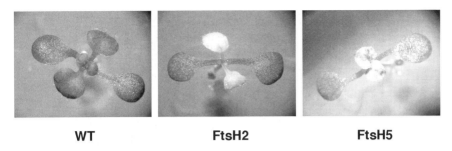

WT   FtsH2   FtsH5

Figure 7.4 *Arabidopsis* FtsH mutants. Seeds of wild type (WT) and mutants in the *ftsH2* and *ftsH5* genes were germinated on plates. Ten-day-old seedlings are presented. Courtesy of W. Sakamoto.

patchy nature of the phenotype suggests that the loss of FtsH2 can be compensated for, at least in the green sectors, probably by other FtsH proteins, and that the compensation mechanism becomes more and more efficient with time. The FtsH5 mutant phenotype is similar, but less pronounced (Sakamoto *et al.*, 2002). Here, the first true leaves are already variegated (Figure 7.4); variegation decreases in subsequent younger leaves; and mature plants cannot be distinguished from wild type. Interestingly, mutations in other FtsH genes, including *ftsH1* and *ftsH8*, whose products are found in thylakoids, have no phenotypic effects (Sakamoto *et al.*, 2003).

The first indication for chloroplast FtsH function came from an *in vitro* study. After import into isolated chloroplasts, unassembled Rieske Fe–S protein that accumulated on the stromal face of the thylakoid membrane, as well as a soluble mutant of this protein that accumulated in the stroma, was rapidly degraded (Ostersetzer and Adam, 1997). Characteristics of the degradation process were reminiscent of FtsH, and indeed, antibodies against the native protease could specifically inhibit degradation *in vitro*. Further support to the link between this proteolytic activity and FtsH came from column chromatography separation of solubilized thylakoid proteins. Activity assays and immunoblot analysis revealed that activity peaks always coincided with the presence of FtsH (Ostersetzer, 2001). Thus, FtsH is likely to play a central role in degradation of unassembled thylakoid proteins.

Another critical role attributed to FtsH protease is involvement in the degradation of oxidatively damaged thylakoid proteins. This has been demonstrated for the D1 protein of PSII reaction center, in the context of repair from photoinhibition. Oxidative damage to the D1 protein is considered the primary reason for photoinhibition of PSII. The repair cycle of PSII involves transcription and translation of new D1 protein copies, but these cannot be incorporated into the damaged complex unless the damaged copies are degraded. Thus, D1 protein degradation represents a key step in the repair of PSII from photoinhibition. Degradation of PSII takes place in at least two steps. The first one is a single proteolytic cleavage in the stromal loop connecting the fourth and the fifth transmembrane helices, to yield 23- and 10-kDa fragments, followed by complete degradation of these. An *in vitro* study suggested that FtsH was involved in the degradation of the 23-kDa fragment of the D1 protein. This degradation could be inhibited by removing FtsH from the membrane, and degradation could be restored by addition of recombinant FtsH (Lindahl *et al.*, 2000). The role of FtsH in the D1 protein degradation was supported by *in vivo* studies as well. The *Arabidopsis* FtsH2 mutant was found to be more susceptible to photoinhibition and demonstrated slower repair, as revealed by chlorophyll fluorescence measurements. In this plant, under conditions where synthesis of chloroplast-encoded proteins, including the D1 protein, was inhibited, damaged D1 protein accumulated in the mutant, whereas it rapidly disappeared in the wild-type plant (Bailey *et al.*, 2002).

7.5.3 DegP

Phenotypic consequences of disruption or inhibition of expression of chloroplast-targeted DegP gene products have not been documented yet. Thus, exploring the roles of DegP protease using mutant or transgenic plants is yet to come. The only reported activity of a chloroplast DegP to date is the *in vitro* study on DegP2 (Haussuhl *et al.*, 2001). Recombinant DegP2 was demonstrated to cleave the D1 protein of photoinhibited PSII, to yield the two typical fragments described above, whereas other thylakoid proteins were insensitive to the presence of protease. Thus, this enzyme is a strong candidate for performing the initial step in the D1 protein degradation.

Although recombinant DegP1 was capable of degrading lumenal proteins such as plastocyanin and OE33 *in vitro* (Chassin *et al.*, 2002), its role *in vivo*, as well as the role of the other two lumenal DegP proteins, DegP5 and DegP8, is yet to be determined. An appealing role could be participation in degradation of integral thylakoid membrane proteins. Degradation of these is poorly understood (see next section). Nevertheless, DegP proteases could theoretically facilitate degradation of these substrates by cleaving and degrading their lumenal exposed loops.

7.6 Degradation of integral membrane proteins

Thylakoid membranes are very rich in proteins, which are prone to light-induced oxidative damage that results in their inactivation. As was shown for photoinhibited PSII, their repair is likely to involve proteolytic degradation. However, the mechanisms involved in degradation of chloroplast integral membrane proteins, just as any other membrane protein, are poorly understood. Two models were suggested to explain this process: the "shedding" and the "pulling" models (Langer, 2000). Both are based on data obtained in studies on FtsH proteases in *E. coli* and yeast mitochondria. The "shedding model" suggests that membrane-bound ATP-dependent proteases degrade the solvent-exposed domains of their membrane substrates, leading to destabilization of hydrophobic domains, and making them accessible to proteolysis on the membrane surface. The "pulling model" suggests that these proteases extract the membrane-spanning helices from the lipid bilayer and degrade them in a hydrophilic environment.

Dislocation of intramembrane domains, as well as domains from the trans-side of the membrane, was demonstrated in *E. coli* and yeast mitochondria (Kihara *et al.*, 1999; Leonhard *et al.*, 2000), where the dislocation is attributed to the ATPase activity of the proteases. In yeast mitochondria, two different FtsH-like proteases (designated m-AAA and i-AAA) are oppositely oriented on the inner membrane (Leonhard *et al.*, 1996). They cooperate in degradation of membrane substrates, and degradation of substrates that span the membrane only once was shown to occur on either side of the membrane (Leonhard *et al.*, 2000). However, in the *E. coli* cytoplasmic membrane, as well as in thylakoid membranes, FtsH proteases are facing only one side of the membrane. This implies that pulling of substrates out of the membrane is likely to be unidirectional. Such a pulling could be facilitated by ATP-independent cleavage of hydrophilic loops on the opposite side of the membrane, as mentioned above, but the details of such cooperation have not been demonstrated yet. Moreover, cleavage of transmembrane helices within the lipid bilayer could theoretically also facilitate complete degradation of membrane substrates, but there is no experimental evidence to support this idea either.

A possible candidate for cleavage within the membrane is the Rhomboid protease. It is an intramembrane serine protease with substrates that are usually cleaved within a membrane-spanning segment. It was first described in *Drosophila*, where it cleaves the membrane-bound growth factor Spitz, allowing it to activate the EGF receptor (Urban and Freeman, 2002, 2003; Urban *et al.*, 2002a, b). *Drosophila* Rhomboid is an integral ER membrane protein, with six transmembranes helices (Urban *et al.*, 2001). Mutation analysis revealed that its catalytic triad is composed of Ser217, His281, and Asn169. (In typical serine proteases, Asp residue is found instead of Asn.) Interestingly, all these residues are located within transmembrane domains, suggesting that Rhomboid is an unusual serine protease with an active site within the membrane bilayer. This feature places Rhomboid within the framework of "regulated intramembrane proteolysis," a new paradigm of signal transduction, together with Presenilin-1, Notch, and S2P (cf. Koonin *et al.*, 2003).

Bioinformatics studies have revealed that Rhomboid proteases are ubiquitous, present in nearly all the sequenced genomes of archaea, bacteria, and eukaryotes (Koonin *et al.*, 2003). They all share six transmembrane helices and the catalytic triad Ser-His-Asn within the lipid bilayer. Whereas most prokaryotic species have a single gene coding for a Rhomboid protein, eukaryotes have evolved extended families, with seven members in *Drosophila* for instance, and at least eight in *Arabidopsis* (Koonin *et al.*, 2003).

The function of Rhomboid proteases has been shown very recently to extend beyond signal transduction. Yeast mitochondria contain two Rhomboids, one of them (Rbd1) is involved in processing the bipartite signal peptides of cytochrome *c* peroxidase (Ccp1) and a dynamin-like GTPase (Mgm1), both located in the intermembrane space of the mitochondria (Esser *et al.*, 2002; Herlan *et al.*, 2003; McQuibban *et al.*, 2003; van der Bliek and Koehler, 2003). Whereas typical residents of the intermembrane space such as cyt b_2 are cleaved first by the matrix processing peptidase (MPP) and then by the intermembrane space processing peptidase, Ccp1 and Mgm1 are cleaved first by the mAAA protease and MPP, respectively, and then by Rbd1. Thus, in addition to mediating signal transduction and processing, Rhomboids may have additional biological functions. A very plausible one would be to facilitate degradation of membrane proteins by cleaving their transmembrane α-helices.

The eight predicted Rhomboid proteins in *Arabidopsis* show homology in the regions of the three residues comprising the catalytic triad, and these residues are predicted to reside within the transmembrane α-helices. Out of the eight proteins, four appear to have chloroplast targeting sequences and two could be targeted to mitochondria. Thus, it is possible that chloroplasts contain intramembrane proteases that could function in precursor processing, like the aforementioned examples from mitochondria, but maybe to participate in the degradation of chloroplast integral membrane proteins as well. This suggestion will have to be tested experimentally.

7.7 Evolutionary aspects

The prokaryotic origin of the chloroplast is also manifested in its proteolytic machinery, as all components of this machinery are orthologs of bacterial proteases and peptidases. However, it is striking that whereas the great majority of bacterial proteases are encoded by single genes, chloroplast proteases belong to gene families. It appears that this multiplication is associated, at least partially, with the evolution of photosynthesis. Comparison of the proteolytic machinery in *E. coli* to those found in the photosynthetic cyanobacterium *Synechocystis* sp. PCC 6803 and the higher plant *Arabidopsis* reveals an interesting trend. Whereas ClpP is encoded by a single gene in *E. coli*, three such genes are found in *Synechocystis* and six in *Arabidopsis*. Its nonproteolytic homolog ClpR, which is absent from *E. coli*, is found in *Synechocystis* and *Arabidopsis* in one and four copies, respectively. ClpC and ClpX are found in single copies in *E. coli* and *Synechocystis*, whereas *Arabidopsis* has two copies of each. *E. coli* has a single copy of FtsH, *Synechocystis* has 4, and *Arabidopsis* 12. DegP has three related proteins in *E. coli*, and the same number is kept in *Synechocystis*. However, *Arabidopsis* has 14 homologs of this protease. It is not clear what the driving forces were that led to this multiplication of genes. A plausible explanation could be that the photosynthetic process is accompanied by an increasing risk of oxidative damage, generating a demand for more efficient and versatile proteolytic machinery to dispose of those proteins that are damaged beyond repair.

As already mentioned above, homologs of several chloroplast proteases are found in mitochondria as well. Phylogenetic analysis of *Arabidopsis* FtsH proteins, together with *Synechocystis* and yeast mitochondrial ones, reveals an interesting phenomenon. Seven out of the nine FtsHs that can be targeted to chloroplasts, FtsH1, FtsH2, and FtsH5–9 (Sakamoto *et al.*, 2003), are more related to the cyanobacterial proteins than to the yeast mitochondrial ones (see Figure 7.5),

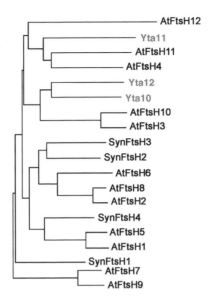

Figure 7.5 A phylogenetic tree of FtsH proteins. *Arabidopsis* (At) FtsHs, *Synechocystis* sp. PCC 6803 (Syn) FtsHs, and the yeast mitochondrial FtsH-like proteins (Yta10–12) are shown.

suggesting that they are all descendents of the cyanobacterial progenitor of the chloroplast. Apparently, FtsHs already started to diverge in this progenitor and higher plants inherited three different forms of these. In almost all cases the higher plant ones were duplicated later on. The three FtsHs that were found in mitochondria, FtsH3, FtsH4, and FtsH10 (Sakamoto *et al.*, 2003), are more related to those found in yeast mitochondria, suggesting that these have derived from α-proteobacteria, the progenitor of nowadays mitochondria. However, retargeting might also have occurred in plants, as FtsH11 and FtsH12, which are more related to the mitochondrial isozymes, are found in chloroplasts (Sakamoto *et al.*, 2003).

Another interesting case is that of Lon protease. Whereas Lon is found in bacteria such as *E. coli*, it is absent from the *Synechocystis* genome. Nevertheless, it is found both in plant mitochondria and in chloroplasts. Here again, the enzyme might have been inherited from α-proteobacteria, duplicated in plants, and then at least one of these gene products might have been redirected to chloroplasts instead of mitochondria. Such a redirection might have occurred for ClpP2 and the ClpXs as well.

7.8 Future prospects

In the past decade, we have witnessed the identification of several chloroplast proteases and peptidases. The plant sequences that have accumulated throughout these years, and the completion of the *Arabidopsis* and rice genome sequencing projects, suggest that the identity of most, if not all, chloroplast proteases and peptidases is now known. It appears that the upcoming challenge is to assign specific functions to each one of the identified proteins. The function of two peptidases, SPP and TPP, is apparently clear – they are responsible for the maturation of imported proteins in the stroma and the lumen, respectively. CtpA also functions in protein maturation. However, the function of a number of other peptidase, especially aminopeptidases, which were identified primarily by analysis of sequence data, is not known. One can assume that they are involved in degradation of short peptides, the products of ATP-dependent proteolysis, but such a function still needs to be demonstrated. Whether they have other more specific substrates and additional functions is another open question.

The degradation of the D1 protein of PSII reaction center, in the context of repair from photoinhibition, has been the focal point of numerous studies for more than 20 years. Only in recent years has the identity of the proteases involved started to unravel. Although the details of this process are still far from being fully understood, there is now experimental evidence for the involvement of two proteases, FtsH and DegP2, in this process. Another challenge in the field will be to link characterized proteolytic processes, such as degradation of LHCII or ELIP during acclimation to changes in light intensities, to specific proteases. In this case, a reverse genetics approach might be useful. Specific protease mutants will have to be tested for their ability to degrade specific proteins under given conditions.

The multiplication of genes within the different protease families is now evident. However, what is the functional significance of this is not clear. Two alternative possibilities should be considered. Within each gene family, the products that are targeted to the same subcompartment fulfill the same functions. The evolutionary driving force in this case might have been the essential functions fulfilled by a given protease that led to this redundancy. In this case, the differential importance of different gene products, as revealed by mutant analysis of FtsHs, is a result of differential expression levels, and not necessarily of different functions. Alternatively, different gene products within a family may have acquired specialized functions during evolution. It is possible that each one of these two alternatives describes better the relationship within a different family. A conclusive explanation will have to await a detailed examination of each gene family. Such studies are likely to shed additional light on the different functions of chloroplasts and the role of proteases and peptidases in controlling and maintaining these functions.

References

Adam, Z. (1996) Protein stability and degradation in chloroplasts. *Plant Mol. Biol.*, 32, 773–783.

Adam, Z. (2000) Chloroplast proteases: possible regulators of gene expression? *Biochimie*, 82, 647–654.

Adam, Z., Adamska, I., Nakabayashi, K. *et al.* (2001) Chloroplast and mitochondrial proteases in *Arabidopsis*. A proposed nomenclature. *Plant Physiol.*, 125, 1912–1918.

Adam, Z., and Clarke, A.K. (2002) Cutting edge of chloroplast proteolysis. *Trends Plant Sci.*, 7, 451–456.

Allen, J.F. (1992) How does protein phosphorylation regulate photosynthesis? *Trends Biochem. Sci.*, 17, 12–17.

Andersson, B. and Aro, E.-M. (1997) Proteolytic activities and proteases of plant chloroplasts. *Physiol. Plant*, 100, 780–793.

Andersson, B. and Aro, E.-M. (2001) Photodamage and D1 protein turnover in photosystem II. In *Regulation of Photosynthesis*, Kluwer Academic Publishers, Dordrecht, pp. 377–393.

Apel, K. and Kloppstech, K. (1980) The effect of light in the biosynthesis of the light-harvesting chlorophyll *a/b* protein. Evidence for the stabilization of the apoprotein. *Planta*, 150, 426–430.

Bailey, S., Thompson, E., Nixon, P.J. *et al.* (2002) A critical role for the Var2 FtsH homologue of *Arabidopsis thaliana* in the photosystem II repair cycle *in vivo*. *J. Biol. Chem.*, 277, 2006–2011.

Barber, J. and Andersson, B. (1992) Too much of a good thing: light can be bad for photosynthesis. *Trends Biochem. Sci.*, 17, 61–66.

Bennett, J. (1981) Biosynthesis of the light-harvesting chlorophyll *a/b* protein. Polypeptide turnover in darkness. *Eur. J. Biochem.*, 118, 61–70.

Bhushan, S., Lefebvre, B., Stahl, A. *et al.* (2003) Dual targeting and function of a protease in mitochondria and chloroplasts. *EMBO Rep.*, 4, 1073–1078.

Chaal, B.K., Mould, R.M., Barbrook, A.C., Gray, J.C. and Howe, C.J. (1998) Characterization of a cDNA encoding the thylakoidal processing peptidase from *Arabidopsis thaliana*. Implications for the origin and catalytic mechanism of the enzyme. *J. Biol. Chem.*, 273, 689–692.

Chassin, Y., Kapri-Pardes, E., Sinvany, G., Arad, T. and Adam, Z. (2002) Expression and characterization of the thylakoid lumen protease DegP1 from *Arabidopsis thaliana*. *Plant Physiol.*, 130, 857–864.

Chen, M., Choi, Y., Voytas, D.F. and Rodermel, S. (2000) Mutations in the *Arabidopsis* VAR2 locus cause leaf variegation due to the loss of a chloroplast FtsH protease. *Plant J.*, 22, 303–313.

Clarke, A.K. (1999) ATP-dependent Clp proteases in photosynthetic organisms – a cut above the rest! *Ann. Bot.*, 83, 593–599.

Clausen, T., Southan, C. and Ehrmann, M. (2002) The HtrA family of proteases: implications for protein composition and cell fate. *Mol. Cell*, 10, 443–455.

Dalbey, R.E. and Robinson, C. (1999) Protein translocation into and across the bacterial plasma membrane and the plant thylakoid membrane. *Trends Biochem. Sci.*, 24, 17–22.

Depamphilis, C.W. and Palmer, J.D. (1990) Loss of photosynthetic and chlororespiratory genes from the plastid genome of a parasitic flowering plant. *Nature*, 348, 337–339.

Esser, K., Tursun, B., Ingenhoven, M., Michaelis, G. and Pratje, E. (2002) A novel two-step mechanism for removal of a mitochondrial signal sequence involves the mAAA complex and the putative rhomboid protease Pcp1. *J. Mol. Biol.*, 323, 835–843.

Fanning, A.S. and Anderson, J.M. (1996) Protein–protein interactions: PDZ domain networks. *Curr. Biol.*, 6, 1385–1388.

Gottesman, S. (1996) Proteases and their targets in *Escherichia coli*. *Annu. Rev. Genet.*, 30, 465–506.

Gray, C.W., Ward, R.V., Karran, E. *et al.* (2000) Characterization of human HtrA2, a novel serine protease involved in the mammalian cellular stress response. *Eur. J. Biochem.*, 267, 5699–5710.

Halperin, T. and Adam, Z. (1996) Degradation of mistargeted OEE33 in the chloroplast stroma. *Plant Mol. Biol.*, 30, 925–933.

Halperin, T., Ostersetzer, O. and Adam, Z. (2001a) ATP-dependent association between subunits of Clp protease in pea chloroplasts. *Planta*, 213, 614–619.

Halperin, T., Zheng, B., Itzhaki, H., Clarke, A.K. and Adam, Z. (2001b) Plant mitochondria contain proteolytic and regulatory subunits of the ATP-dependent Clp protease. *Plant Mol. Biol.*, 45, 461–468.

Haussuhl, K., Andersson, B. and Adamska, I. (2001) A chloroplast DegP2 protease performs the primary cleavage of the photodamaged D1 protein in plant photosystem II. *EMBO J.*, 20, 713–722.

Herlan, M., Vogel, F., Bornhovd, C., Neupert, W. and Reichert, A.S. (2003) Processing of Mgm1 by the Rhomboid-type protease Pcp1 is required for maintenance of mitochondrial morphology and of mitochondrial DNA. *J. Biol. Chem.*, 278, 27781–27788.

Huang, C., Wang, S., Lemieux, C., Otis, C., Turmel, M. and Liu, X.Q. (1994) The *Chlamydomonas* chloroplast *clpP* gene contains translated large insertion sequences and is essential for cell growth. *Mol. Gen. Genet.*, 244, 151–159.

Inagaki, N., Yamamoto, Y., Mori, H. and Satoh, K. (1996) Carboxyl-terminal processing protease for the D1 precursor protein: cloning and sequencing of the spinach cDNA. *Plant Mol. Biol.*, 30, 39–50.

Itzhaki, H., Naveh, L., Lindahl, M., Cook, M. and Adam, Z. (1998). Identification and characterization of DegP, a serine protease associated with the luminal side of the thylakoid membrane. *J. Biol. Chem.*, 273, 7094–7098.

Jarvis, P. and Soll, J. (2002) Toc, tic, and chloroplast protein import. *Biochim. Biophys. Acta*, 1590, 177–189.

Kihara, A., Akiyama, Y. and Ito, K. (1999) Dislocation of membrane proteins in FtsH-mediated proteolysis. *EMBO J.*, 18, 2970–2981.

Kim, J., Eichacker, L.A., Rudiger, W. and Mullet, J.E. (1994) Chlorophyll regulates accumulation of the plastid-encoded chlorophyll proteins P700 and D1 by increasing apoprotein stability. *Plant Physiol.*, 104, 907–916.

Kolmar, H., Waller, P.R.H. and Sauer, R.T. (1996) The DegP and DegQ periplasmic endoproteases of *Escherichia coli*: specificity for cleavage sites and substrate conformation. *J. Bacteriol.*, 178, 5925–5929.

Koonin, E.V., Makarova, K.S., Rogozin, I.B., Davidovic, L., Letellier, M.C. and Pellegrini, L. (2003) The rhomboids: a nearly ubiquitous family of intramembrane serine proteases that probably evolved by multiple ancient horizontal gene transfers. *Genome Biol.*, 4, R19.

Krojer, T., Garrido-Franco, M., Huber, R., Ehrmann, M. and Clausen, T. (2002) Crystal structure of DegP (HtrA) reveals a new protease-chaperone machine. *Nature*, 416, 455–459.

Krzywda, S., Brzozowski, A.M., Verma, C., Karata, K., Ogura, T. and Wilkinson, A.J. (2002) The crystal structure of the AAA domain of the ATP-dependent protease FtsH of *Escherichia coli* at 1.5 Å resolution. *Structure*, 10, 1073–1083.

Kuroda, H. and Maliga, P. (2003) The plastid *clpP1* protease gene is essential for plant development. *Nature*, 425, 86–89.

Kyle, D.J., Ohad, I. and Arntzen, C.J. (1984) Membrane protein demage and repair: selective loss of a quinone–protein function in chloroplast membranes. *Proc. Natl. Acad. Sci. U.S.A.*, 81, 4070–4074.

Langer, T. (2000) AAA proteases: cellular machines for degrading membrane proteins. *Trends Biochem. Sci.*, 25, 247–251.

Lensch, M., Herrmann, R.G. and Sokolenko, A. (2001) Identification and characterization of SppA, a novel light-inducible chloroplast protease complex associated with thylakoid membranes. *J. Biol. Chem.* 276, 33645–33651.

Leonhard, K., Guiard, B., Pellecchia, G., Tzagoloff, A., Neupert, W. and Langer, T. (2000) Membrane protein degradation by AAA proteases in mitochondria: extraction of substrates from either membrane surface. *Mol. Cell*, 5, 629–638.

Leonhard, K., Herrmann, J.M., Stuart, R.A., Mannhaupt, G., Neupert, W. and Langer, T. (1996) AAA proteases with catalytic sites on opposite membrane surfaces comprise a proteolytic system for the ATP-dependent degradation of inner membrane proteins in mitochondria. *EMBO J.*, 15, 4218–4229.

Leto, K.J., Bell, E. and McIntosh, L. (1985) Nuclear mutation leads to an accelerated turnover of chloroplast-encoded 48 kd and 34.5 kd polypeptides in thylakoids lacking photosystem II. *EMBO J.*, 4, 1645–1653.

Levchenko, I., Smith, C.K., Walsh, N.P., Sauer, R.T. and Baker, T.A. (1997) PDZ-like domains mediate binding specificity in the Clp/Hsp100 family of chaperones and protease regulatory subunits. *Cell*, 91, 939–947.

Levy, M., Bachmair, A. and Adam, Z. (2004) A single recessive mutation in the proteolytic machinery of *Arabidopsis* chloroplasts impairs photoprotection and photosynthesis upon cold stress. *Planta*, 218, 396–405.

Li, H.H. and Merchant, S. (1995) Degradation of plastocyanin in copper-deficient *C. reinhardtii*–evidence for a protease-susceptible conformation of the apoprotein and regulated proteolysis. *J. Biol. Chem.*, 270, 23504–23510.

Li, W., Srinivasula, S.M., Chai, J. *et al.* (2002) Structural insights into the pro-apoptotic function of mitochondrial serine protease HtrA2/Omi. *Nat. Struct. Biol.*, 9, 436–441.

Lindahl, M., Spetea, C., Hundal, T., Oppenheim, A.B., Adam, Z. and Andersson, B. (2000) The thylakoid FtsH protease plays a role in the light-induced turnover of the photosystem II D1 protein. *Plant Cell*, 12, 419–431.

Lindahl, M., Tabak, S., Cseke, L., Pichersky, E., Andersson, B. and Adam, Z. (1996) Identification, characterization, and molecular cloning of a homologue of the bacterial FtsH protease in chloroplasts of higher plants. *J. Biol. Chem.*, 271, 29329–29334.

Lindahl, M., Yang, D.H. and Andersson, B. (1995) Regulatory proteolysis of the major light-harvesting chlorophyll *a/b* protein of photosystem II by a light-induced membrane-associated enzymic system. *Eur. J. Biochem.*, 231, 503–509.

Lipinska, B., Zylicz, M. and Georgopoulos, C. (1990) The HtrA (DegP) protein, essential for *Escherichia coli* survival at high temperatures, is an essential endopeptidase. *J. Bacteriol.*, 172, 1791–1797.

Majeran, W., Olive, J., Drapier, D., Vallon, O. and Wollman, F.A. (2001). The light sensitivity of ATP synthase mutants of *Chlamydomonas reinhardtii*. *Plant Physiol.*, 126, 421–433.

Majeran, W., Wollman, F.-A. and Vallon, O. (2000). Evidence for a role of ClpP in the degradation of the chloroplast cytochrome b_6f complex. *Plant Cell*, 12, 137–149.

Matile, P. (2001) Senescence and cell death in plant development: chloroplast senescence and its regulation. In *Regulation of Photosynthesis*, Kluwer Academic Publishers, Dordrecht, pp. 277–296.

Mattoo, A.K., Hoffman-Falk, H., Marder, J.B. and Edelman, M. (1984). Regulation of protein metabolism: coupling of photosynthetic electron transport to *in vivo* degradation of the rapidly metabolized 32-kilodalton protein of the chloroplast membranes. *Proc. Natl. Acad. Sci. U.S.A.*, 81, 1380–1384.

McQuibban, G.A., Saurya, S. and Freeman, M. (2003) Mitochondrial membrane remodelling regulated by a conserved rhomboid protease. *Nature*, 423, 537–541.

Melis, A. (1999) Photosystem-II damage and repair cycle in chloroplasts: what modulates the rate of photodamage? *Trends Plant Sci.*, 4, 130–135.

Merchant, S. and Bogorad, L. (1986) Rapid degradation of apoplastocyanin in Cu(II)-deficient cells of *Chlamydomonas reinhardtii*. *J. Biol. Chem.*, 261, 15850–15853.

Mullet, J.E., Klein, P.G. and Klein, R.R. (1990). Chlorophyll regulates accumulation of the plastid-encoded chlorophyll apoprotein-CP43 and apoprotein-D1 by increasing apoprotein stability. *Proc. Natl. Acad. Sci. U.S.A.*, 87, 4038–4042.

Oelmuller, R., Herrmann, R.G. and Pakrasi, H.B. (1996) Molecular studies of CtpA, the carboxyl-terminal processing protease for the D1 protein of the photosystem II reaction center in higher plants. *J. Biol. Chem.*, 271, 21848–21852.

Ostersetzer, O. (2001) *Characterization of the Proteolytic Machinery in Thylakoid Membranes of Higher Plants*. Ph.D. thesis, The Hebrew University.

Ostersetzer, O. and Adam, Z. (1997) Light-stimulated degradation of an unassembled Rieske FeS protein by a thylakoid-bound protease: the possible role of the FtsH protease. *Plant Cell*, 9, 957–965.

Ostersetzer, O., Tabak, S., Yarden, O., Shapira, R. and Adam, Z. (1996) Immunological detection of proteins similar to bacterial proteases in higher plant chloroplasts. *Eur. J. Biochem.*, 236, 932–936.

Pallen, M.J. and Wren, B.W. (1997) The HtrA family of serine proteases. *Mol. Microbiol.*, 26, 209–221.

Peltier, J.-B., Emanuelsson, O., Kalume, D.E. *et al.* (2002) Central functions of the lumenal and peripheral thylakoid proteome of *Arabidopsis* determined by experimentation and genome-wide prediction. *Plant Cell*, 14, 211–236.

Peltier, J.-B., Ripoll, D.R., Friso, G. *et al.* (2004) Clp protease complexes from photosynthetic and non-photosynthetic plastids and mitochondria of plants, their predicted 3-D structures and functional implications. *J. Biol. Chem.*, 279, 4768–4781.

Peltier, J.-B., Ytterberg, J., Liberles, D.A., Roepstorff, P. and van Wijk, K.J. (2001) Identification of a 350 kDa ClpP protease complex with 10 different Clp isoforms in chloroplasts of *Arabidopsis thaliana*. *J. Biol. Chem.*, 276, 16318–16327.

Ponting, C.P. (1997) Evidence for PDZ domains in bacteria, yeast, and plants. *Protein Sci.*, 6, 464–468.

Porankiewicz, J., Wang, J. and Clarke, A.K. (1999) New insights into the ATP-dependent Clp protease: *Escherichia coli* and beyond. *Mol. Microbiol.*, 32, 449–458.

Prasil, O., Adir, N. and Ohad, I. (1992) Dynamics of photosystem II: mechanism of photoin-hibition and recovery processes. In *Topics in Photosynthesis*, Elsevier Science Publishers, Amsterdam, pp. 295–348.

Ramesh, H., Srinivasula, S.M., Zhang, Z. *et al.* (2002) Identification of Omi/HtrA2 as a mitochondrial apoptotic serine protease that disrupts inhibitor of apoptosis protein–caspase interaction. *J. Biol. Chem.*, 277, 432–438.

Richter, S. and Lamppa, G.K. (1998) A chloroplast processing enzyme functions as the general stromal processing peptidase. *Proc. Natl. Acad. Sci. U.S.A.*, 95, 7463–7468.

Richter, S. and Lamppa, G.K. (1999) Stromal processing peptidase binds transit peptides and initiates their ATP-dependent turnover in chloroplasts. *J. Cell Biol.*, 147, 33–43.

Richter, S. and Lamppa, G.K. (2003) Structural properties of the chloroplast stromal processing peptidase required for its function in transit peptide removal. *J. Biol. Chem.*, 278, 39497–39502.

Sakamoto, W. (2003) Leaf-variegated mutations and their responsible genes in *Arabidopsis thaliana*. *Genes Genet. Syst.*, 78, 1–9.

Sakamoto, W., Tamura, T., Hanba-Tomita, Y. and Sodmergen M. Murata (2002) The VAR1 locus of *Arabidopsis* encodes a chloroplastic FtsH and is responsible for leaf variegation in the mutant alleles. *Genes Cells*, 7, 769–780.

Sakamoto, W., Zaltsman, A., Adam, Z. and Takahashi, Y. (2003) Coordinated regulation and complex formation of VAR1 and VAR2, chloroplastic FtsH metalloproteases involved in the repair cycle of photosystem II in *Arabidopsis* thylakoid membranes. *Plant Cell.*, 15, 2843–2855.

Sarria, R., Lyznik, A., Vallejos, C.E. and Mackenzie, S.A. (1998) A cytoplasmic male sterility-associated mitochondrial peptide in common bean is post-translationally regulated. *Plant Cell*, 10, 1217–1228.

Sassoon, N., Arie, J.P. and Betton, J.M. (1999) PDZ domains determine the native oligomeric structure of the DegP (HtrA) protease. *Mol. Microbiol.*, 33, 583–589.

Sato, S., Nakamura, Y., Kaneko, T., Asamizu, E. and Tabata, S. (1999) Complete structure of the chloroplast genome of *Arabidopsis* thaliana. *DNA Res.*, 6, 283–290.

Schmidt, G.W. and Mishkind, M.L. (1983) Rapid degradation of unassembled ribulose 1,5-biphosphate carboxylase small subunit in chloroplasts. *Proc. Natl. Acad. Sci. U.S.A.*, 80, 2632–2636.

Schubert, M., Petersson, U.A., Haas, B.J., Funk, C., Schroder, W.P. and Kieselbach, T. (2002) Proteome map of the chloroplast lumen of *Arabidopsis thaliana*. *J. Biol. Chem.*, 277, 8354–8365.

Shanklin, J., Dewitt, N.D. and Flanagan, J.M. (1995) The stroma of higher plant plastids contain ClpP and ClpC, functional homologs of *Escherichia coli* ClpP and ClpA: an archetypal two-component ATP-dependent protease. *Plant Cell*, 7, 1713–1722.

Shestakov, S.V., Anbudurai, P.R., Stanbekova, G.E., Gadzhiev, A., Lind, L.K. and Pakrasi, H.B. (1994) Molecular cloning and characterization of the ctpA gene encoding a carboxyl-terminal processing protease – analysis of a spontaneous photosystem II-deficient mutant strain of the cyanobacterium *Synechocystis* sp. PCC 6803. *J. Biol. Chem.*, 269, 19354–19359.

Shikanai, T., Shimizu, K., Ueda, K., Nishimura, Y., Kuroiwa, T. and Hashimoto, T. (2001) The chloroplast clpP gene, encoding a proteolytic subunit of ATP-dependent protease, is indispensable for chloroplast development in tobacco. *Plant Cell Physiol.*, 42, 264–273.

Sinvany-Villalobo, G., Davydoy, O., Ben-Ari, G., Zaltsman, A., Raskind, A. and Adam, Z. (in press) Expression in multi-gene families: analysis of chloroplast and mitochondrial proteases. *Plant Physiol.*

Skorko-Glonek, J., Wawrzynow, A., Krzewski, K., Kurpierz, K. and Lipinska, B. (1995) Site-directed mutagenesis of the HtrA(DegP) serine protease, whose proteolytic activity is indispensable for escherichia coli survival at elevated temperatures. *Gene*, 163, 47–52.

Sokolenko, A., Pojidaeva, E., Zinchenko, V. *et al.* (2002) The gene complement for proteolysis in the cyanobacterium *Synechocystis* sp. PCC 6803 and *Arabidopsis thaliana* chloroplasts. *Curr. Genet.*, 41, 291–310.

Spiess, C., Beil, A. and Ehrmann, M. (1999). A temperature-dependent switch from chaperone to protease in a widely conserved heat shock protein. *Cell*, 97, 339–347.

Strauch, K.L., Johnson, K. and Beckwith, J. (1989) Characterization of *degP*, a gene required for proteolysis in the cell envelope and essential for growth of *Escherichia coli* at high temperatures. *J. Bacteriol.*, 171, 2689–2696.

Takechi, K., Sodmergen, M. Murata, Motoyoshi, F. and Sakamoto, W. (2000) The YELLOW VARIEGATED (VAR2) locus encodes a homologue of FtsH, an ATP-dependent protease in *Arabidopsis*. *Plant Cell Physiol.*, 41, 1334–1346.

The *Arabidopsis* Genome Initiative (2000) Analysis of the genome sequence of the flowering plant *Arabidopsis thaliana*. *Nature*, 408, 796–815.

Tjus, S.E., Moller, B.L. and Scheller, H.V. (1998) Photosystem I is an early target of photoinhibition in barley illuminated at chilling temperatures. *Plant Physiol.*, 116, 755–764.

Trost, J.T., Chisholm, D.A., Jordan, D.B. and Diner, B.A. (1997). The D1 C-terminal processing protease of photosystem II from *Scenedesmus obliquus*. Protein purification and gene characterization in wild type and processing mutants. *J. Biol. Chem.*, 272, 20348–20356.

Urban, S. and Freeman, M. (2002) Intramembrane proteolysis controls diverse signalling pathways throughout evolution. *Curr. Opin. Genet. Dev.*, 12, 512–518.

Urban, S. and Freeman, M. (2003) Substrate specificity of rhomboid intramembrane proteases is governed by helix-breaking residues in the substrate transmembrane domain. *Mol. Cell*, 11, 1425–1434.

Urban, S., Lee, J.R. and Freeman, M. (2001) Drosophila rhomboid-1 defines a family of putative intramembrane serine proteases. *Cell*, 107, 173–182.

Urban, S., Lee, J.R. and Freeman, M. (2002a) A family of Rhomboid intramembrane proteases activates all Drosophila membrane-tethered EGF ligands. *EMBO J.*, 21, 4277–4286.

Urban, S., Schlieper, D. and Freeman, M. (2002b) Conservation of intramembrane proteolytic activity and substrate specificity in prokaryotic and eukaryotic rhomboids. *Curr. Biol.*, 12, 1507–1512.

van der Bliek, A.M. and Koehler, C.M. (2003) A mitochondrial rhomboid protease. *Dev. Cell*, 4, 769–70.

Vandervere, P.S., Bennett, T.M., Oblong, J.E. and Lamppa, G.K. (1995) A chloroplast processing enzyme involved in precursor maturation shares a zinc-binding motif with a recently
recognized family of metalloendopeptidases. *Proc. Natl. Acad. Sci. U.S.A.*, 92, 7177–7181.

Wada, M., Kagawa, T. and Sato, Y. (2003) Chloroplast movement. *Annu. Rev. Plant Biol.*, 54,
455–468.

Waller, P.R.H. and Sauer, R.T. (1996) Characterization of degQ and degS, *Escherichia coli* genes
encoding homologs of the DegP protease. *J. Bacteriol.*, 178, 1146–1153.

Wang, J., Hartling, J.A. and Flanagan, J.M. (1997) The structure of ClpP at 2.3 Å resolution
suggests a model for ATP-dependent proteolysis. *Cell*, 91, 447–456.

Wolfe, K.H., Morden, C.W. and Palmer, J.D. (1992) Function and evolution of a minimal plastid
genome from a nonphotosynthetic parasitic plant. *Proc. Natl. Acad. Sci. U.S.A.*, 89, 10648–
10652.

Yamamoto, Y., Inagaki, N. and Satoh, K. (2001) Overexpression and characterization of carboxyl-
terminal processing protease for precursor D1 protein: regulation of enzyme–substrate interaction by molecular environments. *J. Biol. Chem.*, 276, 7518–7525.

Yang, D.H., Webster, J., Adam, Z., Lindahl, M. and Andersson, B. (1998) Induction of acclimative
proteolysis of the light-harvesting chlorophyll a/b protein of photosystem II in response to
elevated light intensities. *Plant Physiol.*, 118, 827–834.

Yu, F., Park, S. and Rodermel, S.R. (2004) The *Arabidopsis* FtsH metalloprotease gene family:
interchangeability of subunits in chloroplast oligomeric complexes. *Plant J.*, 37, 864–876.

Zheng, B., Halperin, T., Hruskova-Heidingsfeldova, O., Adam, Z. and Clarke, A.K. (2002).
Characterization of chloroplast Clp proteins in *Arabidopsis*: localization, tissue specificity
and stress responses. *Physiol. Plant*, 114, 92–101.

Zhong, R., Wan, J., Jin, R. and Lamppa, G. (2003) A pea antisense gene for the chloroplast stromal
processing peptidase yields seedling lethals in *Arabidopsis*: survivors show defective GFP
import *in vivo*. *Plant J.*, 34, 802–812.

8 Regulation of nuclear gene expression by plastid signals

John C. Gray

8.1 Introduction

Plastid biogenesis requires the coordinated expression of genes distributed between the nuclear and plastid genomes. The plastid genome of most higher plants consists of circular, double-stranded DNA molecules of 120–160 kbp, encoding approximately 80 proteins necessary for photosynthesis or the plastid genetic system. However, these plastid-encoded proteins make up a very small proportion of the proteins required to produce functional chloroplasts. It has been estimated that the nuclear genome of *Arabidopsis* encodes nearly 3000 proteins targeted to plastids (Abdallah *et al.*, 2000), and it can therefore be envisaged that control systems will be necessary to ensure coordinated gene expression. All known plastid gene products are components of multi-subunit complexes containing at least one nuclear gene product. This includes Rubisco (ribulose 1,5-bisphosphate carboxylase), photosystems I and II, the cytochrome *bf* complex, ATP synthase, the NADH dehydrogenase complex, acetyl-CoA carboxylase, RNA polymerase, ribosomes and the Clp protease (Herrmann *et al.*, 1992). Each of these complexes has a distinct stoichiometry of plastid- and nuclear-encoded subunits, each of which has to be synthesised in sufficient amounts for complex assembly. Initially, it was suggested that this was determined exclusively at the level of assembly, with the amount of the complex assembled being determined by the amount of the nuclear-encoded, cytoplasmically synthesised subunits (Ellis, 1977). However, evidence soon accumulated showing that the functional or developmental state of the plastids was able to influence the expression of nuclear genes encoding plastid proteins. This retrograde signalling from plastids is the subject of this chapter.

However, it must be recognised that retrograde signalling occurs in a background of anterograde signalling from the nucleus to the plastids. Anterograde signalling takes many forms: from the provision of regulatory components of the plastid genetic system (such as sigma factors for RNA polymerase, and proteins regulating transcript stability and translation initiation) to components of the signal-generating systems in the plastids. The integration of retrograde and anterograde signalling systems is likely to be essential for the assembly and maintenance of functional plastids. There is now considerable evidence for multiple plastid signals operating at different stages of plant development. This chapter is organised in sections attempting to answer specific questions relating to plastid signalling: What is the evidence for plastid signalling? Which genes are regulated by plastid signals? What are plastid signals?

and How do plastid signals work? I hope the answers will illuminate our current knowledge of plastid signalling and identify areas where more research is needed. The chapter has benefited from several illuminating reviews and commentary articles that have appeared in the last few years (Rodermel, 2001; Surpin *et al.*, 2002; Jarvis, 2003; Pfannschmidt, 2003; Rodermel and Park, 2003).

8.2   What is the evidence for plastid signalling?

The first evidence for an effect of plastids on the expression of nuclear genes for plastid components was obtained in the late 1970s using barley mutants (Bradbeer *et al.*, 1979). In the following 10 years, studies using mutants or inhibitors whose primary site of action affected plastid function produced a sufficient body of evidence to be addressed by two influential reviews (Oelmüller, 1989; Taylor, 1989). Subsequently, a large body of evidence has supported the concept of retrograde signalling and the recognition that there must be multiple plastid signals.

8.2.1   Evidence from mutants

Bradbeer *et al.* (1979) observed that the activities of phosphoribulokinase and NADP[+]-glyceraldehyde 3-phosphate dehydrogenase in white leaf tissue of the barley mutants *albostrians* and *Saskatoon* were much lower than in green leaf tissue. The white mutant tissue contained plastids deficient in 70S ribosomes and was unable to synthesise and assemble the plastid-encoded components of the photosynthetic apparatus. This defect was proposed to result in decreased expression of the nuclear genes encoding the Calvin cycle enzymes. Subsequent studies with the barley *albostrians* mutant, using RNA gel blots, have shown that transcripts of a large number of genes encoding photosynthesis-related proteins are depleted in the white tissue, compared to the green tissue (Hess *et al.*, 1991, 1994).

Similar results have been obtained with other mutants affecting plastid processes. Mutants of maize, barley, tomato and *Arabidopsis* affecting carotenoid biosynthesis show dramatic effects on the expression of photosynthesis-related nuclear genes. The carotenoid-deficient maize mutants *lw*, *w3*, *w1-47*, *vp2* and *vp7* were shown to have decreased amounts of *Lhcb* transcripts encoding the light-harvesting chlorophyll proteins of photosystem II in light-grown seedlings (Harpster *et al.*, 1984; Mayfield and Taylor, 1984, 1987; Burgess and Taylor, 1988; La Rocca *et al.*, 2000a). These seedlings bleached rapidly in the light as a result of the photo-oxidation of chlorophyll in the absence of photoprotective carotenoids. However, chlorophyll-deficient seedlings, such as *oy-1040*, *113* and *l-Blandy4*, which are defective in the conversion of protoporphyrin to Mg-protoporphyrin, accumulated *Lhcb* transcripts (Mayfield and Taylor, 1984; Burgess and Taylor, 1988), indicating that it was not the absence of chlorophyll *per se* that affected nuclear gene expression. Similar results were obtained with barley mutants (Batschauer *et al.*, 1986). Carotenoid-deficient *alb-f*[17] seedlings failed to accumulate *Lhcb* transcripts, whereas a range of mutants affecting chlorophyll synthesis, including *xan-f* and *xan-g* mutants (which were later shown to encode subunits of Mg-chelatase; Jensen *et al.*, 1996), accumulated *Lhcb* transcripts in 6-day-old seedlings (Batschauer *et al.*, 1986).

White regions of variegated leaves of tomato *ghost* and *Arabidopsis immutans* mutants also showed decreased accumulation of *Lhcb* and *RbcS* transcripts (Giuliano and Scolnik, 1988; Wetzel *et al.*, 1994). These white regions were unable to complete carotenoid biosynthesis because of mutation of genes encoding quinol oxidase, which is required for the oxidation of plastoquinol, the cofactor for carotenoid desaturation (Carol *et al.*, 1999; Wu *et al.*, 1999; Josse *et al.*, 2000). The albino *Arabidopsis* mutant *cla1* (*chloroplastos alteredos1*) also showed decreased accumulation of *Lhcb* transcripts, most probably due to photobleaching in the absence of carotenoids (Mandel *et al.*, 1996). The *cla1* mutant is defective in 1-deoxyxylulose 5-phosphate synthase, a key plastid-located enzyme of isoprenoid biosynthesis (Estevez *et al.*, 2000).

Several other *Arabidopsis* mutants with defects in plastid processes have been shown to have altered expression of photosynthesis-related nuclear genes. The *Arabidopsis cue* (*Cab underexpressed*) mutants were selected for decreased expression of the β-glucuronidase (GUS) reporter gene from the *Arabidopsis Lhcb1*2* promoter on transfer of dark-grown seedlings to the light (Li *et al.*, 1995; López-Juez *et al.*, 1998). *CUE1* has been shown to encode a plastid inner-envelope-located phosphoenolpyruvate (PEP)/phosphate translocator, which is involved in the import of PEP for synthesis of aromatic compounds via the shikimate pathway (Streatfield *et al.*, 1999). However, it has been shown recently that the *cue1* phenotype is not simply caused by a decrease in the shikimate pathway (Voll *et al.*, 2003). Array technology has shown that transcripts of ~1500 genes encoding plastid-targeted proteins are decreased in the *cue1* mutant (Richly *et al.*, 2003). The mechanism by which decreased PEP/phosphate translocator activity is signalled to the nucleus is not known.

The *Arabidopsis laf6* (*long after far-red6*) mutant shows reduced responsiveness to far-red light and fails to accumulate normal amounts of transcripts of *Lhcb*, *PetH* (ferredoxin–NADP$^+$ oxidoreductase) and *CHS* (chalcone synthase) genes (Møller *et al.*, 2001). *LAF6* was reported to encode a plastid inner-envelope-associated protein, similar to an ATP-binding cassette (ABC) protein, which was predicted to be involved in transporting protoporphyrin IX from the envelope into the stroma. The mutant showed a twofold accumulation of protoporphyrin IX, and it was suggested this was involved in plastid signalling to the nucleus (Jarvis, 2001; Møller *et al.*, 2001). However, *CHS* expression is not down-regulated by plastid signals in mustard (Oelmüller *et al.*, 1986), barley (Hess *et al.*, 1994) or *Arabidopsis* (Strand *et al.*, 2003), and negative signalling by tetrapyrroles has been shown to be due to Mg-protoporphyrin, not protoporphyrin IX (Strand *et al.*, 2003). *LAF6* is homologous to bacterial *sufB*, which is involved in iron–sulphur centre assembly (Wilson *et al.*, 2003), suggesting that an alternative explanation for the effect of the *laf6* mutation on nuclear gene expression is required.

The application of transcript profiling using array technology has demonstrated that mutations affecting a wide range of plastid processes have dramatic effects on nuclear gene expression in *Arabidopsis* (Kurth *et al.*, 2002; Kubis *et al.*, 2003; Maiwald *et al.*, 2003; Richly *et al.*, 2003). The *Arabidopsis prpl11-1* mutant, which has a defect in the nuclear gene encoding plastid ribosomal protein L11, showed decreased plastid protein synthesis, decreased accumulation of Rubisco and photosynthetic electron transfer components, and reduced quantum yield of photosystem II (Pesaresi *et al.*, 2001). Transcriptome analysis, using an array containing probes for 1827 genes encoding chloroplast-targeted proteins, identified 25 genes showing at least two- to fivefold increases in transcript abundance and 42 genes showing two- to fivefold decreases in transcript abundance (Kurth *et al.*, 2002). Subsequent studies with a larger array containing probes for 3292 genes encoding putative plastid-targeted proteins identified additional nuclear genes whose transcript abundance changes in the *prpl11-1* mutant (Richly *et al.*, 2003). These experiments clearly establish that a lesion in plastid ribosome assembly affects the expression of a large number of nuclear genes.

Similar studies have been carried out with the same arrays with a variety of *Arabidopsis* mutants with lesions in nuclear genes encoding plastid proteins (Richly *et al.*, 2003). These include mutants of photosynthesis proteins (*psad1, psae1, psan, psao, atpc1, atpd, hcf145*), the PEP phosphate translocator (*cue1*), the Toc34 component of plastid protein import (*ppi1*) and of chlorophyll synthesis (*gun5, flu*). The overall patterns of responses of the nuclear genes to these changes in plastid function fall into just three categories (Richly *et al.*, 2003). The mutants *psan, psao, atpd, gun5* and *flu* (in the dark) showed increased transcripts from a large number of nuclear genes encoding plastid proteins, whereas *psae, atpc, hcf145, cue1* and *flu* (in the light) showed decreased transcripts from these nuclear genes, compared to wild-type plants. In a third group, the mutant *psad* showed similar numbers of genes with increased and decreased transcripts, as found for the *prpl11-1* mutant. This was interpreted to indicate a two-state master switch that integrates information about the perturbations caused by the mutations and results in one of two major responses. For example, almost all genes up-regulated in the *gun5* mutant were down-regulated in the *cue1* mutant. Similarly, mutants of *FLU*, which encodes a regulator of chlorophyll synthesis (Meskauskiene *et al.*, 2001), showed opposite effects on the same nuclear genes when the plants were either in the light or in the dark (Richly *et al.*, 2003).

These experiments with mutants indicate that perturbations in carotenoid biosynthesis, tetrapyrrole synthesis, plastid protein synthesis and photosynthesis in barley, maize, tomato and *Arabidopsis* have resulted in changes in the expression of nuclear genes encoding plastid proteins.

8.2.2 *Evidence from inhibitors*

A wide range of inhibitors with a primary mechanism of action in plastids has been shown to affect the expression of nuclear genes encoding plastid proteins, providing further evidence for retrograde signalling systems.

8.2.2.1   Inhibitors of carotenoid biosynthesis

Norflurazon, an inhibitor of phytoene desaturase, has been widely used in studies of plastid signalling. Plants treated with norflurazon fail to accumulate carotenoids and are unable to protect chlorophyll from photo-oxidation under high-light conditions. Chlorophyll photo-oxidation results in the eventual destruction of the internal structure of the chloroplasts, with the loss of plastid ribosomes and stromal enzymes (Frosch *et al.*, 1979; Reiss *et al.*, 1983). Norflurazon treatment was first shown to affect the accumulation of transcripts of photosynthesis-related nuclear genes (e.g. *Lhcb*) in maize seedlings grown in daylight for 8 days (Mayfield and Taylor, 1984). Subsequently, norflurazon has been the inhibitor of choice for many studies on plastid signalling. It has been shown to decrease the expression of numerous nuclear genes encoding plastid-related proteins (see section 8.3) in a wide range of plants, including mustard (*Sinapis alba*), barley, *Arabidopsis*, tobacco and pea (Batschauer *et al.*, 1986; Oelmüller and Mohr, 1986; Sagar *et al.*, 1988; Susek *et al.*, 1993; Sullivan and Gray, 2002).

Amitrole, a carotenoid-biosynthesis inhibitor acting on phytoene desaturation and lycopene cyclisation (Young, 1991), has been shown to decrease the accumulation of *Lhcb* and *RbcS* transcripts in light-grown photobleached seedlings of barley and maize (La Rocca *et al.*, 2000a, b, 2001). The inhibitory effects of amitrole on transcript accumulation were not as severe as those produced by norflurazon. Amitrole-treated seedlings retained some chlorophyll and showed the presence of some thylakoid membrane, whereas norflurazon-treated seedlings were completely devoid of chlorophyll and thylakoid membranes (La Rocca *et al.*, 2000a, b). Amitrole also differed from norflurazon in a strong inhibition of accumulation of *RbcS* transcripts in dark-grown barley seedlings, and in inhibition of the accumulation of *Lhcb* and *RbcS* transcripts in dark-grown seedlings treated with a brief pulse of low-intensity light sufficient to induce gene expression but not photo-oxidation (La Rocca *et al.*, 2001). Norflurazon had no effect on *Lhcb* and *RbcS* transcript accumulation under these conditions. These differences suggest that amitrole may be affecting processes other than carotenoid biosynthesis. Both amitrole and norflurazon are known to affect the lipid composition of plastids in dark-grown barley seedlings (Di Baccio *et al.*, 2002), and amitrole has been shown to stimulate tetrapyrrole synthesis in dark-grown barley seedlings, unlike norflurazon (La Rocca *et al.*, 2001).

8.2.2.2   Inhibitors of plastid gene expression

Treatment of mustard seedlings with chloramphenicol, an inhibitor of plastid protein synthesis, during the first 48 h of germination resulted in decreased accumulation of transcripts of *Lhcb* and *RbcS* genes (Oelmüller *et al.*, 1986). Chloramphenicol treatment also decreased *Lhcb1* expression in *Arabidopsis* seedlings (Susek *et al.*, 1993). Similar results were obtained with lincomycin, erythromycin and streptomycin, leading to decreased expression of photosynthesis-related nuclear genes in pea, tobacco and rice (Adamska, 1995; Gray *et al.*, 1995, 2003; Yoshida *et al.*, 1998; Sullivan and Gray, 1999). Lincomycin treatment was also able to decrease expression of photosynthesis-related nuclear genes in the dark in wild-type tobacco seedlings (Gray *et al.*, 2003) and in the pea *lip1* and *Arabidopsis cop1* mutants, which show photomorphogenic development in the dark (Sullivan and Gray, 1999).

Inhibition of plastid transcription by tagetitoxin, an inhibitor of the plastid-encoded RNA polymerase, resulted in decreased *Lhcb* and *RbcS* transcripts in barley seedlings (Rapp and Mullet, 1991). Nalidixic acid, a prokaryotic DNA gyrase inhibitor that affects plastid transcription (Gray *et al.*, 2003), also decreased expression from promoters of the *RbcS* and *PetH* genes in tobacco seedlings (Gray *et al.*, 1995). Nalidixic acid was effective in preventing inhibition of nuclear gene expression only when applied during the first 2–3 days of seedling development (Gray *et al.*, 1995), similar to the time frame shown for inhibition of nuclear gene expression by chloramphenicol and lincomycin (Oelmüller *et al.*, 1986; Bajracharya *et al.*, 1987; Gray *et al.*, 1995). These observations suggest that inhibition of plastid gene expression results in decreased expression of photosynthesis-related nuclear genes.

8.2.2.3   Inhibitors of photosynthesis

An effect of the photosynthesis inhibitors DCMU, an inhibitor of the Q_B site of photosystem II, and DBMIB, an inhibitor of the Q_O site of the cytochrome *bf* complex, on nuclear gene expression was first observed in the green alga *Dunaliella tertiolecta* (Escoubas *et al.*, 1995). They showed that DCMU treatment resulted in increased accumulation of *Lhcb* transcripts, whereas DBMIB treatment resulted in decreased *Lhcb* transcripts, mimicking light-intensity-dependent changes in *Lhcb* expression and suggesting a role for the plastoquinone redox state in signalling during light acclimation (Escoubas *et al.*, 1995; Durnford and Falkowski, 1997).

DCMU and DBMIB treatments have also indicated that the spinach *PetE* promoter is regulated by redox signals in transgenic tobacco seedlings (Pfannschmidt *et al.*, 2001b). However, expression from the spinach *PsaD* and *PsaF* promoters in transgenic tobacco seedlings did not respond to DCMU and DBMIB treatments in the same way as *PetE* (Pfannschmidt *et al.*, 2001b). The *Arabidopsis PetE* promoter appeared to show a rather complex set of responses to DCMU in *Arabidopsis* cell cultures and in detached leaves (Oswald *et al.*, 2001). DCMU prevented the starvation-induced increase in expression from *PetE* and *Lhcb1* promoters when added at the time of sugar removal, but had no effect when added to cell cultures or leaf discs that were actively expressing *PetE* or *Lhcb1* (Oswald *et al.*, 2001). Petracek *et al.* (1997) had previously reported that DCMU had no effect on the accumulation of *Lhcb* transcripts in illuminated dark-adapted tobacco plants. However, DCMU had a marked inhibitory effect on *Lhcb* transcripts in light-grown tobacco plants (Sullivan and Gray, 2002), and nuclear run-on assays showed that this was due largely, if not exclusively, to a decreased rate of *Lhcb* transcription (Sullivan and Gray, 2002).

In contrast, DCMU treatment had little effect on expression from the *RbcS* or *PetH* promoters in 7-day-old transgenic tobacco seedlings (Gray *et al.*, 1995) or on the accumulation of *Lhcb* or *PetE* transcripts in 7-day-old pea and tobacco seedlings (Sullivan and Gray, 2002). These experiments indicate that the developmental state of the plant material has a marked influence on whether disruption of photosynthetic electron transfer with DCMU has an effect on the expression of photosynthesis-related nuclear genes. Developing plastids without a functional photosynthetic apparatus would not be expected to respond to DCMU treatment.

Bromoxynil, an inhibitor of electron transfer in photosystem II, had a major effect on transcripts of nuclear genes encoding plastid-targeted proteins in 4-week-old *Arabidopsis* plants (Richly *et al.*, 2003). Transcripts of approximately 700 genes, out of 3292 on the array, increased in the presence of bromoxynil, similar to the responses observed in the *psao, psan, atpd, gun5* and *ppi1* mutants (Richly *et al.*, 2003). A similar response was observed on treatment of *Arabidopsis* plants with paraquat (Richly *et al.*, 2003), which acts as an electron acceptor for photosystem I, preventing electron flow to $NADP^+$. These experiments clearly indicate that disrupting photosynthetic electron transfer can have marked effects on nuclear gene expression in older plants.

8.2.3   Evidence from light treatments

The responses to light treatments perceived in the chloroplasts provide additional evidence for signalling from plastids. Light perception by phytochrome, crytochrome or phototrophins is excluded from this discussion because these photoreceptors are not located within plastids. However, light perception by the photosynthetic apparatus can lead to changes in nuclear gene expression (Pfannschmidt *et al.*, 2001a; Pfannschmidt, 2003).

8.2.3.1   Light quality
Optimised photosynthetic electron transfer requires the delivery of equal quanta of light to the reaction centres of photosystems I and II. In the short term, imbalances are perceived by the redox state of the plastoquinone pool leading to protein phosphorylation and changes in the stoichiometry of light-harvesting complexes and the reaction centres. In the longer term, changes in gene expression have been observed. Light treatments preferentially exciting photosystem II lead to increased expression from the spinach *PetE, PsaD* and *PsaF* promoters in transgenic tobacco plants (Pfannschmidt *et al.*, 2001b). This sort of light regime also affected the expression of large numbers of nuclear genes encoding plastid proteins in *Arabidopsis* (Richly *et al.*, 2003). Transcripts of more than 1100 genes were increased, and transcripts of more than 700 genes were decreased, when RNA samples from plants grown under light preferentially exciting photosystem II were compared with RNA samples from plants grown under light preferentially exciting photosystem I (Richly *et al.*, 2003).

8.2.3.2   Light quantity
Escoubas *et al.* (1995) showed that *Lhcb* expression in *D. tertiolecta* was increased on transfer from high-light to low-light conditions, and showed that this was probably a response to alteration of the redox state of the plastoquinone pool. Similar changes in *Lhcb* transcripts were observed in *D. salina* following light intensity and temperature treatments affecting photosystem II excitation pressure and the redox state of Q_A in

photosystem II (Maxwell *et al.*, 1995). The situation in higher plants appears to be more complicated. Transfer of barley plants from low-light to high-light conditions resulted in a large decrease in *Lhcb* transcripts, as observed for *Dunaliella*, but a similar decrease was not observed on transfer of plants to a low CO_2 and O_2 atmosphere, which produced the same change in photosystem II excitation pressure and Q_A redox state (Montané *et al.*, 1998). Pursiheimo *et al.* (2001) concluded that in rye seedlings subjected to short-term shifts in light intensity and temperature, the accumulation of *Lhcb* transcripts was not directly related to the photosystem II excitation pressure or the redox state of the plastoquinone pool. They showed that increased accumulation of *Lhcb* transcripts, due to decreased light intensity or temperature treatments, was correlated with the activity of the light-harvesting chlorophyll-binding proteins of photosystem II (LHCII) kinase, which has been shown to be under the control of the thioredoxin system, reflecting the activity of photosystem I (Rintamäki *et al.*, 2000).

Transcriptome analysis of *Arabidopsis* plants subjected to a $4°C$ treatment under medium-light intensity ($100 \ \mu mol \ m^{-2} \ s^{-1}$), producing severe photoinhibition, showed the down-regulation of ~2000 nuclear genes encoding plastid-located proteins. This response was similar to that obtained with the *psae* and *atpc* mutants (Richly *et al.*, 2003). Treatment of *Arabidopsis* for 1 h at much higher light intensities ($2000 \ \mu mol \ m^{-2} \ s^{-1}$), at normal temperatures, resulted in the opposite response (Richly *et al.*, 2003). Transcripts from approximately 1500 genes encoding plastid-located proteins were increased, whereas transcripts of less than 200 genes were decreased by this treatment.

These experiments with mutants, inhibitors and light point to four processes in plastids that are intimately involved in plastid signalling: carotenoid and tetrapyrrole synthesis, plastid protein synthesis and photosynthesis.

8.3 Which genes are regulated by plastid signals?

The first report to recognise the regulation of nuclear gene expression by plastid signals used enzyme assays to show decreased activities of phosphoribulokinase and $NADP^+$-glyceraldehyde 3-phosphate dehydrogenase in white tissue of the barley *albostrians* mutant (Bradbeer *et al.*, 1979). However, this approach is not valid for plant material obtained following treatments that result in photo-oxidation and the destruction of the internal structure of plastids. In most studies, estimates of gene expression have been obtained by measurements of transcript abundance, initially by translation of polyA-RNA *in vitro* (Oelmüller *et al.*, 1986; Burgess and Taylor, 1987), and subsequently by RNA gel blot hybridisation. More recently, array technology has been used, allowing the simultaneous analysis of large numbers of genes (Kurth *et al.*, 2002; Richly *et al.*, 2003; Strand *et al.*, 2003). In addition, promoter analysis of a small number of genes in transgenic plants has allowed the identification of promoter elements involved in plastid regulation (see Section 8.5.1).

8.3.1   Light reactions of photosynthesis

The *Lhcb* (*Cab*) genes encoding LHCII proteins appear to be among the most sensitive genes to plastid signals, and they have been widely used to examine features of plastid signalling. The *Lhcb* genes are extremely sensitive to chlorophyll photo-oxidation due to defective carotenoid biosynthesis, either in mutants or in plants treated with inhibitors such as norflurazon or amitrole. They are also strongly repressed by the tetrapyrrole intermediate Mg-protoporphyrin (Johanningmeier and Howell, 1984; Strand *et al.*, 2003), which increases in plants treated with inhibitors of carotenoid biosynthesis, such as norflurazon or amitrole (Kittsteiner *et al.*, 1991; La Rocca *et al.*, 2001), or inhibitors of tetrapyrrole biosynthesis, such as dipyridyl or β-thujaplicin (Johanningmeier and Howell, 1984; Oster *et al.*, 1996). *Lhcb* genes are repressed in white tissue of the barley *albostrians* mutant which lacks plastid ribosomes (Hess *et al.*, 1991, 1994), and also by treatments of young seedlings of mustard, *Arabidopsis*, pea and tobacco with plastid protein synthesis inhibitors, such as chloramphenicol, lincomycin and erythromycin (Oelmüller *et al.*, 1986; Susek *et al.*, 1993; Sullivan and Gray, 1999). Expression of *Lhcb* genes is also decreased in barley seedlings treated with tagetitoxin, an inhibitor of plastid-encoded RNA polymerase (Rapp and Mullet, 1991), and in the *Arabidopsis ppi1* mutant defective in chloroplast protein import (Kubis *et al.*, 2003). In contrast, the accumulation of *Lhcb* transcripts is increased three- to fourfold in the *Arabidopsis prpl11-1* mutant, which is defective in plastid ribosomal protein L11 (Kurth *et al.*, 2002). This appears counter-intuitive when compared to the effects of plastid protein synthesis inhibitors, and illustrates the complexities of plastid signalling.

The expression of *Lhcb* genes is sensitive to light intensity, detected via the photosynthetic apparatus, and the genes are up-regulated in low-light intensities and down-regulated in high-light intensities (Escoubas *et al.*, 1995; Maxwell *et al.*, 1995; Montané *et al.*, 1998; Pursiheimo *et al.*, 2001). However, there appear to be differences in signal generation in algae and in higher plants (see Section 8.2.3.2). The redox state of the plastoquinone pool is correlated with *Lhcb* transcription in *Dunaliella* (Escoubas *et al.*, 1995; Maxwell *et al.*, 1996), whereas control by the thioredoxin system appears to be superimposed on this in higher plants (Pursiheimo *et al.*, 2001). *Lhcb* genes appear to be responsive to all treatments known to affect plastid signalling.

In most of the studies described above it is not possible to be precise about which *Lhcb* genes are being examined, because of the possibility of cross-hybridisation between genes in the same family and between genes in different families. However, the Affymetrix microarrays allow sufficient discrimination to distinguish individual family members, and have shown that transcripts of *Lhcb2*, *Lhcb4*, *Lhcb5* and *Lhcb6* genes are decreased in *Arabidopsis* seedlings grown in the presence of norflurazon in the light (Strand *et al.*, 2003).

Genes encoding some of the components of the photosynthetic electron transfer chain have been shown to respond to a range of conditions known to affect plastid signalling. The *PetC*, *PetE*, *PetF* (or *Fed-1*) and *PetH* genes, encoding the Rieske FeS protein, plastocyanin, ferredoxin and ferredoxin-NADP$^+$ oxidoreductase, respectively, have all been shown to be sensitive to chlorophyll photo-oxidation (Burgess and Taylor, 1987; Vorst *et al.*, 1993; Bolle *et al.*, 1994; Gray *et al.*, 1995; Knight *et al.*, 2002; Sullivan and Gray, 2002; Strand *et al.*, 2003). *PetE* has also been shown to be sensitive to lincomycin treatment in pea and tobacco (Sullivan and Gray, 1999, 2002). *PetE* and *PetF* transcripts were decreased in the *Arabidopsis ppi1* mutant (Kubis *et al.*, 2003) and were sensitive to DCMU treatment in *Arabidopsis* and tobacco (Petracek *et al.*, 1998; Oswald *et al.*, 2001; Pfannschmidt *et al.*, 2001b). *PetH* transcripts decreased fourfold in the *prpl11-1* mutant of *Arabidopsis* (Kurth *et al.*, 2002).

Transcripts of most of the nuclear genes encoding components of photosystems I and II show the same patterns of behaviour as the *Lhcb* genes. Transcripts of the *PsbO* and *PsbP* genes, encoding the 33- and 23-kDa polypeptides of the oxygen-evolving complex (OEC) of photosystem II, were shown to decrease in norflurazon-treated maize seedlings, using translation of polyA-RNA *in vitro* (Burgess and Taylor, 1987). Decreased transcripts for these genes, as well as *PsbQ*, which encodes the 17-kDa OEC polypeptide, have also been shown in norflurazon-treated *Arabidopsis* using Affymetrix microarrays (Strand *et al.*, 2003). The potato *PsbR* gene (*ST-LS1*), encoding another lumenal protein, is also repressed by norflurazon treatment (Stockhaus *et al.*, 1987), but is up-regulated in the *Arabidopsis prpl11-1* mutant (Kurth *et al.*, 2002). Transcripts of *PsbS* are decreased by norflurazon treatment of *Arabidopsis* seedlings (Strand *et al.*, 2003) and transcripts of *PsbY* are decreased in the *Arabidopsis ppi1* mutant (Kubis *et al.*, 2003).

Transcripts of most genes encoding subunits of photosystem I are decreased by norflurazon treatment (Bolle *et al.*, 1994, 1996a; Chandok *et al.*, 2001; Strand *et al.*, 2003). Transcripts of *PsaK* and *PsaN* were increased three- and eightfold, respectively, in the *Arabidopsis prpl11-1* mutant (Kurth *et al.*, 2002). Expression of spinach *PsaD* and *PsaF* was sensitive to light quality and to the photosynthetic inhibitors DCMU and DBMIB (Pfannschmidt *et al.*, 2001b). Transcripts of the *PsaD* and *Lhca4* genes were decreased in the *Arabidopsis ppi1* mutant (Kubis *et al.*, 2003). *Lhca* genes were sensitive to treatments, such as β-thujaplicin, which increased Mg-protoporphyrin levels in cress seedlings (Oster *et al.*, 1996).

The synthesis of all three nuclear-encoded subunits of the plastid ATP synthase, the γ and δ subunits of CF$_1$ and the CF$_O$II subunit, is regulated by plastid signals. Transcripts of the tobacco *AtpC*, *AtpD* and *AtpG* genes were decreased in seedlings treated with norflurazon (Kusnetsov *et al.*, 1999). Expression from the promoters of the spinach *AtpC* and *AtpD* genes was sensitive to photo-oxidation caused by norflurazon treatment in transgenic tobacco seedlings (Bolle *et al.*, 1994, 1996b) and transcripts of *AtpC* were decreased in pea seedlings treated with lincomycin (Sullivan and Gray, 1999).

8.3.2 *CO$_2$ fixation and photorespiratory pathways*

The enzymes of the Calvin cycle are highly regulated by plastid signals. The *RbcS* genes encoding the small subunit of Rubisco have been studied most extensively and are sensitive to a range of treatments affecting plastid function and development. Transcripts of *RbcS* genes were decreased in photobleached tissues of mustard, maize, tomato, pea, tobacco and *Arabidopsis* (Oelmüller and Mohr, 1986; Mayfield and Taylor, 1987; Giuliano and Scolnik, 1988; Susek *et al.*, 1993; Bolle *et al.*, 1994; Gray *et al.*, 1995). *RbcS* genes were also down-regulated in white plastid-ribosome-deficient tissue of the barley *albostrians* mutant (Hess *et al.*, 1991, 1994), and by treatment of young seedlings of mustard, *Arabidopsis*, pea and barley with inhibitors of plastid protein synthesis, such as chloramphenicol, lincomycin and streptomycin (Oelmüller *et al.*, 1986; Susek *et al.*, 1993; Yoshida *et al.*, 1998; Sullivan and Gray, 1999). However, in contrast, transcripts of *RbcS* genes were increased fivefold in the *Arabidopsis prpl11-1* mutant (Kurth *et al.*, 2002). *RbcS* genes were down-regulated in the *Arabidopsis ppi1* mutant (Kubis *et al.*, 2003).

Transcripts of the genes encoding phosphoglycerate kinase, glyceraldehyde 3-phosphate dehydrogenase, aldolase, sedoheptulose 1,7-bisphosphatase and phosphoribulokinase are decreased by photobleaching due to norflurazon treatment (Longstaff *et al.*, 1989; Strand *et al.*, 2003). Decreased activities of $NADP^+$-glyceraldehyde 3-phosphate dehydrogenase and phosphoribulokinase (Bradbeer *et al.*, 1979), 3-phosphoglycerate kinase, triosephosphate isomerase and aldolase (Boldt *et al.*, 1992) and decreased transcripts of genes encoding the plastid 3-phosphoglycerate kinase, $NADP^+$-glyceraldehyde 3-phosphate dehydrogenase, fructose 1,6-bisphosphatase and phosphoribulokinase (Hess *et al.*, 1994) have been shown in white tissue of the barley *albostrians* mutant. The activity of $NADP^+$-glyceraldehyde 3-phosphate dehydrogenase was also decreased in mustard seedlings treated with chloramphenicol (Oelmüller *et al.*, 1986). Transcripts of the *Arabidopsis* gene encoding 3-phosphoglycerate kinase were decreased in the *ppi1* mutant (Kubis *et al.*, 2003).

The effect of photobleaching on the expression of some of the genes encoding components of the C_4 pathway has been examined in maize. Translation of polyA-RNA *in vitro* indicated that transcripts of the genes encoding pyruvate phosphate dikinase (PPDK) and $NADP^+$-malic enzyme (NADP-ME) were decreased to 10 and 30%, respectively, of the control amounts in maize leaves treated with norflurazon in high light (Burgess and Taylor, 1987). However, quantification of transcripts by RNA gel blot analysis did not show such large decreases (Tamada *et al.*, 2003). Transcripts encoding PPDK and NADP-ME were reported to be decreased by norflurazon treatment to only 60 and 70%, respectively, of control tissue (Tamada *et al.*, 2003). Transcripts encoding the cytosolic enzyme phosphoenolpyruvate carboxylase (PEPC) were also decreased to 70% of control tissue in these experiments, similar to the relatively small change previously observed in the *lw* carotenoid-deficient mutant of maize (Mayfield and Taylor, 1987).

Plastid signalling was recognised at a relatively early stage to be a feature of the regulation of genes encoding some of the enzymes of the photorespiratory pathway that were not plastid-localised, but located in the peroxisomes. Norflurazon treatment decreased the activity of glycolate oxidase and hydroxypyruvate reductase in rye, mustard and cucumber seedlings (Feierabend and Kemmerich, 1983;

Bajracharya *et al.*, 1987; Schwartz *et al.*, 1992), and decreased the expression of genes encoding these proteins, and NAD$^+$-malate dehydrogenase, in cucumber, tobacco and *Arabidopsis* (Schwartz *et al.*, 1992; Barak *et al.*, 2001; Strand *et al.*, 2003). The activities of enzymes of the glycolate pathway are also decreased in white leaf tissue of the barley *albostrians* mutant (Boldt *et al.*, 1997).

The effect of plastid signalling on genes encoding enzymes of starch synthesis and degradation has not been studied directly, but norflurazon treatment has recently been shown to decrease transcripts encoding β-amylase and both subunits of ADPglucose pyrophosphorylase in *Arabidopsis* (Strand *et al.*, 2003). In contrast, transcripts encoding a starch-synthase-like protein were increased in the *Arabidopsis ppi1* mutant (Kubis *et al.*, 2003).

8.3.3   Tetrapyrrole and other biosynthetic pathways

Although the disruption of tetrapyrrole and carotenoid biosynthesis pathways has been shown to generate plastid signals (see Section 8.2), there is evidence that some of the genes encoding enzymes of these pathways are regulated by plastid signals. The pathway of tetrapyrrole biosynthesis from glutamate is shown in Figure 8.1. Norflurazon treatment has been shown to decrease expression from the promoters of the genes encoding glutamyl-tRNA reductase (*HEMA*) and ferrochelatase II in *Arabidopsis* (McCormac *et al.*, 2001; Singh *et al.*, 2002) and to decrease transcripts of the *Arabidopsis* genes encoding glutamyl-tRNA reductase (Kumar *et al.*, 1999), glutamate 1-semialdehyde aminotransferase (*GSA*), hydroxymethylbilane synthase, uroporphyrinogen decarboxylase, protochlorophyllide reductase (*PORB*) and chlorophyll synthase (Strand *et al.*, 2003). Reduced expression of the tobacco *ChlH* gene, encoding the H subunit of Mg-chelatase, by antisense RNA (Papenbrock *et al.*, 2000) or by virus-induced gene silencing (Hiriart *et al.*, 2002), resulted in decreased transcripts encoding HEMA, GSA and ALA (ε-aminolaevulinate) dehydratase (Papenbrock *et al.*, 2000) and ChlD, the D subunit of Mg-chelatase, and chlorophyll synthase (Hiriart *et al.*, 2002). Increased transcripts encoding Mg-protoporphyrin methyltransferase and haem oxygenase were detected in the *Arabidopsis ppi1* mutant (Kubis *et al.*, 2003).

Genes encoding enzymes of carotenoid synthesis are induced by photo-oxidative stress produced in the plastids. Transcripts encoding phytoene synthase and phytoene desaturase were higher in the white leaf tissue of the tomato *ghost* mutant (Giuliano *et al.*, 1993), and in tomato and tobacco tissue treated with norflurazon (Giuliano *et al.*, 1993; Corona *et al.*, 1996). However, norflurazon treatment was shown to decrease transcripts encoding phytoene synthase in *Arabidopsis* (Strand *et al.*, 2003). Transcripts encoding lycopene ε-cyclase increase in the *Arabidopsis ppi1* mutant (Kubis *et al.*, 2003).

The expression of a large number of genes encoding a wide range of enzymes in a variety of metabolic pathways is affected by plastid signals, as shown by analysis of transcripts by array technology (Kurth *et al.*, 2002; Kubis *et al.*, 2003; Strand *et al.*, 2003). However, further analysis of the data is required before meaningful patterns of regulation can be recognised.

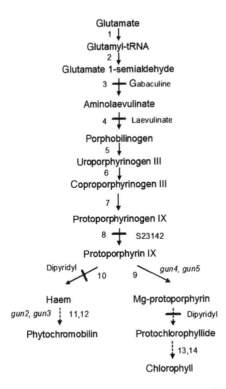

Figure 8.1 Pathway of tetrapyrrole biosynthesis in chloroplasts. The intermediates in the biosynthesis of chlorophyll, haem and phytochromobilin are shown, together with inhibitors (shown next to the black bars) and *gun* mutants (in italics). The enzymes are (1) glutamyl-tRNA synthase, (2) glutamyl-tRNA reductase (HEMA), (3) glutamate 1-semialdehyde reductase (GSA), (4) 4-aminolaevulinate dehydratase (ALAD), (5) porphobilinogen deaminase, or hydroxymethylbilane synthase, (6) uroporphyrinogen decarboxylase, (7) coproporphyrinogen oxidase, (8) protoporphyrinogen oxidase, (9) Mg-chelatase, (10) ferrochelatase, (11) haem oxygenase, (12) phytochromobilin synthase, (13) protochlorophyllide oxidoreductase (POR) and (14) chlorophyll synthase. The enzymes converting Mg-protoporphyrin to protochlorophyllide are not listed.

8.3.4 Plastid genetic system, protein import and chaperones

Array technology has identified numerous nuclear genes encoding components of the plastid genetic system that are regulated by plastid signals. Norflurazon treatment of *Arabidopsis* seedlings resulted in decreased transcripts of the *Rpl11*, *Rpl35*, *Rps5* and *Rps17* genes encoding plastid ribosomal subunits (Strand *et al.*, 2003). However, decreased expression of *Rpl11* in the *Arabidopsis prpl11-1* mutant resulted in increased transcripts of the *Rpl15* and *Rpl18* genes (Kurth *et al.*, 2002). In addition, transcripts of *Rpl11*, *Rpl12* and *Rps9* were increased in the *Arabidopsis ppi1* mutant (Kubis *et al.*, 2003), indicating the complexity of the regulatory network involved in plastid ribosome biogenesis. Transcripts encoding a variety of plastid genetic system proteins were increased in the *ppi1* mutant; these include a putative DNA polymerase, a sigma factor, the RNA-binding protein cp29 and translation initiation and release factors (Kubis *et al.*, 2003). Expression of a barley RNA-binding protein cp31A was decreased in white tissue of the barley *albostrians* mutant and in norflurazon-treated wild-type seedlings, although expression of two other RNA-binding proteins, cp31B and cp33, was not affected (Churin *et al.*, 1999). Transcripts encoding the *Arabidopsis* RNA-binding protein cp33 were decreased in norflurazon-treated seedlings (Strand *et al.*, 2003).

Transcripts encoding several components of the plastid protein import machinery in the plastid envelope, including Toc159, Toc75, Tic110, Tic40, Tic22 and Tic20, were increased in the *Arabidopsis ppi1* mutant, which is defective in Toc34 (Kubis *et al.*, 2003). Increased transcripts encoding components of the thylakoid protein import systems, including Tha1 (hcf106), Tha4, SecY and SRP54, were detected in the *prpl11-1* or *ppi1* mutants (Kurth *et al.*, 2002; Kubis *et al.*, 2003).

The expression of various genes encoding molecular chaperones has been shown to be regulated by plastid signals in *Chlamydomonas* and *Arabidopsis*. The *Chlamydomonas HSP70* genes, encoding cytosolic and chloroplast heat-shock proteins, are up-regulated by Mg-protoporphyrin (Kropat *et al.*, 1997, 2000), whereas transcripts of *Arabidopsis HSP70* and *DnaJ* genes are increased in the *ppi1* mutant (Kubis *et al.*, 2003). The *ppi1* mutant also contains increased transcripts encoding the cpn60β and cpn21 subunits of the plastid chaperonin complex and a ClpC protein (Kubis *et al.*, 2003).

The sections above indicate the large number of nuclear genes that are influenced by the developmental or functional state of plastids. In particular, the use of arrays containing probes for large numbers of genes encoding plastid proteins (Kurth *et al.*, 2002) has indicated that a large proportion of these genes are regulated by signals from plastids (Richly *et al.*, 2003).

8.4 What are plastid signals?

Although the evidence outlined above provides a compelling case for retrograde signalling from plastids to regulate the expression of nuclear genes encoding plastid proteins, the identity of components of the signalling pathways has remained elusive, until recently (Kropat *et al.*, 1997, 2000; Strand *et al.*, 2003). This section considers our current knowledge of the identity of plastid signals.

8.4.1 Positive or negative signals?

The initial descriptions of effects of the state of plastids on nuclear gene expression were placed in a conceptual framework of functional plastids producing positive signals, which were disrupted by damage to the plastids (Oelmüller *et al.*, 1986; Taylor, 1989). The original proposal was for a single plastid signal with a short half-life that is required continuously for optimal gene expression (Oelmüller *et al.*, 1986; Taylor, 1989). However, the recessive nature of the *Arabidopsis gun* (*genomes uncoupled*) mutants (see Section 8.5), which express photosynthesis-related nuclear genes in the presence of norflurazon, suggested that negative signalling was likely (Susek *et al.*, 1993) and this has been confirmed by the identification of Mg-protoporphyrin as a negative signalling component (Strand *et al.*, 2003).

However, there is growing evidence that plastid signals may have opposite effects on different genes. For example, Mg-protoporphyrin appears to act as a positive regulator of *HSP70* genes and a negative regulator of *Lhcb* genes in *Chlamydomonas* (Johanningmeier and Howell, 1984; Kropat *et al.*, 1997, 2000). Transcriptome analysis of a range of *Arabidopsis* mutants provides further evidence that plastid signalling can have positive and negative effects simultaneously (Kurth *et al.*, 2003).

8.4.2 Tetrapyrrole signals

A large amount of evidence has accumulated implicating tetrapyrrole biosynthesis intermediates in plastid signalling, culminating in the demonstration that Mg-protoporphyrin IX acts as positive signal for *HSP70* expression in *Chlamydomonas* (Kropat *et al.*, 1997, 2000) and as a negative signal for *Lhcb1* expression in *Arabidopsis* (Strand *et al.*, 2003). The principal approaches have been the use of inhibitors or mutants affecting the tetrapyrrole biosynthesis pathway, or the direct application of tetrapyrrole biosynthesis intermediates to plants (see Sections 8.4.2.1 and 8.4.2.2). The pathway of tetrapyrrole synthesis from glutamate is located exclusively in plastids (Figure 8.1; Cornah *et al.*, 2003) and is a possible target for many treatments affecting chloroplast function, leading to altered expression of nuclear genes.

8.4.2.1 Inhibitors and application of intermediates

The first effects of inhibitors of tetrapyrrole biosynthesis on nuclear gene expression were shown in *Chlamydomonas* (Johanningmeier and Howell, 1984; Johanningmeier, 1988). The application of 2,2′-bipyridyl, an iron chelator believed to inhibit formation of the isocyclic ring of chlorophyll, resulted in decreased amounts of *Lhcb* transcripts. The cells accumulated large amounts of Mg-protoporphyrin methyl ester, leading to the suggestion that this tetrapyrrole biosynthesis intermediate acts as a negative effector of *Lhcb* expression. Nuclear run-on experiments showed that the 2,2′-bipyridyl treatment resulted in decreased transcription of *Lhcb* genes (Jasper *et al.*, 1991). Similar observations were made with cress seedlings treated with 2,2′-bipyridyl or 8-hydroxyquinoline, another iron chelator (Kittsteiner *et al.*, 1991). The seedlings showed decreased amounts of *Lhcb* transcripts by RNA gel blot hybridisation, and decreased *Lhcb* transcription in nuclear run-on assays (Kittsteiner *et al.*, 1991). Treatment of cress seedlings with β-thujaplicin also resulted in decreased transcription of *Lhcb1*, *Lhcb2* and *Lhca1*, and was correlated with increased amounts of Mg-protoporphyrin methyl ester (Oster *et al.*, 1996). Barley seedlings treated with amitrole, normally regarded as a carotenoid biosynthesis inhibitor, showed increased amounts of Mg-protoporphyrin in the dark, and showed decreased accumulation of *Lhcb* and *RbcS* transcripts when transferred to light (La Rocca *et al.*, 2001). These experiments have been interpreted to suggest that Mg-protoporphyrin and its methyl ester are negative signalling components for the regulation of photosynthesis-related nuclear genes.

An opposite effect of Mg-protoporphyrin and its methyl ester on nuclear gene expression has been observed in *Chlamydomonas* (Kropat *et al.*, 1997, 2000). Addition of Mg-protoporphyrin to *Chlamydomonas* cultures in the dark resulted in the induction of the nuclear *HSP70A* and *HSP70B* genes, encoding cytosolic and plastid-localised heat-shock proteins, respectively (Kropat *et al.*, 1997). The light induction of expression of these genes was absent in the *brs-1* mutant (Kropat *et al.*, 1997), which has been shown to result in a frameshift mutation in the *ChlH* gene encoding the H subunit of Mg-chelatase (Chekounova *et al.*, 2001). This mutant accumulates protoporphyrin IX, which it cannot convert to Mg-protoporphyrin (Kropat *et al.*, 2000). The addition of 4 μM protoporphyrin IX to wild-type *Chlamydomonas* cultures in the dark was unable to induce expression of *HSP70A* (Kropat *et al.*, 1997), even though it was rapidly converted to Mg-protoporphyrin and its methyl ester (Kropat *et al.*, 2000). It was suggested that exogenously added protoporphyrin IX was converted to Mg-protoporphyrin and retained in the chloroplast, the location of Mg-chelatase, whereas exogenously added Mg-protoporphyrin was available in the cytosol or nucleus to interact with other components of the signalling pathway (Kropat *et al.*, 2000).

A direct demonstration of the negative effects of Mg-protoporphyrin on nuclear gene expression was provided by Strand *et al.* (2003). They showed that the addition of 50 μM Mg-protoporphyrin to *Arabidopsis* leaf protoplasts resulted in decreased amounts of *Lhcb* transcripts within 5 h. The addition of 50 μM porphobilinogen, protoporphyrin IX or haem had no effect on *Lhcb* transcripts, indicating the specific inhibitory effect of Mg-protoporphyrin (Strand *et al.*, 2003). Norflurazon treatment of *Arabidopsis* plants was shown to result in $\sim$15-fold increase in Mg-protoporphyrin (to 6380 pmol/g fresh weight) and a 65% decrease in Mg-protoporphyrin methyl ester (to 24 pmol/g fresh weight), suggesting that Mg-protoporphyrin is the key negative regulator.

However, it is not clear to what extent Mg-protoporphyrin acts as a negative regulator of nuclear gene expression under ambient conditions in wild-type plants. Inhibitors of the early steps of tetrapyrrole biosynthesis do not appear to affect the expression of photosynthesis-related nuclear genes in *Chlamydomonas* or tobacco seedlings under standard growth conditions. Laevulinic acid and 4,6-dioxoheptanoic acid, competitive inhibitors of aminolaevulinate dehydratase, did not affect *Lhcb* expression in *Chlamydomonas* (Johanningmeier and Howell, 1984; Johanningmeier, 1988), and gabaculine, an inhibitor of glutamate semialdehyde aminotransferase, had little effect on expression from *RbcS* and *PetH* promoters in transgenic tobacco seedlings (Gray *et al.*, 1995). These treatments might be expected to decrease the amounts of all tetrapyrrole biosynthesis intermediates, including Mg-protoporphyrin, potentially relieving any inhibitory effects and possibly resulting in increased expression of target genes. On the other hand, treatment of seedlings with ALA, a tetrapyrrole precursor, has been reported to decrease expression of *Lhcb* genes in cress and *Arabidopsis* (Kittsteiner *et al.*, 1991; Vinti *et al.*, 2000), possibly due to increased synthesis of a repressive intermediate. It will be important to establish under what conditions repressive concentrations of Mg-protoporphyrin accumulate in plants. There have been few measurements of Mg-protoporphyrin levels in higher plants. Pöpperl *et al.* (1998) measured changes in tetrapyrrole intermediates during the diurnal light/dark cycle in young tobacco leaves and showed a 1000-fold increase in Mg-protoporphyrin as the leaves entered the light phase. The maximum concentration observed ($\sim$5 nmol/g dry weight) was similar to the concentration reported for green *Arabidopsis* leaves ($\sim$0.4 nmol/g fresh weight; Strand *et al.*, 2003). This concentration appears to be at least an order of magnitude lower than required for repression of nuclear gene expression. Dark-grown leaves contain much lower amounts of Mg-protoporphyrin ($\sim$0.2 nmol/g dry weight for barley; La Rocca *et al.*, 2001).

8.4.2.2   *Mutants*

A genetic approach to identify components of the plastid signalling pathway was initiated by Susek *et al.* (1993) and has recently contributed to the identification of Mg-protoporphyrin as a signalling component. A collection of *gun* mutants of *Arabidopsis* was generated by ethylmethane sulphonate mutagenesis of seeds of an *Arabidopsis* line containing a chimeric gene encoding the GUS reporter and the hygromycin phosphotransferase selectable marker both under the control of identical *Lhcb1**2 promoters (Susek *et al.*, 1993). *gun* mutants were identified by the presence of GUS activity when seedlings were grown in the light in the presence of norflurazon, which results in repression of expression from the *Lhcb1* promoter in wild-type plants. The mutants were placed in five complementation groups (Susek *et al.*, 1993) and the identity of four of the five *gun* loci have now been established (Mochizuki *et al.*, 2001; Larkin *et al.*, 2003). *gun2* and *gun3* were shown to be allelic to the long hypocotyl photomorphogenic mutants *hy1* and *hy2*, which have mutations in the genes encoding haem oxygenase (Davis *et al.*, 1999; Muramoto *et al.*, 1999) and phytochromobilin synthase (Kohchi *et al.*, 2001), respectively. These enzymes are required for the synthesis of the phytochrome chromophore from haem in plastids. The *gun5-1* mutant was shown to result from a point mutation in the *ChlH* gene encoding the H subunit of Mg-chelatase, which introduces Mg^{2+} into protoporphyrin IX as the first committed step of chlorophyll synthesis (Mochizuki *et al.*, 2001). A mutation in the same gene in maize is found in the *l-Blandy4* mutant, which also expresses *Lhcb* in photobleached leaves in the presence of norflurazon (Burgess and Taylor, 1988). *GUN4* encodes another protein involved in Mg-protoporphyrin formation (Larkin *et al.*, 2003), although its exact role *in vivo* has not been fully defined. GUN4 has been shown to bind protoporphyrin IX and Mg-protoporphyrin and to associate with the H subunit of Mg-chelatase, resulting in stimulation of Mg-chelatase activity (Larkin *et al.*, 2003). However, there appears to be no absolute requirement for GUN4 for chlorophyll synthesis (Larkin *et al.*, 2003).

Genetic analysis of these *gun* mutants indicated they were all recessive and none of the mutants completely inactivated plastid signalling (Mochizuki *et al.*, 2001). Analysis of the phenotypes of *gun* double mutants suggested that *gun2, gun3, gun4* and *gun5* all affected the same plastid signalling pathway, whereas *gun1* appeared to affect a separate signalling pathway (Vinti *et al.*, 2000; Mochizuki *et al.*, 2001). The recognition that four of the five *gun* mutants contained mutations in genes encoding enzymes of tetrapyrrole biosynthesis led to the demonstration of a *gun* phenotype of *Arabidopsis* mutants of genes encoding porphobilinogen deaminase, coproporphyrinogen oxidase and the D subunit of Mg-chelatase (Strand *et al.*, 2003). In addition, McCormac and Terry (2002) showed that *Arabidopsis* lines overexpressing *PORA* or *PORB* (genes encoding protochlorophyllide oxidoreductase) have a *gun* phenotype in the presence of norflurazon, perhaps suggesting that these lines contain decreased amounts of Mg-protoporphyrin.

Treatment of *gun2* and *gun5* with norflurazon resulted in much lower levels of Mg-protoporphyrin than in wild-type seedlings (Strand *et al.*, 2003). *Lhcb1* expression in norflurazon-treated *gun2* and *gun5* seedlings was decreased by 2,2′-dipyridyl, and this effect could be reversed by the addition of S23142, an inhibitor of protoporphyrinogen oxidase (Strand *et al.*, 2003). These observations are all consistent with a role for Mg-protoporphyrin as a negative regulator of *Lhcb1* expression in *Arabidopsis*.

There are, however, some inconsistencies that need to be resolved. Mochizuki *et al.* (2001) reported that the *Arabidopsis cs* and *ch-42* mutants, with mutations in the *ChlI* gene encoding the I subunit of Mg-chelatase, did not show a *gun* phenotype in the presence of norflurazon. This led to a model in which the H subunit of Mg-chelatase acted as an indicator of the protoporphyrin/Mg-protoporphyrin ratio, leading to signal transduction to the nucleus (Mochizuki *et al.*, 2001). This model has apparently been abandoned in favour of a direct action of cytosolic Mg-protoporphyrin (Strand *et al.*, 2003).

The use of the *Chlamydomonas brs-1* mutant in studies of light induction of gene expression has also produced some results that need further resolution. The *brs-1* mutant is defective in the H subunit of Mg-chelatase (Chekounova *et al.*, 2001) and fails to accumulate Mg-protoporphyrin on transfer from dark to light, unlike wild-type cultures (Kropat *et al.*, 2000). The *brs-1* mutant shows decreased light induction of both *HSP70* genes (Kropat *et al.* 1997) and *Lhcb* genes (Johanningmeier and Howell, 1984), although the *HSP70* genes are positively regulated by Mg-protoporphyrin, whereas the *Lhcb* genes appear to be negatively regulated (see Section 8.4.2.1).

Most *gun* mutants do not show large changes in chlorophyll content or altered expression of photosynthesis-related nuclear genes when grown under normal conditions (Susek *et al.* 1993; Mochizuki *et al.*, 1996, 2001). However, subtle differences in de-etiolation following transfer of dark-grown seedlings to the light have been observed for *gun1* (Mochizuki *et al.*, 1996). The mutant seedlings showed delayed greening and expression of *Lhcb* and *RbcS* genes when dark-grown seedlings were transferred to light (Mochizuki *et al.*, 1996). This suggests the importance of *GUN1* for coordinating nuclear and plastid gene expression during the very early stages of the transition from heterotrophic to photoautotrophic growth.

gun1 has been shown to be distinct from the other *gun* mutants, on the basis of phenotypes of double *gun* mutants (Vinti *et al.*, 2000; Mochizuki *et al.*, 2003). It has also been shown to express *Lhcb* and *RbcS* in seedlings treated with lincomycin, a specific inhibitor of plastid translation, unlike wild-type seedlings or the other *gun* mutants (J.H. Wang, J.A. Sullivan and J.C. Gray, unpublished data). This suggests that *GUN1* encodes a component of a signalling pathway affected by inhibition of plastid protein synthesis. This pathway appears to be important for regulating nuclear gene expression during the early stages of seedling development (see Section 8.2.2.2).

8.4.3 Protein phosphorylation/dephosphorylation

There have been suggestions that protein phosphorylation/dephosphorylation plays a role in the transduction of signals from plastids (Escoubas *et al.*, 1995; Chandok *et al.*, 2001). Escoubas *et al.* (1995) showed that the cytoplasmic protein phosphatase inhibitors, okadaic acid, microcystin-LR and tautomycin, prevented chlorophyll *a* accumulation on transfer of *D. tertiolecta* from high-light to low-light conditions. The change in light intensity stimulates *Lhcb* expression via the redox status of the plastoquinone pool in the chloroplasts (Escoubas *et al.*, 1995).

The plastoquinone redox state is known to regulate the activity of a protein kinase associated with the cytochrome *bf* complex in the thylakoid membrane (Vener *et al.*, 1997). Escoubas *et al.* (1995) suggested that the phosphorylation of a chloroplast protein was linked to the activation of a transcriptional repressor by phosphorylation in the cytosol. Inhibition of cytosolic protein phosphatase activity was proposed to increase the proportion of the active phosphorylated state of the repressor, leading to decreased expression of *Lhcb* genes. Unfortunately, none of the proposed protein targets for the phosphorylation/dephosphorylation system has been identified in *Dunaliella*.

Chandok *et al.* (2001) have shown that autophosphorylation of a 70-kDa polypeptide immunoprecipitated with antibodies to protein kinase C was increased in soluble extracts of tobacco seedlings treated with norflurazon in the light. It was suggested that protein kinase C may be a component coupling nuclear gene expression to the functional state of the plastids. However, phorbolmyristate acetate (PMA), which activates protein kinase C and stimulates *PsaF* expression in tobacco seedlings, had no effect on plastid-regulated expression from short promoter regions of the tobacco *PsaF* gene (Chandok *et al.*, 2001). Further work is needed to establish a role for cytosolic protein phosphorylation/dephosphorylation in plastid signalling.

8.5 How do plastid signals work?

Until the advent of microarray analysis, relatively few genes had been identified as targets of plastid signalling, and even fewer had been studied in detail to determine the ways in which plastid signals regulated gene expression. Unfortunately, most studies report single time-point measurements of transcript abundance, by RNA gel blot analysis or by microarray analysis, which do not provide information on the mechanism of regulation of gene expression. Experiments using nuclear run-on assays or transgenic plants containing promoter–reporter gene constructs have been used to establish whether plastid regulation is operating at transcription or at post-transcriptional events. These studies have shown that a majority of the genes examined are regulated transcriptionally by plastid signals, although evidence for post-transcriptional regulation has been obtained for a few genes.

8.5.1   Transcriptional regulation

Decreased accumulation of transcripts of the *Lhcb* and *RbcS* genes in norflurazon-treated seedlings of barley, maize, rye and pea was shown to be due to decreased rates of transcription in run-on assays with isolated nuclei (Batschauer *et al.*, 1986; Burgess and Taylor, 1988; Ernst and Schefbeck, 1988; Sagar *et al.*, 1988). Decreased rates of transcription of *Lhcb* and *RbcS* have also been demonstrated in nuclei isolated from photobleached tissue of the carotenoid-deficient *alb-f*[17] mutant of barley (Batschauer *et al.*, 1986) and *ghost* mutant of tomato (Giuliano and Scolnik, 1988). Run-on transcription assays with isolated nuclei were also used to show decreased transcription of *Lhcb* in the *albostrians* mutant of barley (Hess *et al.*, 1994) and in DCMU-treated tobacco leaves (Sullivan and Gray, 2002). Transcription of the *Lhcb* genes in *D. tertiolecta* was shown, using run-on transcription assays, to increase threefold on transfer from high-light to low-light conditions (Escoubas *et al.*, 1995). These experiments clearly demonstrate the transcriptional regulation of the *Lhcb* genes in various plants in response to treatments known to be mediated by plastid signals.

Similar conclusions on transcriptional regulation can be drawn from the effects of inhibitor treatments on transgenic plants containing reporter genes under the control of promoters of photosynthesis-related nuclear genes. Decreased expression of the neomycin phosphotransferase (*nptII*) reporter gene under the control of the pea *Lhcb1* and *RbcS* promoters was observed in transgenic tobacco seedlings treated with norflurazon (Simpson *et al.*, 1986). Subsequently, the promoters of *Arabidopsis Lhcb1* (Susek *et al.*, 1993), *Arabidopsis PetE* (Vorst *et al.*, 1993), spinach *Lhcb1*, *PetE*, *PetH*, *PsaF*, and *AtpC* and *AtpD* (Bolle *et al.*, 1994, 1996b), tobacco *RbcS* and pea *PetH* (Gray *et al.*, 1995), *Arabidopsis HemA* (McCormac *et al.*, 2001), and *PetC* (Knight *et al.*, 2002) have all been shown to direct norflurazon-sensitive expression of the GUS reporter gene in transgenic tobacco or *Arabidopsis*. In addition, transgenic tobacco seedlings expressing the GUS reporter gene under the control of the promoter of a tobacco gene encoding peroxisomal glycolate oxidase showed decreased GUS activity in the presence of norflurazon (Barak *et al.*, 2001).

Inhibitors of chloroplast gene expression, such as lincomycin and nalidixic acid, have also been shown to decrease expression of the GUS reporter gene under the control of the tobacco *RbcS* and pea *PetH* genes in transgenic tobacco seedlings

(Gray *et al.*, 1995), demonstrating transcriptional responses of these genes to plastid signals responding to the state of plastid gene expression. The photosynthetic electron transfer inhibitors, DCMU and DBMIB, resulted in decreased expression from the spinach *PetE*, *PsaD* and *PsaF* promoters in 14-day-old tobacco seedlings (Pfannschmidt *et al.*, 2001b). These seedlings also showed light-quality-dependent changes in expression of GUS reporter gene activity. The spinach *PetE*, *PsaD* and *PsaF* promoters showed increased activity on transfer from light predominantly exciting photosystem I to light predominantly exciting photosystem II; however, only the *PetE* promoter showed decreased activity on transfer from photosystem II light to photosystem I light (Pfannschmidt *et al.*, 2001b). These experiments demonstrate that these spinach genes respond to plastid signals derived from photosynthesis at the transcriptional level.

These experiments with transgenic plants clearly demonstrate the transcriptional responses of promoters of photosynthesis-related nuclear genes following treatments known to be mediated by plastid signals. They also provide a means of identifying promoter elements responsive to plastid signals. Deletion analysis of many of the promoters described above has identified shorter regions that are able to direct similar patterns of reporter-gene expression in norflurazon-treated seedlings (Vorst *et al.*, 1993; Bolle *et al.*, 1994; Lübberstedt *et al.*, 1994; Gray *et al.*, 1995, Bolle *et al.*, 1996a, b; Kusnetsov *et al.*, 1996; Hahn and Kück, 1999). However, in none of the promoters was it possible to separate the regulatory elements responding to norflurazon treatment from light-regulatory elements. This suggests that light regulation and plastid regulation of photosynthesis-related nuclear genes may operate through the same transcription factors and regulatory elements.

Gain-of-function assays using combinations of multimers of light-regulatory *cis*-elements in *Arabidopsis* have shown that all combinations showing high levels of light regulation also show plastid regulation (Puente *et al.*, 1996). GUS reporter gene expression from combinations of GT1, G-box, Z-DNA and GATA elements fused to the minimal *nos* promoter was sensitive to photobleaching caused by norflurazon treatment (Puente *et al.*, 1996). A 52 bp region of the *Nicotiana plumbaginifolia RbcS 8B* promoter, containing single copies of a GATA element (I box) and a G box (CACGTG), was sufficient to confer norflurazon-sensitive expression on the minimal CaMV 35S -46 promoter in *Arabidopsis* (Martinez-Hernandez *et al.*, 2002). The involvement of G-box elements in plastid regulation is also indicated by studies on expression from the *Arabidopsis Lhcb1*2* promoter in the *gun5* mutant (Strand *et al.*, 2003). Mutation of the binding site for the *CAB* upstream factor 1 (CUF1), which binds to the G-box-related CUF1 element (CACGTA) in the *Lhcb1*2* promoter, resulted in decreased expression in the *gun5* mutant in the presence of norflurazon.

A separate element involved in plastid-regulated expression has been identified in the spinach *AtpC* promoter. Bolle *et al.* (1996b) showed that replacement of the nucleotides AAAAT within the sequence CCTCCAAAATCAAT by GGGGC in the *AtpC* promoter resulted in constitutive expression of the GUS reporter gene in the dark and in photobleached seedlings following norflurazon treatment. The mutation of the central A nucleotide in this sequence to G (CCTCCAAGATCAAT) also resulted in a loss of plastid-responsive expression (Kusnetsov *et al.*, 1999). Screening an *Arabidopsis* cDNA expression library resulted in the identification of a CAAT-box binding factor (Kusnetsov *et al.*, 1999) and of ATPC-2, a protein with DNA helicase domains (Bezhani*et al.*, 2001). ATPC-2 failed to bind to the mutated sequence containing the G nucleotide and competed for binding with the CAAT-box binding factor *in vitro*, leading to the suggestion that ATPC-2 acts as a repressor for light- and plastid-regulated expression of *AtpC* (Bezhani *et al.*, 2001).

In none of these studies is it clear how the plastid signal affects the binding of transcription factors to the plastid-response elements. Strand *et al.* (2003) suggested that the simplest model for the regulation of *Lhcb1*2* by Mg-protoporphyrin in *Arabidopsis* is a direct interaction of Mg-protoporphyrin with the transcription factor CUF1, modifying its DNA-binding activity (see Figure 8.2). However, in the absence of information on whether CUF1 is an activator or repressor, Mg-protoporphyrin binding might either inactivate a transcriptional activator (Figure 8.2A) or activate a transcriptional repressor (Figure 8.2B). There is as yet no evidence to determine whether these simple models describe plastid regulation of the *Lhcb1*2* gene in *Arabidopsis*.

8.5.2 *Post-transcriptional regulation*

Although a majority of plastid-regulated nuclear genes studied show control at transcriptional initiation, there are a few genes whose expression has been shown to be regulated at post-transcriptional stages. Sequences downstream of the transcription start site have been shown to be involved in light- and plastid-regulated expression of the pea *Fed-1* (Petracek *et al.*, 1997, 1998), spinach *PsaD* (Bolle *et al.*, 1996b; Sherameti *et al.*, 2002) and pea *PetE* genes (Helliwell *et al.*, 1997; Sullivan and Gray, 2002). An intron in the spinach *PsaD* gene has been shown to be required for plastid regulation in transgenic tobacco seedlings (Bolle *et al.*, 1996b).

Photosynthetic electron transfer controls the polyribosome association of transcripts of the pea *Fed-1* and spinach *PsaD*, *PsaF* and *PsaL* transcripts. DCMU treatment prevented the polyribosome association of pea *Fed-1*, tobacco *Lhcb1* and spinach *PsaD*, *PsaF and PsaL* transcripts in transgenic tobacco seedlings or in spinach seedlings (Petracek *et al.*, 1997; Sherameti *et al.*, 2002), resulting in the destabilisation of the *Fed-1*, but not the *Lhcb1* or *PsaD*, transcripts. The destabilisation of pea *PetE* transcripts in DCMU-treated transgenic tobacco plants has also been reported (Sullivan and Gray, 2002). The correct 5′ untranslated region (5′UTR) and part of the coding region were necessary for regulated expression of the pea *Fed-1* and pea *PetE* genes (Dickey *et al.*, 1998; Hansen *et al.*, 2001; Sullivan and Gray, 2002). However, only the 5′UTR is required for polyribosome association of spinach *PsaD* transcripts, and the lack of ribosome association does not appear to result in destabilisation of the transcripts (Sherameti *et al.*, 2002).

The 5′UTR and the coding region of the pea *PetE* gene are also required for regulation in response to norflurazon and lincomycin treatment of transgenic tobacco seedlings (Sullivan and Gray, 2002). So far, it has not been possible to separate the

A Mg-protoporphyrin inhibits an activator

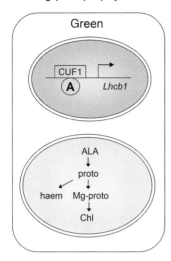
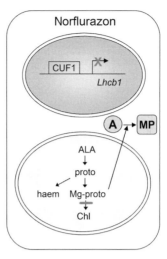

B Mg-protoporphyrin activates a repressor

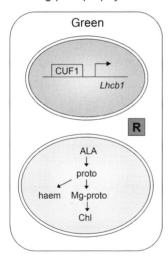
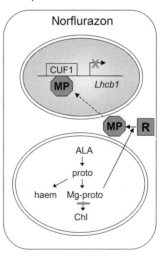

Figure 8.2 Models of transcriptional regulation of the *Lhcb1* promoter by Mg-protoporphyrin. (A) Mg-protoporphyrin binding to a transcriptional activator causes a conformational change, preventing binding to the CUF1 element in the *Lhcb1* promoter. (B) Mg-protoporphyrin binding to an inactive repressor protein causes a conformational change, allowing binding to the CUF1 element in the *Lhcb1* promoter. In each part, the diagrams show a nucleus and a chloroplast in a single cell. The cell on the left is in green tissue, whereas the cell on the right has been photobleached by treatment with norflurazon in the light. Norflurazon treatment leads to increased Mg-protoporphyrin which escapes from the plastid and interacts with the activator or repressor in the cytosol. Based on the model in Strand *et al.* (2003).

elements in the *PetE* transcript that respond to light or plastid signals affected by norflurazon or lincomycin treatment (N. Brown and J.C. Gray, unpublished data). Mechanisms by which signals from three apparently separate signalling pathways regulate post-transcriptional events are currently unknown.

8.6 Conclusions

This review set out to answer a series of questions concerning the control of nuclear gene expression by retrograde signals from plastids. I believe the following series of statements provide an overview of our current understanding of plastid signalling. A majority of the nuclear genes encoding plastid-located proteins are regulated by plastid signals. These signals appear to be generated by three processes within plastids: tetrapyrrole synthesis, plastid protein synthesis and photosynthesis. Signals produced by tetrapyrrole synthesis and plastid protein synthesis are likely to be involved in regulating gene expression during plastid development, whereas photosynthetic signals derived from changes in photosynthetic electron transfer and the redox state of electron carriers are involved in acclimatory changes to altered environmental conditions. Plastid signals may act positively or negatively. Mg-protoporphyrin, the best characterised signalling intermediate, is a negative signal for *Lhcb* gene expression, and a positive signal for *HSP70* gene expression. Plastid-response elements for transcriptional regulation cannot be separated from light-response elements. Plastid and light signals appear to converge at or before transcription factor binding. A small number of genes show post-transcriptional plastid regulation via changes in polyribosome association or transcript stability.

References

Abdallah, F., Salamini, F. and Leister, D. (2000) A prediction of the size and evolutionary origin of the proteome of chloroplasts of *Arabidopsis*. *Trends Plant Sci.*, 5, 141–142.

Adamska, I. (1995) Regulation of early light-inducible protein gene expression by blue and red light in etiolated seedlings involves nuclear and plastid factors. *Plant Physiol.*, 107, 1167–1175.

Bajracharya, D., Bergfeld, R., Hatzfeld, W.D., Klein, S. and Schopfer, P. (1987) Regulatory involvement of plastids in the development of peroxisomal enzymes in the cotyledons of mustard (*Sinapis alba* L.) seedlings. *J. Plant Physiol.*, 126, 421–436.

Barak, S., Nejidat, A., Heimer, Y. and Volokita, M. (2001) Transcriptional and posttranscriptional regulation of the glycolate oxidase gene in tobacco seedlings. *Plant Mol. Biol.*, 45, 399–407.

Batschauer, A., Mösinger, E., Krenz, K., Dörr, I. and Apel, K. (1986) The implication of a plastid-derived factor in the transcriptional control of nuclear genes encoding the light-harvesting chlorophyll *a/b* protein. *Eur. J. Biochem.*, 154, 625–634.

Bezhani, S., Sherameti, I., Pfannschmidt, T. and Oelmüller, R. (2001) A repressor with similarities to prokaryotic and eukaryotic DNA helicases control the assembly of the CAAT box binding complex at a photosynthesis gene promoter. *J. Biol. Chem.*, 276, 23785–23789.

Boldt, R., Börner, T. and Scharrenberger, C. (1992) Repression of the plastidic isoenzymes of aldolase, 3-phosphoglycerate kinase, and triosephosphate isomerase in the barley mutant "albostrians." *Plant Physiol.*, 99, 895–900.

Boldt, R., Koshuchowa, S., Gross, W., Börner, T. and Schnarrenberger, C. (1997) Decrease in glycolate pathway enzyme activities in plastids and peroxisomes of the albostrians mutant of barley (*Hordeum vulgare* L.). *Plant Sci.*, 124, 33–40.

Bolle, C., Herrmann, R.G. and Oelmüller, R. (1996a) Intron sequences are involved in the plastid- and light- dependent expression of the spinach *PsaD* gene. *Plant J.*, 10, 919–924.

Bolle, C., Kusnetsov, V.V., Herrmann, R.G. and Oelmüller, R. (1996b) The spinach *AtpC* and *AtpD* genes contain elements for light-regulated, plastid-dependent and organ-specific expression in the vicinity of the transcription start sites. *Plant J.*, 9, 21–30.

Bolle, C., Sopory, S., Lubberstedt, T., Klosgen, R., Herrmann, R. and Oelmüller, R. (1994) The role of plastids in the expression of nuclear genes for thylakoid proteins studied with chimeric β-glucuronidase gene fusions. *Plant Physiol.*, 105, 1355–1364.

Bradbeer, J.W., Atkinson, Y.E., Börner, T. and Hagemann, R. (1979) Cytoplasmic synthesis of plastid polypeptides may be controlled by plastid-synthesized RNA. *Nature*, 279, 816–817.

Burgess, D.G. and Taylor, W.C. (1987) Chloroplast photooxidation affects the accumulation of cytosolic messenger RNA encoding chloroplast proteins in maize. *Planta*, 170, 520–527.

Burgess, D.G. and Taylor, W.C. (1988) The chloroplast affects the transcription of a nuclear gene family. *Mol. Gen. Genet.*, 214, 89–96.

Carol, P., Stevenson, D., Bisanz, C. *et al.* (1999) Mutations in the *Arabidopsis* gene *IMMUTANS* cause a variegated phenotype by inactivating a chloroplast terminal oxidase associated with phytoene desaturation. *Plant Cell*, 11, 57–68.

Chandok, M.R., Sopory, S.K. and Oelmüller, R. (2001) Cytoplasmic kinase and phosphatase activities can induce *PsaF* gene expression in the absence of functional plastids: evidence that phosphorylation/dephosphorylation events are involved in interorganellar crosstalk. *Mol. Gen. Genet.*, 264, 819–826.

Chekounova, E., Voronetskaya, V., Papenbrock, J., Grimm, B. and Beck, C.F. (2001) Characterization of *Chlamydomonas* mutants defective in the H subunit of Mg-chelatase. *Mol. Gen. Genet.*, 266, 363–373.

Churin, Y., Hess, W.R. and Börner, T. (1999) Cloning and characterization of three cDNAs encoding chloroplast RNA-binding proteins from barley (*Hordeum vulgare* L.): differential regulation of expression by light and plastid development. *Curr. Genet.*, 36, 173–181.

Cornah, J.E., Terry, M.J. and Smith, A.G. (2003) Green or red: what stops the traffic in the tetrapyrrole pathway? *Trends Plant Sci.*, 8, 224–230.

Corona, V., Aracri, B., Kosturkova, G. *et al.* (1996) Regulation of a carotenoid biosynthesis gene promoter during plant development. *Plant J.*, 9, 505–512.

Davis, S.J., Kurepa, J. and Vierstra, R.D. (1999) The *Arabidopsis thaliana HY1* locus, required for phytochrome- chromophore biosynthesis, encodes a protein related to heme oxygenase. *Proc. Natl. Acad. Sci. U.S.A.*, 96, 6541–6546.

Di Baccio, D., Quartacci, M.F., Dalla Vecchia, F., La Rocca, N., Rascio, N. and Navari-Izzo, F. (2002) Bleaching herbicide effects on plastids of dark-grown plants: lipid composition of etioplasts in amitrole and norflurazon-treated barley leaves. *J. Exp. Bot.*, 53, 1857–1865.

Dickey, L.F., Petracek, M.E., Nguyen, T.T., Hansen, E.R. and Thompson, W.F. (1998) Light regulation of *Fed1* mRNA requires an element in the 5′ untranslated region and correlates with differential polyribosome association. *Plant Cell*, 10, 475–484.

Durnford, D.G. and Falkowski, P.G. (1997) Chloroplast redox regulation of nuclear gene transcription during photoacclimation. *Photosynth. Res.*, 53, 229–241.

Ellis, R.J. (1977) Protein synthesis by isolated chloroplasts. *Biochim. Biophys. Acta*, 463, 185–215.

Ernst, D. and Schefbeck, K. (1988) Photooxidation of plastids inhibits transcription of nuclear encoded genes in rye (*Secale cereale*). *Plant Physiol.*, 88, 255–258.

Escoubas, J.-M., Lomas, M., LaRoche, J. and Falkowski, P.G. (1995) Light intensity regulation of *cab* gene transcription is signaled by the redox state of the plastoquinone pool. *Proc. Natl. Acad. Sci. U.S.A.*, 92, 10237–10241.

Estevez, J.M., Cantero, A., Romero, C. *et al.* (2000) Analysis of the expression of *CLA1*, a gene that encodes the 1-deoxyxylulose 5-phosphate synthase of the 2-*C*-methyl-D-erythritol -4-phosphate pathway in *Arabidopsis*. *Plant Physiol.*, 124, 95–103.

Feierabend, J. and Kemmerich, P. (1983) Mode of interference of chlorosis-inducing herbicides with peroxisomal enzyme activities. *Physiol. Plant*, 57, 346–357.

Frosch, S., Jabben, M, Bergfeld, R., Kleinig, H. and Mohr, H. (1979) Inhibition of carotenoid biosynthesis by the herbicide SAN9789 and its consequences for the action of phytochrome on plastidogenesis. *Planta*, 145, 497–505.

Giuliano, G., Bartley, G.E. and Scolnik, P.A. (1993) Regulation of carotenoid biosynthesis during tomato development. *Plant Cell*, 5, 379–387.

Giuliano, G. and Scolnik, P.A. (1988) Transcription of two photosynthesis-associated nuclear gene families correlates with the presence of chloroplasts in leaves of the variegated tomato *ghost* mutant. *Plant Physiol.*, 86, 7–9.

Gray, J.C., Sornarajah, R., Zabron, A.A., Duckett, C.M. and Khan, M.S. (1995) Chloroplast control of nuclear gene expression. In *Photosynthesis, From Light to Biosphere*, Vol. 3 (ed. P. Mathis), Kluwer, Dordrecht, pp. 543–550.

Gray, J.C., Sullivan, J.A., Wang, J.-H., Jerome, C.A. and MacLean, D. (2003) Coordination of plastid and nuclear gene expression. *Philos. Trans. R. Soc. Lond. B*, 358, 135–145.

Hahn, D. and Kück, U. (1999) Identification of DNA sequences controlling light- and chloroplast-dependent expression of the *Lhcb1* gene from *Chlamydomonas reinhardtii*. *Curr. Genet.*, 34, 459–466.

Hansen, E.R., Petracek, M.E., Dickey, L.F. and Thompson, W.F. (2001) The 5′ end of the pea ferredoxin-1 mRNA mediates rapid and reversible light-directed changes in translation in tobacco. *Plant Physiol.*, 125, 770–778.

Harpster, M.H., Mayfield, S.P. and Taylor, W.C. (1984) Effects of pigment-deficient mutants on the accumulation of photosynthetic proteins in maize. *Plant Mol. Biol.*, 3, 59–71.

Helliwell, C.A., Webster, C.I. and Gray, J.C. (1997) Light-regulated expression of the pea plastocyanin gene is mediated by elements within the transcribed region of the gene. *Plant J.*, 12, 499–506.

Herrmann, R.G., Westhoff, P. and Link, G. (1992) Biogenesis of plastids in higher plants. In *Cell Organelles* (ed. R.G. Herrmann), Springer, Wien, pp. 276–349.

Hess, W.R., Müller, A., Nagy, F. and Börner, A. (1994) Ribosome-deficient plastids affect transcription of light- induced nuclear genes: genetic evidence for a plastid-derived signal. *Mol. Gen. Genet.*, 242, 505–512.

Hess, W.R., Schendel, R., Börner, T. and Rüdiger, W. (1991) Reduction of mRNA level for two nuclear encoded light regulated genes in the barley mutant *albostrians* is not correlated with phytochrome content and activity. *J. Plant Physiol.*, 138, 292–298.

Hiriart, J.-B., Lehto, K., Tyystjärvi, E., Junttila, T. and Aro, E.-M. (2002) Suppression of a key gene involved in chlorophyll biosynthesis by means of virus-induced gene silencing. *Plant Mol. Biol.*, 50, 213–224.

Jarvis, P. (2001) Intracellular signalling: the chloroplast talks! *Curr. Biol.*, 11, R307–R310.

Jarvis, P. (2003) Intracellular signalling: the language of the chloroplast. *Curr. Biol.*, 13, R314–R316.

Jasper, F., Quednau, B., Kortenjann, M. and Johanningmeier, U. (1991) Control of *cab* gene expression in synchronized *Chlamydomonas reinhardtii* cells. *J. Photochem. Photobiol. B: Biol.*, 11, 139–150.

Jensen, P.E., Willows, R.D., Petersen, B.L. *et al.* (1996) Structural genes for Mg-chelatase subunits in barley: *Xantha-f, -g* and *-h*. *Mol. Gen. Genet.*, 250, 383–394.

Johanningmeier, U. (1988) Possible control of transcript levels by chlorophyll precursors in *Chlamydomonas*. *Eur. J. Biochem.*, 177, 417–424.

Johanningmeier, U. and Howell, S. (1984) Regulation of light-harvesting chlorophyll binding protein mRNA accumulation in *Chlamydomonas reinhardtii* – possible involvement of chlorophyll synthesis precursors. *J. Biol. Chem.*, 259, 13541–13549.

Josse, E.-M., Simkin, A.J., Gaffé, J., Labouré, A.-M., Kuntz, M. and Carol, P. (2000) Plastid terminal oxidase associated with carotenoid desaturation during chromoplast differentiation. *Plant Physiol.*, 123, 1427– 1436.

Kittsteiner, U., Brunner, H. and Rüdiger, W. (1991) The greening process in cress seedlings, I: Complexing agents and 5-aminolevulinate inhibit accumulation of *cab* mRNA coding for the light-harvesting chlorophyll *a/b* protein. *Physiol. Plant*, 81, 190–196.

Knight, J.S., Duckett, C.M., Sullivan, J.A., Walker, A.R. and Gray, J.C. (2002) Tissue-specific, light-regulated and plastid-regulated expression of the single-copy nuclear gene encoding the chloroplast Rieske FeS protein of *Arabidopsis thaliana*. *Plant Cell Physiol.*, 43, 522–531.

Kohchi, T., Mukougawa, K., Frankenberg, N., Masuda, M., Yokata, A. and Lagarias, J.C. (2001) The Arabidopsis *HY2* gene encodes phytochromobilin synthase, a ferredoxin-dependent biliverdin reductase. *Plant Cell*, 13, 425–436.

Kropat, J., Oster, U., Rüdiger, W. and Beck, C.F. (1997) Chlorophyll precursors are signals of chloroplast origin involved in light induction of nuclear heat-shock genes. *Proc. Natl. Acad. Sci. U.S.A.*, 94, 14168–14172.

Kropat, J., Oster, U., Rüdiger, W. and Beck, C.F. (2000) Chloroplast signalling in the light induction of nuclear *HSP70* genes requires the accumulation of chlorophyll precursors and their accessibility to cytoplasm/nucleus. *Plant J.*, 24, 523–531.

Kubis, S., Baldwin, A., Patel, R. *et al.* (2003) The Arabidopsis *ppi1* mutant is specifically defective in the expression, chloroplast import, and accumulation of photosynthetic proteins. *Plant Cell*, 15, 1859–1871.

Kumar, A.M., Chaturvedi, S and Söll, D. (1999) Selective inhibition of *HEMA* gene expression by photooxidation in *Arabidopsis thaliana*. *Phytochemistry*, 51, 847–850.

Kurth, J., Varotto, C., Pesaresi, P. *et al.* (2002). Gene-sequence-tag expression analyses of 1800 genes related to chloroplast functions. *Planta*, 215, 101–109.

Kusnetsov, V., Bolle, C., Lübberstedt, T., Sopory, S., Herrmann, R.G. and Oelmüller, R. (1996) Evidence that the plastid signal and light operate via the same *cis*-acting elements in the promoters of nuclear genes for plastid proteins. *Mol. Gen. Genet.*, 252, 631–639.

Kusnetsov, V., Landsberger, M., Meurer, J. and Oelmüller, R. (1999) The assembly of the CAAT-box binding complex at a photosynthesis gene promoter is regulated by light, cytokinin, and the stage of the plastids. *J. Biol. Chem.*, 274, 36009–36014.

Larkin, R.M., Alonso, J.M., Ecker, J.R. and Chory, J. (2003) GUN4, a regulator of chlorophyll synthesis and intracellular signalling. *Science*, 299, 902–906.

La Rocca, N.L., Barbato, R., Dalla Vecchia, F. and Rascio, N. (2000a) *Cab* gene expression in bleached leaves of carotenoid-deficient maize. *Photosynth. Res.*, 64, 119–126.

La Rocca, N.L., Dalla Vecchia, F., Barbato, R., Bonora, A., Bergantino, E. and Rascio, N. (2000b) Plastid photodamage and *Cab* gene expression in barley leaves. *Physiol. Plant*, 109, 51–57.

La Rocca, N.L., Rascio, N., Oster, U. and Rüdiger, W. (2001) Amitrole treatment of etiolated barley seedlings leads to deregulation of tetrapyrrole synthesis and to reduced expression of *Lhc* and *RbcS* genes. *Planta*, 213, 101–108.

Li, H., Culligan, K., Dixon, R.A. and Chory, J. (1995) *CUE1*: a mesophyll cell-specific positive regulator of light-controlled gene expression in *Arabidopsis*. *Plant Cell*, 7, 1599–1610.

Longstaff, M., Raines, C.A., McMorrow, E.M., Bradbeer, J.W. and Dyer, T.A. (1989) Wheat phosphoglycerate kinase – evidence for recombination between the genes for the chloroplastic and cytosolic enzymes. *Nucleic Acids Res.*, 17, 6569–6580.

López-Juez, E., Jarvis, R.P., Yakeuchi, A., Page, A.M. and Chory, J. (1998) New Arabidopsis *cue* mutants suggest a close connection between plastid and phytochrome regulation of nuclear gene expression. *Plant Physiol.*, 118, 803–815.

Lübberstedt, T., Oelmüller, R., Wanner, G. and Herrmann, R.G. (1994) Interacting *cis* elements in the plastocyanin promoter from spinach ensure regulated high-level expression. *Mol. Gen. Genet.*, 242, 602–613.

Maiwald, D., Dietzmann, A., Jahns, P. *et al.* (2003) Knock-out of the genes coding for the Rieske protein and the ATP-synthase δ-subunit of Arabidopsis. Effects on photosynthesis, thylakoid protein composition, and nuclear chloroplast gene expression. *Plant Physiol.*, 133, 191–202.

Mandel, M.A., Feldman, F.A., Herrera-Estrella, L., Rocha-Sosa, M. and Leon, P. (1996) *CLA1*, a novel gene required for chloroplast development, is highly conserved in evolution. *Plant J.*, 9, 649–658.

Martinez-Hernandez, A., Lopez-Ochoa, L., Argüelo-Astorga, G. and Herrera-Estrella, L. (2002) Functional properties and regulatory complexity of a minimal *RBCS* light-responsive unit activated by phytochrome, cryptochrome, and plastid signals. *Plant Physiol.*, 128, 1223–1233.

Maxwell, D.P., Laudenbach, D.E. and Huner, N.P.A. (1995) Redox regulation of light-harvesting complex II and *Cab* mRNA abundance in *Dunaliella salina*. *Plant Physiol.*, 109, 787–795.

Mayfield, S. and Taylor, W. (1984) Carotenoid-deficient maize seedlings fail to accumulate light harvesting chlorophyll *a/b* binding protein (LHCP) mRNA. *Eur. J. Biochem.*, 144, 79–84.

Mayfield, S. and Taylor, W. (1987) Chloroplast photooxidation inhibits the expression of a set of nuclear genes. *Mol. Gen. Genet.*, 208, 309–314.

McCormac, A.C., Fischer, A., Kumar, A.M., Söll, D. and Terry, M.J. (2001) Regulation of *HEMA1* expression by phytochrome and a plastid signal during de-etiolation in *Arabidopsis thaliana*. *Plant J.*, 25, 549–561.

McCormac, A.C. and Terry, M.J. (2002) Loss of nuclear gene expression during the phytochrome A-mediated far-red block of greening response. *Plant Physiol.*, 130, 402–414.

Meskauskiene, R., Nater, M., Goslings, D., Kessler, F., op den Camp, R. and Apel, K. (2001) FLU: a negative regulator of chlorophyll biosynthesis in *Arabidopsis thaliana*. *Proc. Natl. Acad. Sci. U.S.A.*, 98, 12826–12831.

Mochizuki, N., Brusslan, J., Larkin, J., Nagatani, A. and Chory, J. (2001) Arabidopsis *genomes uncoupled 5 (GUN5)* mutant reveals the involvement of Mg chelatase H subunit in plastid-to-nucleus signal transduction. *Proc. Natl. Acad. Sci. U.S.A.*, 98, 2053–2058.

Mochizuki, N., Susek, R.E. and Chory, J. (1996) An intracellular signal transduction pathway between the chloroplast and nucleus is involved in de-etiolation. *Plant Physiol.*, 112, 1465–1469.

Montané, M.-H., Tardy, F., Kloppstech, K. and Havaux, M. (1998) Differential control of xanthophylls and light-induced stress proteins, as opposed to light-harvesting chlorophyll *a/b* proteins, during photosynthetic acclimation of barley leaves to light irradiance. *Plant Physiol.*, 118, 227–235.

Møller, S.G., Kunkel, T. and Chua, N.-H. (2001) A plastidic ABC protein involved in intercompartmental communication of light signaling. *Genes Dev.*, 15, 90–103.

Muramoto, T., Kohchi, T., Yokota, A., Hwang, I. and Goodman, H.M. (1999) The *Arabidopsis* photomorphogenic mutant *hy1* is deficient in phytochrome chromophore biosynthesis as a result of a mutation in a plastid heme oxygenase. *Plant Cell*, 11, 335–347.

Oelmüller, R. (1989) Photooxidative destruction of chloroplasts and its effects on nuclear gene expression and extraplastidic enzyme levels. *Photochem. Photobiol.*, 49, 229–239.

Oelmüller, R., Levitan, I., Bergfeld, R., Rajasekhar, V.K. and Mohr, H. (1986) Expression of nuclear genes as affected by treatments acting on plastids. *Planta*, 168, 482–492.

Oelmüller, R. and Mohr, H. (1986) Photooxidative destruction of chloroplasts and its consequences for expression of nuclear genes. *Planta*, 167, 106–113.

Oster, U., Brunner, H. and Rüdiger, W. (1996) The greening process in cress seedlings. Possible interference of chlorophyll precursors, accumulated after thujiplacin treatment with light-regulated expression of *Lhc* genes. *J. Photochem. Photobiol. B: Biol.*, 36, 255–261.

Oswald, O., Martin, T., Dominy, P.J. and Graham, I. (2001) Plastid redox state and sugars: interactive regulators of nuclear-encoded photosynthetic gene expression. *Proc. Natl. Acad. Sci. U.S.A.*, 98, 2047–2052.

Papenbrock, J., Pfündel, E., Mock, H.-P. and Grimm, B. (2000) Decreased and increased expression of the subunit CHLI diminishes Mg chelatase activity and reduces chlorophyll synthesis in transgenic tobacco plants. *Plant J.*, 22, 155–164.

Pesaresi, P., Varotto, C., Meurer, J., Jahns, P., Salamini, F. and Leister, D. (2001) Knock-out of the plastid ribosomal protein L11 in *Arabidopsis*: effects on mRNA translation and photosynthesis. *Plant J.*, 27, 179–189.

Petracek, M.E., Dickey, L.F., Nguyen, T.T. *et al.* (1998) Ferredoxin-1 mRNA is destabilized by changes in photosynthetic electron transport. *Proc. Natl. Acad. Sci. U.S.A.*, 95, 9009–9013.

Petracek, M.E., Dickey, L.F. and Thompson, W.F. (1997) Light-regulated changes in abundance and polyribosome association of ferredoxin are dependent on photosynthesis. *Plant Cell*, 9, 2291–2300.

Pfannschmidt, T. (2003) Chloroplast redox signals: how photosynthesis controls its own genes. *Trends Plant Sci.*, 8, 33–41.

Pfannschmidt, T., Allen, J.F. and Oelmüller, R. (2001a) Principles of redox control of photosynthesis gene expression. *Physiol. Plant*, 112, 1–9.

Pfannschmidt, T., Schütze, K., Brost, M. and Oelmüller, R. (2001b) A novel mechanism of nuclear photosynthesis gene regulation by redox signals from the chloroplast during photosystem stoichiometry adjustment. *J. Biol. Chem.*, 276, 36125–36130.

Pöpperl, G., Oster, U. and Rüdiger, W. (1998) Light-dependent increase in chlorophyll precursors during day-night cycle in tobacco and barley seedlings. *J. Plant Physiol.*, 153, 40–45.

Puente, P., Wei, N. and Deng, X.-W. (1996) Combinatorial interplay of promoter elements constitutes the minimal determinants for light and developmental control of gene expression in Arabidopsis. *EMBO J.*, 15, 3732–3743.

Pursiheimo, S., Mulo, P., Rintamäki, E. and Aro, E-M. (2001) Coregulation of light-harvesting complex II phosphorylation and *lhcb* mRNA accumulation in winter rye. *Plant J.*, 26, 317–327.

Rapp, J.C. and Mullet, J.E. (1991) Chloroplast transcription is required to express the nuclear genes *RbcS* and *Cab*. Plastid DNA copy number is regulated independently. *Plant Mol. Biol.*, 17, 813–823.

Reiss, T., Bergfeld, R., Link, G., Thien, W. and Mohr, H. (1983) Photooxidative destruction of chloroplasts and its consequences to cytosolic enzyme levels and plant development. *Planta*, 159, 518–528.

Richly, E., Dietzmann, A., Biehl, A. *et al.* (2003) Covariations in the nuclear chloroplast transcriptome reveal a regulatory master-switch. *EMBO Rep.*, 4, 491–498.

Rintamäki, E., Matinsuo, P., Pursiheimo, S and Aro, E-M. (2000) Cooperative regulation of light-harvesting complex II phosphorylation via the plastoquinol and ferredoxin-thioredoxin system in chloroplasts. *Proc. Natl. Acad. Sci. U.S.A.*, 97, 11644–11649.

Rodermel, S. (2001) Pathways of plastid-to-nucleus signaling. *Trends Plant Sci.*, 6, 471–478.

Rodermel, S. and Park, S. (2003) Pathways of intracellular communication: tetrapyrroles and plastid-to-nucleus signalling. *BioEsssays*, 25, 631–636.

Sagar, A.D. and Briggs, W.R. (1990) Effects of high light stress on carotenoid-deficient chloroplasts in *Pisum sativum*. *Plant Physiol.*, 94, 1663–1670.

Sagar, A.D., Horwitz, B.A., Elliott, R.C., Thompson, W.F. and Briggs, W.R. (1988) Light effects on several chloroplast components in norflurazon-treated pea seedlings. *Plant Physiol.*, 88, 340–347.

Schwartz, B., Daniel, S.G. and Becker, W. (1992) Photooxidative destruction of chloroplasts leads to reduced expression of peroxisomal NADH-dependent hydroxypyruvate reductase in developing cucumber cotyledons. *Plant Physiol.*, 99, 681–685.

Sherameti, I., Nakamura, M., Yamamoto, Y., Pfannschmidt, T., Obokata, J. and Oelmüller, R. (2002) Polyribosome loading of spinach mRNAs for photosystem I subunits is controlled by photosynthetic electron transport. *Plant J.*, 32, 631–639.

Simpson, J., van Montagu, M. and Herrera-Estrella, L. (1986) Photosynthesis-associated gene families: differences in response to tissue-specific and environmental factors. *Science*, 233, 34–38.

Singh, D.P., Cornah, J.E., Hadingham, S. and Smith, A.G. (2002) Expression analysis of the two ferrochelatase genes in *Arabidopsis* in different tissues and under stress conditions reveal their different roles in haem biosynthesis. *Plant Mol. Biol.*, 50, 773–788.

Stockhaus, J., Eckes, P., Blau, A., Schell, J. and Willmitzer, L. (1987) Organ-specific and dosage-dependent expression of a leaf/stem specific gene from potato after tagging and transfer into potato and tobacco plants. *Nucleic Acids Res.*, 15, 3479–3491.

Strand, A., Asami, T., Alonso, J., Ecker, J.R. and Chory, J. (2003) Chloroplast to nucleus communication triggered by accumulation of Mg-protoporphyrin IX. *Nature*, 421, 79–83.

Streatfield, S.J., Weber, A., Kinsman E.A. *et al.* (1999) The phosphoenolpyruvate/phosphate translocator is required for phenolic metabolism, palisade cell development, and plastid-dependent nuclear gene express. *Plant Cell*, 11, 1609–1621.

Sullivan, J.A. and Gray, J.C. (1999) Plastid translation is required for the expression of nuclear photosynthesis genes in the dark and in roots of the pea *lip 1* mutant. *Plant Cell*, 11, 901–911.

Sullivan, J.A. and Gray, J.C. (2002) Multiple plastid signals regulate the expression of the pea plastocyanin gene in pea and transgenic tobacco plants. *Plant J.*, 32, 763–774.

Surpin, M., Larkin, R.M. and Chory, J. (2002) Signal transduction between the chloroplast and the nucleus. *Plant Cell*, 13, S327–S338.

Susek, R.E., Ausubel, F.M. and Chory, J. (1993) Signal transduction mutants of *Arabidopsis* uncouple nuclear *CAB* and *RBCS* expression from chloroplast development. *Cell*, 74, 787–799.

Tamada, Y., Imanari, E., Kurotani, K.-I., Nakai, M., Andreo, C.S. and Izui, K. (2003) Effect of photooxidative destruction of chloroplasts on the expression of nuclear genes for C4 photosynthesis and for chloroplast biogenesis in maize. *J. Plant Physiol.*, 160, 3–8.

Taylor, W.C. (1989) Regulatory interactions between nuclear and plastid genomes. *Ann. Rev. Plant Physiol. Plant Mol. Biol.*, 40, 211–233.

Vener, A.V., van Kim, P.J.M., Rich, P.R., Ohad, I. and Andersson, B. (1997) Plastoquinol at the quinol oxidation site of reduced cytochrome bf mediates signal transduction between light and protein phosphorylation: thylakoid protein kinase deactivation by a single-turnover flash. *Proc. Natl. Acad. Sci. U.S.A.*, 94, 1585–1590.

Vinti, G., Hills, A., Campbell, S. *et al.* (2000) Interactions between *hy1* and *gun* mutants of *Arabidopsis*, and their implications for plastid/nuclear signalling. *Plant J.*, 24, 883–894.

Voll, L., Häusler, R.E., Hecker, R. *et al.* (2003) The phenotype of the *Arabidopsis cue1* mutant is not simply caused by a general restriction of the shikimate pathway. *Plant J.*, 36, 301–317.

Vorst, O., Kock, P., Lever, A. *et al.* (1993) The promoter of the *Arabidopsis thaliana* plastocyanin gene contains a far upstream enhancer-like element involved in chloroplast-dependent expression. *Plant J.*, 4, 933–945.

Wetzel, C.M., Jiang, C.-Z., Meehan, L.J., Voytas, D.F. and Rodermel, S.R. (1994) Nuclear-organelle interactions: the *immutans* variegation mutant of *Arabidopsis* is plastid autonomous and impaired in carotenoid biosynthesis. *Plant J.*, 6, 161–175.

Wilson, R.J.M., Rangachi, A., Saldanha, J.W., Rickman, L., Buxton, R.S. and Eccleston, J.F. (2003) Parasite plastids: maintenance and functions. *Philos. Trans. R. Soc. Lond. B*, 358, 155–164.

Wu, D., Wright, D.A., Wetzer, C., Voytas, D.F. and Rodermel, S. (1999) The *IMMUTANS* variegation locus of *Arabidopsis* defines a mitochondrial alternative oxidase homolog that functions during early chloroplast development. *Plant Cell*, 11, 43–55.

Yoshida, R., Sato, T., Kanno, A. and Kameya, T. (1998) Streptomycin mimics the cool temperature response in rice plants. *J. Exp. Bot.*, 49, 221–227.

Young, A.J. (1991) Inhibition of carotenoid biosynthesis. In *Herbicides* (eds N.R. Baker and M.P. Percival), Elsevier, Amsterdam, pp. 131–171.

9 Chloroplast avoidance movement

Masahiro Kasahara and Masamitsu Wada

9.1 Introduction

Although plants are static organisms, their organelles move actively inside cells and take up their adequate position to optimize metabolic activities. The intracellular positions of organelles are therefore affected by various environmental stimuli. Light-directed chloroplast movement is one of the well-characterized organelle movements induced by environmental changes. Under low-light conditions, chloroplasts spread over the cell surface perpendicular to the direction of light to harvest sufficient light and to maximize photosynthetic activity (Figure 9.1), whereas under high-light conditions, chloroplasts situate along the cell sides parallel to the direction of incident light so as to minimize potential photodamage (Figure 9.1). When a part of a cell is irradiated by light during observation under the microscope, it can be further realized that chloroplast movement is dynamic and well organized. Figure 9.2 demonstrates chloroplast accumulation movement in a dark-adapted cell, where chloroplasts move toward a spot irradiated for 1 min with low-intensity light. In contrast, chloroplasts move away from a high-light spot irradiated continuously by chloroplast avoidance movement (Figure 9.3). It takes approximately 30–60 min to accomplish the relocation in gametophytes of fern *Adiantum capillus-veneris* (Wada and Kagawa, 2001) and in mesophyll cells of *Arabidopsis* leaves (Trojan and Gabrys, 1996; Kagawa and Wada, 2000). Light-directed chloroplast movement has been observed in most of plant cells tested so far, including algae, mosses, ferns, and seed plants. This implies that chloroplast movement is most probably indispensable for such sessile organisms to live under sunlight.

Chloroplast movement was already known at the beginning of the twentieth century (Senn, 1908). Since then, many plant scientists have been interested in identifying the photoreceptor controlling chloroplast movement, undertaking studies of light quality versus movement. In seed plants, chloroplast movement is exclusively controlled by blue light (Inoue and Shibata, 1974). Red light as well as blue light induces chloroplast movement in a fern, *A. capillus-veneris* (Yatsuhashi *et al.*, 1985), a moss, *Physcomitrella patens* (Kadota *et al.*, 2000; Sato *et al.*, 2001a), and algae, *Mougeotia scalaris* (Haupt and Scheuerlein, 1990), and *Mesotaenium caldariorum* (Haupt and Scheuerlein, 1990). The effect of chloroplast movement on the photosynthetic activity has also been vigorously studied (Zurzycki, 1955, 1957; Haupt and Scheuerlein, 1990; Park *et al.*, 1996).

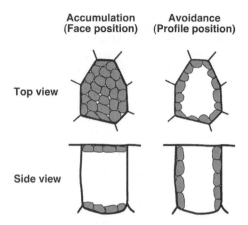

Figure 9.1 Chloroplast arrangements in accumulation and avoidance responses in a cell. Chloroplasts change their position depending on the light conditions. They accumulate along periclinal cell walls and take the face position under low-light condition (left) whereas they move to sidewalls and take the profile position under strong light condition (right). Top and side views are shown.

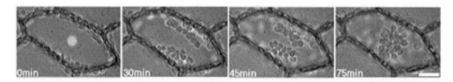

Figure 9.2 Serial images of chloroplast accumulation movement. A microbeam of red light (30 W m^{-2}, 10 μm in diameter) is shown as a white spot (0 min) on a dark-adapted prothallial cell of *A. capillus-veneris*. The microbeam was turned on at 0 min and turned off after 1 min of irradiation. Chloroplasts moved toward the area where the microbeam had been irradiated (30–75 min). Bar: 20 μm. Reproduced from Wada and Kagawa (2001).

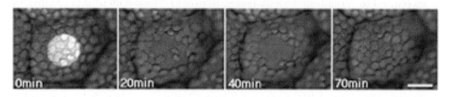

Figure 9.3 Serial images of chloroplast avoidance movement. A microbeam of strong blue light (30 W m^{-2}, 27 μm in diameter) was applied continuously to a weak-light-adapted prothallial cell of *A. capillus-veneris* from 0 to 40 min. The cell was then kept in the dark up to 70 min. Note that weak light signals are raised simultaneously at the irradiated area even under strong light and remained for a while. Consequently, chloroplasts move inside the irradiated area even after the strong light was turned off (70 min). On the other hand, strong light signals have a relatively short life because as soon as the strong light was turned off chloroplasts moved into the area where the light had been irradiated and weak light signals were still present. Bar: 20 μm. Reproduced from Wada and Kagawa (2001).

After years of studies, the blue-light photoreceptor controlling chloroplast movement was identified from *Arabidopsis thaliana* in 2001 (Jarillo *et al.*, 2001; Kagawa *et al.*, 2001). In this review, we will mainly introduce the recent progress in this field. Because chloroplast avoidance movement and accumulation movement are closely related to each other, we will describe both movements. For a more detailed analysis of classic experiments and information about several aspects, refer to earlier reviews (Schönbohm, 1980; Zurzycki, 1980; Haupt and Scheuerlein, 1990; Wada and Kagawa, 2001; Kagawa and Wada, 2002; Wada *et al.*, 2003).

9.2 Photoreceptors controlling chloroplast movement

Because of the simplicity of observation, chloroplast movement was studied in detail using cells with simple structures such as algae, moss, and fern protonemata, and *Lemna trisulca*. Chloroplast movement in land seed plants was less well understood because the multilayered cell organization in leaves made detailed analysis difficult. Both chloroplast accumulation and avoidance movements in *Arabidopsis* have been detected photometrically (Trojan and Gabrys, 1996) and observed microscopically (Kagawa and Wada, 2000), and together with the power of *Arabidopsis* genetics it was realized that *Arabidopsis* would actually become a good system to elucidate chloroplast movement.

9.2.1 Phototropin

Kagawa *et al.* (2001) developed a simple screening method applicable for large-scale mutant isolation (Figure 9.4). Approximately 100,000 ethylmethane sulfonate (EMS) mutagenized lines and 15,000 T-DNA-tagged lines were screened using the method and more than 10 mutants that do not show avoidance movement were isolated. After confirmation of the phenotype microscopically, these mutants were named *cav* (defective in *c*hloroplast *av*oidance movement). Four EMS-mutagenized lines (*cav1-1* to -4) and one T-DNA-tagged line (*cav1-5*) were allelic to each other. In these *cav1* mutants, chloroplast accumulation movement was normal. The mutation sites of the *cav1* mutants were all found in the *PHOT2* gene (encoding phototropin 2 and formerly known as *NPL1*). At the same time, Jarillo *et al.* (2001) showed photometrically that a reverse-genetically obtained *phot2* mutant of *Arabidopsis* lacked chloroplast avoidance movement. *PHOT2* is the paralog of *PHOT1*, phototropin 1 (formerly known as *NPH1*), which has been identified as a blue-light receptor for phototropism under low-light condition (Briggs and Huala, 1999; Sakai *et al.*, 2001; Briggs and Christie, 2002). Thus, it was strongly suggested that *PHOT2* encoded a blue-light receptor for chloroplast avoidance movement.

In the *phot1-5* mutant, chloroplast avoidance movement was normal, whereas chloroplast accumulation movement was slightly less sensitive to the light intensity (Kagawa and Wada, 2000). Microbeam irradiation with blue light at an intensity of $0.1 \ W \ m^{-2}$ induced accumulation movement in wild-type cells, while five times higher blue-light condition ($0.5 \ W \ m^{-2}$) was required to induce accumulation

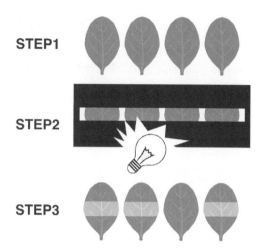

Figure 9.4 Large-scale mutant screening method. STEP1: Leaves of mutagenized *Arabidopsis* are aligned on an agar plate (approximately 200 pieces on a 10 × 14 cm plate). STEP2: The leaves are covered with a non-light-transmitting plate with open slits and are irradiated with strong light for about 1 h. An overhead projector is helpful as the strong light source. STEP3: The irradiated areas become pale green in wild-type leaves because of the increase of light transmittance caused by chloroplast avoidance movement. No band appears in mutant leaves.

movement in *phot1-5* cells. This fact shows that phot1 is involved in accumulation movement; however, another blue-light photoreceptor mediating the accumulation response must still exist. Sakai *et al.* (2001) studied both movements using a *phot1* and *phot2* double mutant and it was very clear that accumulation movement as well as avoidance movement was not observed at any light intensity tested under blue-light microbeam irradiation. This result shows that phot1 and phot2 redundantly function as blue-light receptors for chloroplast accumulation movement and that phot2 solely functions for chloroplast avoidance movement. Thus, phot2 mediates accumulation movement at low-intensity light and avoidance movement at high-intensity light. One interesting question that arises is how does phot2 sense light intensity and switch its function.

So far, it has also been shown that phot1 and phot2 mediate other responses in addition to chloroplast movement, including phototropism, stomatal opening and leaf expansion.

9.2.2 Characteristics of phototropins

Phototropin is a receptor kinase with two LOV (*l*ight, *o*xygen or *v*oltage) domains at the N-terminal region and a Ser/Thr-kinase domain at the C-terminal region (Figure 9.5). In the *Arabidopsis* genome, there are two genes, *PHOT1* and *PHOT2*, encoding proteins consisting of 996 and 915 amino acids, respectively. The

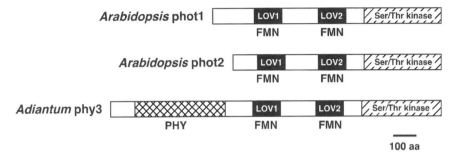

Figure 9.5 Domain organization of *Arabidopsis* phot1 and phot2 and *Adiantum* phy3. *Arabidopsis* phot1 and phot2 have two LOV domains as the binding sites for flavin mononucleotide (FMN) cofactors. *Adiantum* phy3 consists of full domains of phototropin and the N-terminal extension is homologous to the phytochrome chromophore binding domain (PHY; phytochromobilin). Scale bar: 100 amino acids. Note that phot1, PHOT1, *PHOT1* and *phot1* are used to indicate holoprotein, apoprotein, wild-type gene and mutated gene, respectively (Briggs *et al.*, 2001).

N-terminus of PHOT2 is shorter than that of PHOT1. Although hydropathy analysis using the deduced amino acid sequences predicts that phototropin should be a soluble protein, it is shown to be localized at the cytoplasmic membrane (Christie *et al.*, 2002; Sakamoto and Briggs, 2002). The localization of phototropin is consistent with the prediction that the photoreceptor for chloroplast movement should be working at or near the cytoplasmic membrane (Wada *et al.*, 1983; Haupt and Scheuerlein, 1990).

PAS domains are reported to function as protein–protein interaction domains and cofactor binding sites (Taylor and Zhulin, 1999). A subset of PAS domains, which are predicted to be domains for signal sensing such as light, oxygen or voltage, is closely related to the PAS domain of phototropin (Huala *et al.*, 1997). LOV domains of PHOT1 and PHOT2 were expressed in *Escherichia coli* and the purified LOV protein fractions were yellow in color because of flavin mononucleotide binding (Christie *et al.*, 1999; Kasahara *et al.*, 2002b). The absorption spectrum of the LOV domain resembles the action spectrum for chloroplast movement in *L. trisulca* (Figure 9.6). This good correlation is also consistent with the conclusion that phototropin is the blue-light receptor for chloroplast movement.

Since its discovery, the phototropin has been actively studied by biophysicists in addition to plant physiologists. The crystal structure of a LOV domain has already been solved (Crosson and Moffat, 2001, 2002) and the photochemical reactions of LOV domains have been analyzed in detail (Salomon *et al.*, 2000; Swartz *et al.*, 2001; Crosson *et al.*, 2003). cDNA sequences encoding phototropin have been isolated from algae, moss, fern, and various seed plants (Briggs *et al.*, 2001). For more details on phototropin, see other reviews (Briggs and Huala, 1999; Briggs *et al.*, 2001; Briggs and Christie, 2002).

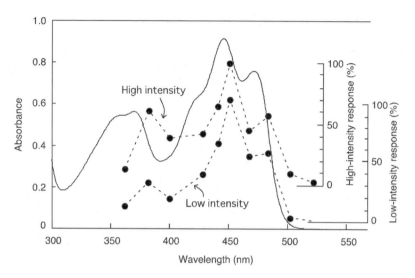

Figure 9.6 The absorption spectrum of the *Arabidopsis* phot2 LOV2 domain with the action spectra of high- and low-intensity chloroplast responses of *L. trisulca*. The *Arabidopsis* phot2 LOV2 domain was expressed in *E. coli* and purified. The absorption spectrum of the purified LOV domain was plotted with the action spectra of the high-intensity response (avoidance movement) and the low-intensity response (accumulation movement) of *L. trisulca*. The action spectra were redrawn from Zurzycki (1980).

9.2.3 A unique photoreceptor-mediating red-light-dependent chloroplast movement

In the gametophytes of the fern *A. capillus-veneris*, both chloroplast accumulation and avoidance movements are controlled by red light as well as blue light (Yatsuhashi *et al.*, 1985). Recently, phytochrome 3 (phy3) has been shown to mediate the red-light-dependent chloroplast movement response (Kawai *et al.*, 2003). The phy3 is a hybrid of phytochrome and phototropin (Figure 9.5) and binds phycocyanobilin as a cofactor in the phytochrome domain exhibiting red/far-red reversibility similar to conventional phytochrome (Nozue *et al.*, 1998). The phytochrome domain and the phototropin-related kinase domain of phy3 are probably working as the red-light perception and the output to downstream signaling components, respectively. In the protonemata of the moss *P. patens*, both movements are controlled by red and blue light (Kadota *et al.*, 2000; Sato *et al.*, 2001a). A phy3-type receptor has not been found in *P. patens*. A phototropin knockout mutant of *P. patens* removed and highly reduced the light sensitivity for both red- and blue-light-dependent avoidance movements and accumulation movements, respectively (Kasahara, *et al.*, in press). This result shows that phototropins are most likely downstream components of phytochrome in red-light-dependent signaling pathway for chloroplast movements in *P. patens* and that phototropins have a crucial role in the red-light-dependent chloroplast movement signaling pathway.

9.3   Downstream signaling from the photoreceptors

Signaling pathways for light-induced chloroplast movement can be divided into three steps: (i) light perception, (ii) signal transduction, and (iii) motile systems. The discovery of phototropin has helped in elucidating the beginning of the light perception step. The other two steps are less well understood at the molecular level. However, calcium and actin-based cytoskeleton have long been thought to be key components in the signal transduction step and the motility of chloroplasts, respectively. Recently, calcium was shown to be a possible downstream component of the phototropin signaling pathways, a finding providing clues on how calcium functions in chloroplast movement (Baum *et al.*, 1999; Babourina *et al.*, 2002; Harada *et al.*, 2003; Stoelzle *et al.*, 2003). We have recently identified a novel protein required for proper chloroplast positioning and movement, named CHUP1 (Oikawa *et al.*, 2003). Since CHUP1 has actin-binding properties, the analysis of this protein should provide clues toward understanding the last movement step.

9.3.1   Calcium

The effect of calcium on chloroplast movement has been studied by manipulating the cytoplasmic calcium levels using calcium-channel inhibitors and calcium ionophores (for review, see Haupt and Scheuerlein, 1990; Wada and Kagawa, 2001; Sato *et al.*, 2003). For example, by increasing cytoplasmic calcium concentrations by addition of a calcium ionophore (A23187) and having high calcium in the medium results in chloroplasts moving to the sidewall, similar to avoidance movement induced by strong light in *L. trisulca* (Tlalka and Fricker, 1999). Internal storage of calcium may be more critical for chloroplast movement in this organism rather than external calcium because the block of calcium influx has little effect on movement (Tlalka and Gabrys, 1993; Sato *et al.*, 2001b).

Sato *et al.* (2001b) have found that chloroplast movement is induced not only by light but also by mechanical stimulation in *A. capillus-veneris*, and clearly shown that distinct calcium storages are involved in the response. The mechanical-stimulation-induced chloroplast movement is suppressed by the addition of plasma membrane calcium-channel blockers, lanthanum (La^{3+}) or gadolinium (Gd^{3+}), whereas chloroplast movement is induced by light stimulation even in the presence of La^{3+} or Gd^{3+} (Sato *et al.*, 2001b). These facts suggest that the influx of external calcium and the release of intracellular calcium storage are required for the mechanical-stimulation-induced movement and the light-induced movement, respectively, assuming that a similar system is involved in both movements. However, the finding that external calcium is required for the mechanical-stimulation-induced movement makes the situation more confusing because it is difficult to judge which movement is induced by artificially manipulating intracellular calcium levels. Whether calcium functions as a signaling molecule for "light-induced" chloroplast movement is still an open question. It should also be kept in mind that calcium is involved in many fundamental physiological activities.

Despite this seemingly complicated situation, there has been an accumulation of data showing possible functions of calcium as a downstream signal for phototropins (Baum *et al.*, 1999; Babourina *et al.*, 2002; Harada *et al.*, 2003; Stoelzle *et al.*, 2003). Baum *et al.* (1999) studied changes in cytosolic calcium levels induced by a blue-light pulse, using transgenic plants expressing aequorin, a calcium indicator. A transient increase (lasting about 80 s) occurs in wild-type *Arabidopsis* seedlings, but the level of transient increase is considerably reduced in *phot1* mutant seedlings. Using the same techniques, Harada *et al.* (2003) showed the occurrence of a similar transient increase in wild-type leaves but reduced transient levels in both *phot1* and *phot2* mutants in a light-intensity-dependent manner. Stoelzle *et al.* (2003) showed that voltage-dependent calcium channels in the plasma membrane are activated by blue-light irradiation, using *Arabidopsis* mesophyll protoplasts and the patch-clamp technique. The channel activation is completely abolished in a *phot1phot2* double mutant. The patch-clamp data suggest that the calcium influx through the plasma membrane is probably an early signaling component of the phototropin signaling pathway. Since phot1 and phot2 probably localize to the periphery of the plasma membrane (Christie *et al.*, 2002; Sakamoto and Briggs, 2002; Harada *et al.*, 2003), the phototropins could directly regulate the activity of calcium channels. It is suggested that phot2 could induce a phospholipase C mediated phosphoinositide signaling pathway and promote the release of calcium from internal calcium storages, such as ER and vacuole, through inositol 1,4,5-triphosphate (Harada *et al.*, 2003). The phototropin-induced transient calcium increase may function in the early signaling steps of light-induced chloroplast movement. Further biochemical analysis and isolation of key signaling components are however required.

9.3.2 Actin-based cytoskeleton

The cytoskeleton acts as the effector system in the signaling pathway for chloroplast movement. Many reports show that cytoskeletal drugs disrupting the function of actin filaments inhibit chloroplast movement while drugs disrupting the function of microtubules have no effect (Wagner *et al.*, 1972; Kadota and Wada, 1992a; Malec *et al.*, 1996; Gorton *et al.*, 1999). In addition, the observation of chloroplasts in cryo-fixed and freeze-substituted leaf cells of *Arabidopsis*, which is a good technique to preserve intact actin filaments, shows that chloroplasts are closely associated with actin filaments but not with microtubules (Kandasamy and Meagher, 1999). Actin filaments are therefore key components in the motile chloroplast movement system. However, it has recently been shown that chloroplasts in the moss *P. patens* cells use microtubules as well as actin filaments (Sato *et al.*, 2001a) but this case seems to be an exception rather than the rule. Interestingly, it has been reported that circular structures composed of actin filaments are visible around chloroplasts in *A. capillus-veneris* (Kadota and Wada, 1992b). These circular structures appear when chloroplasts settle at their destination by accumulation movement, and disappear before the chloroplasts disperse from the area after transfer to darkness (Kadota and Wada, 1992b). The circular structures are proposed to function as anchors for chloroplasts. Similar actin structures have also been seen in *Arabidopsis* (Kandasamy and

Meagher, 1999). This type of chloroplast-specific actin structure may play crucial roles in chloroplast movement.

9.3.3   CHUP1 required for proper chloroplast positioning and movement

During the mutant screening program for chloroplast movement as shown in Figure 9.4, an alternative mutant line was isolated from *Arabidopsis* in addition to the *phot2* mutants (Oikawa *et al.*, 2003). Although *phot2* mutants are defective only in chloroplast avoidance movement, the alternative mutant lacks both chloroplast avoidance and accumulation movements. Microscopic observation revealed that almost all chloroplasts gathered at the bottom of the abaxial side of leaf cells (Figure 9.7A). The mutant locus was named *chup1*, standing for *ch*loroplast *u*nusual *p*ositioning 1. The *CHUP1* gene encodes a protein (1004 amino acids; 112-kDa molecular mass) having an actin-binding domain as well as other multiple putative functional domains, including a hydrophobic signal peptide-like domain, a coiled-coil domain, a proline-rich domain, and two leucine-zipper domains (Figure 9.7B). The actin-binding domain has been shown to bind to F-actin (filamentous actin) *in vitro*. Green fluorescent protein fusion with the hydrophobic signal peptide-like domain was localized at the periphery of chloroplasts, suggesting that CHUP1 may be localized at the chloroplast outer membrane. Orthologs of *CHUP1* have been isolated from the moss *P. patens* and the fern *A. capillus-veneris* (M. Kasahara, unpublished data, 2004), but no similar genes have been found in the sequence databases of organisms without chloroplasts, such as yeast, *Caenorhabditis elegans*, and *Drosophilla melanogaster*. Chloroplast-associated CHUP1 probably has a crucial function,

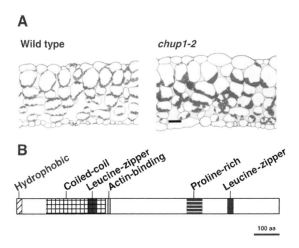

Figure 9.7   (A) Distribution of chloroplasts in the mesophyll cells of wild-type and *chup1-2* leaves. Cross-sections of wild-type and *chup1-2* leaves grown under low-light condition are shown. Bar: 30 μm. (B) Domain organization of CHUP1. Hydrophobic region, coiled-coil domain, actin-binding domain, proline-rich region and two leucine-zipper domains are indicated. Reproduced from Oikawa *et al.* (2003).

utilizing actin filaments required for chloroplast movement. Further analysis of this interesting protein should reveal the precise mechanism of the motile system for chloroplast movement.

9.4   Physiological significance of chloroplast movement

Plants growing under low-light condition, such as at the forest floor, carry out maximal light harvesting for photosynthesis, whereas under high-light condition, such as during daytime in the summer, the photosynthetic apparatus absorbs excess light. The excess light means that the light energy absorbed in the photosynthetic apparatus is beyond that utilized for the photosynthetic reaction. Reactive oxygen species are generated in chloroplasts by the extinction of the excess absorbed energy, which then causes oxidative damage to biological molecules. Since the excess light levels vary according to the environmental conditions, plants have developed numerous adaptation mechanisms at the molecular, subcellular, and organ level toward strong light, including scavenging enzymes of reactive oxygen species, antioxidant molecules and thermal dissipation system, chloroplast avoidance movement, and leaf movement (Demmig-Adams and Adams, 1992; Asada, 1999; Niyogi, 1999).

9.4.1   Chloroplast accumulation movement

Under low-light condition, chloroplasts move toward the illuminated area and take the face position (Figure 9.1). Zurzycki (1955) examined the relationship between the rate of photosynthesis and chloroplast arrangement and showed that when light is the limiting factor, the rate of photosynthesis increases as chloroplasts move to the cell surface from the sidewall. A similar relationship was observed in experiments using leaves of land plants having multilayered organ architecture (Lechowski, 1974). Chloroplast accumulation movement is effective in maximizing photosynthetic activity under light-limiting conditions.

9.4.2   Chloroplast avoidance movement

Chloroplasts move away from the area illuminated with strong light and take the profile position (Figure 9.1). There are several reports investigating the ecological significance of chloroplast avoidance movement. Zurzycki (1957) wondered if chloroplast avoidance movement modifies the destructive effect of strong light on the photosynthetic apparatus and studied the correlation between the inactivation kinetics of photosynthesis and chloroplast arrangement. Brugnoli and Björkman (1992) showed that chloroplast movements can have an important effect on the efficiency of light utilization in photosynthesis. Park et al. (1996) further found that photosystem II (PSII) of Tradescantia albiflora, a shade plant, is more resistant to high-light stress than that of pea. This was an unexpected result because PSII of shade plants is likely to be more susceptible to strong light than that of pea. They investigated some factors affecting PSII inactivation from both plants and suggested that the best explanation for the greater resistance of T. albiflora PSII was the difference in light transmittance of their leaves after strong light irradiation, which is caused by their different capacity of chloroplast avoidance movement. A more marked chloroplast avoidance movement seems to help decrease the light absorption of PSII and contribute to the strong light tolerance. Jeong *et al.* (2002) used an *FtsZ* mutant of tobacco, which has a few large chloroplasts in cells because of impaired chloroplast division. They suggested that growth retardation of the *FtsZ* mutant can be attributed to reduced changes of light transmittance through leaves in response to light intensity. Consequently, the mutant failed to harvest a sufficient amount of light and to decrease light absorption under low- and high-light growth conditions, respectively.

To assess the physiological significance of chloroplast avoidance movement, *Arabidopsis* mutants are now available, representing the best experimental system, because it is easier to equalize physiological conditions and minimize influence to other physiological phenomena. Kasahara *et al.* (2002a) used two *Arabidopsis* mutants, *phot2-1* and *chup1-2*, for this purpose. Chloroplasts in *phot2-1* cells take the face position even under strong light condition because phot1 induces accumulation movement, regardless of the light intensity (Figure 9.8). In *chup1-2* cells, chloroplasts gathered at the bottom surface because of the deficiency in the motile system, as described above (Figure 9.8). The amount of light transmittance through leaves

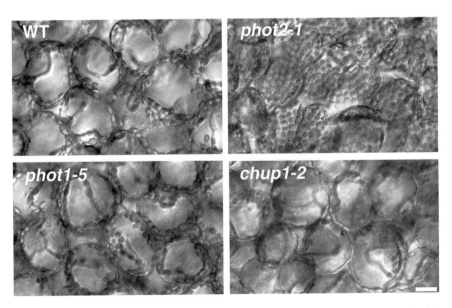

Figure 9.8 Phenotypes of chloroplast movement mutants. Leaves were treated with 500 μmol m^{-2} s^{-1} white light for 1 h before taking the micrographs. Chloroplasts in wild-type (WT) and *phot1-5* cells locate along the sidewalls, whereas those in *phot2-1* align along the cell surface. Chloroplasts in *chup1-2* cell are hardly visible because all the micrographs were focused on the adaxial surface of the palisade cell layers, although the chloroplasts of *chup1-2* stay at the bottom of the cells. Reproduced from Kasahara *et al.* (2002a).

of these mutants is relatively constant under increasing light intensity, indicating that they are unable to decrease light absorption by avoidance movement.

Since PSII is well known to be the primary target of photodamage, F_v/F_m, a parameter indicating the maximal quantum yield of PSII, was measured during a strong light treatment. Plants grown under low-light condition (100 μmol m^{-2} s^{-1}) were treated with strong light (1400 μmol m^{-2} s^{-1}). F_v/F_m values for wild type and *phot1-5*, having normal avoidance movement, declined rapidly within 1 h and kept on declining gradually for the following 5 h. In contrast, F_v/F_m values for *phot2-1* and *chup1-2* continued to decline steeply even after the rapid decrease during the first hour (Kasahara *et al.*, 2002a). Under the same light treatment, mature leaves of *phot2-1* and *chup1-2* showed some signs of bleaching at 10 h and became necrotic after 22 h, whereas those of wild type and *phot1-5* withstood the strong light stress over 31 h (Figure 9.9). Palisade cell layers of *phot2-1* and *chup1-2* were

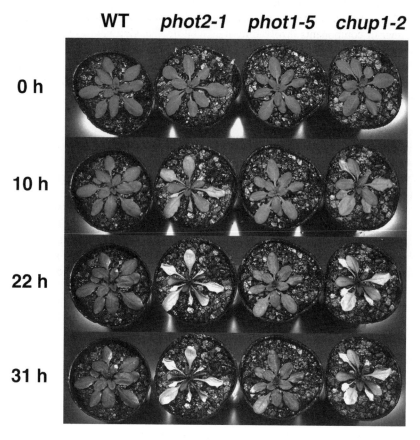

Figure 9.9 Phenotypes of plants exposed to continuous strong light. All plants were grown under weak light condition (100 μmol m^{-2} s^{-1}) for 4 weeks and then irradiated with strong white light (1400 μmol m^{-2} s^{-1}). The photographs were taken before (0 h) and after 10, 22, 31 h. Reproduced from Kasahara *et al.* (2002a).

severely damaged within 12 h. The capacities of other photoprotective mechanisms including thermal dissipation, activity of main reactive oxygen scavenging enzymes (catalase, ascorbate peroxidase, and superoxide dismutase) and the levels of antioxidants (ascorbate and gultathione) are comparable among these plants. The enhanced photodamage of *phot2-1* and *chup1-2* seems to be solely attributed to the deficiency in chloroplast avoidance movement. The photoprotection research field is currently very active and many mechanisms have been elucidated to be effectively working to mitigate photodamage in plants. It is evident that chloroplast avoidance movement has a significant role in photoprotection, together with other mechanisms.

9.5 Conclusion

The identification of the photoreceptors involved in chloroplast movement is unambiguously the major breakthrough in this research field. However, to fully understand this phenomenon at the molecular level, it is undoubtedly necessary to isolate a number of components involved in the signal transduction pathway, starting from the phototropins and the regulation of the actin-based motile systems.

In addition, the availability of mutants provides opportunities to examine the physiological significance of chloroplast movement. Chloroplast avoidance movement is shown to be significantly important to reduce the absorption of excess light, which is one aspect of the avoidance response. The profile arrangement of chloroplasts in cell layers of the irradiated side allows light to penetrate into the deeper layers of the cells and may increase the total photosynthetic activity in whole leaves (Brugnoli and Bjökman, 1992; Terashima and Hikosaka, 1995). Further, chloroplasts in profile arrangement under strong light tend to sit at the cell wall facing the intercellular air space, meaning that CO_2 can reach easily and directly to the chloroplasts through the cell wall (Terashima and Hikosaka, 1995; Gorton *et al.*, 2003). Taken together, chloroplast avoidance response under strong light may be effective not only for avoiding the photodamage of chloroplasts but also for increasing the total activity of photosynthesis by deep light penetration and utilizing air space efficiently. The examination of these hypotheses will be possible using chloroplast movement mutants.

Acknowledgements

We thank Dr Takatoshi Kagawa for providing the micrographs of Figures 9.2 and 9.3.

References

Asada, K. (1999) The water–water cycle in chloroplasts: scavenging of active oxygen and dissipation of excess photons. *Annu. Rev. Plant Physiol. Plant Mol. Biol.*, 50, 601–639.
Babourina, O., Newman, I. and Shabala, S. (2002) Blue light-induced kinetics of H^+ and Ca^{2+} fluxes in etiolated wild-type and phototropin-mutant *Arabidopsis* seedlings. *Proc. Natl. Acad. Sci. U.S.A.*, 99, 2433–2438.

Baum, G., Long, J.C., Jenkins, G.I. and Trewavas, A.J. (1999) Stimulation of the blue light phototropic receptor NPH1 causes a transient increase in cytosolic Ca^{2+}. *Proc. Natl. Acad. Sci. U.S.A.*, 96, 13554–13559.

Briggs, W.R., Beck, C.F., Cashmore, A.R. *et al.* (2001) The phototropin family of photoreceptors. *Plant Cell*, 13, 993–997.

Briggs, W.R. and Christie, J.M. (2002) Phototropins 1 and 2: versatile plant blue-light receptors. *Trends Plant Sci.*, 7, 204–210.

Briggs, W.R. and Huala, E. (1999) Blue-light photoreceptors in higher plants. *Annu. Rev. Cell Dev. Biol.*, 15, 33–62.

Brugnoli, E. and Bjökman, O. (1992) Chloroplast movements in leaves: influence on chlorophyll fluorescence and measurements of light-induced absorbance changes related to 'pH and zeaxanthin formation. *Photosynth. Res.*, 32, 23–35.

Christie, J.M., Salomon, M., Nozue, K., Wada, M. and Briggs, W.R. (1999) LOV (light, oxygen, or voltage) domains of the blue-light photoreceptor phototropin (nph1): binding sites for the chromophore flavin mononucleotide. *Proc. Natl. Acad. Sci. U.S.A.*, 96, 8779–8783.

Christie, J.M., Swartz, T.E., Bogomolni, R.A. and Briggs, W.R. (2002) Phototropin LOV domains exhibit distinct roles in regulating photoreceptor function. *Plant J.*, 32, 205–219.

Crosson, S. and Moffat, K. (2001) Structure of a flavin-binding plant photoreceptor domain: insights into light-mediated signal transduction. *Proc. Natl. Acad. Sci. U.S.A.*, 98, 2995–3000.

Crosson, S. and Moffat, K. (2002) Photoexcited structure of a plant photoreceptor domain reveals a light-driven molecular switch. *Plant Cell*, 14, 1067–1075.

Crosson, S., Rajagopal, S. and Moffat, K. (2003) The LOV domain family: photoresponsive signaling modules coupled to diverse output domains. *Biochemistry*, 42, 2–10.

Demmig-Adams, B. and Adams, W.W.I. (1992) Photoprotection and other responses of plants to high light stress. *Annu. Rev. Plant Physiol. Plant Mol. Biol.*, 43, 599–626.

Gorton, H.L., Herbert, S.K. and Vogelmann, T.C. (2003) Photoacoustic analysis indicates that chloroplast movement does not alter liquid-phase CO_2 diffusion in leaves of *Alocasia brisbanensis*. *Plant Physiol.*, 132, 1529–1539.

Gorton, H.L., Williams, W.E. and Vogelmann, T.C. (1999) Chloroplast movement in *Alocasia macrorrhiza*. *Physiol. Plant.*, 106, 421–428.

Harada, A., Sakai, T. and Okada, K. (2003) phot1 and phot2 mediate blue light-induced transient increases in cytosolic Ca^{2+} differently in *Arabidopsis* leaves. *Proc. Natl. Acad. Sci. U.S.A.*, 100, 8583–8588.

Haupt, W. and Scheuerlein, R. (1990) Chloroplast movement. *Plant Cell Environ.*, 13, 595–614.

Huala, E., Oeller, P.W., Liscum, E., Han, I.S., Larsen, E. and Briggs, W.R. (1997) *Arabidopsis* NPH1: a protein kinase with a putative redox-sensing domain. *Science*, 278, 2120–2123.

Inoue, Y. and Shibata, K. (1974) Comparative examination of terrestrial plant leaves in terms of light-induced absorption changes due to chloroplast rearrangements. *Plant Cell Physiol.*, 17, 717–721.

Jarillo, J.A., Gabrys, H., Capel, J., Alonso, J.M., Ecker, J.R. and Cashmore, A.R. (2001) Phototropin-related NPL1 controls chloroplast relocation induced by blue light. *Nature*, 410, 952–954.

Jeong, W.J., Park, Y.I., Suh, K., Raven, J.A., Yoo, O.J. and Liu, J.R. (2002) A large population of small chloroplasts in tobacco leaf cells allows more effective chloroplast movement than a few enlarged chloroplasts. *Plant Physiol.*, 129, 112–121.

Kadota, A., Sato, Y. and Wada, M. (2000) Intracellular chloroplast photorelocation in the moss *Physcomitrella patens* is mediated by phytochrome as well as by a blue-light receptor. *Planta*, 210, 932–937.

Kadota, A. and Wada, M. (1992a) Photoorientation of chloroplasts in protonemal cells of the fern *Adiantum* as analyzed by use of a video-tracking system. *Bot. Mag. Tokyo*, 105, 265–279.

Kadota, A. and Wada, M. (1992b) Photoinduction of formation of circular structures by microfilaments on chloroplast during intracellular orientation in protonemal cells of the fern *Adiantum capillus-veneris*. *Protoplasma*, 167, 97–107.

Kagawa, T., Sakai, T., Suetsugu, N. *et al.* (2001) *Arabidopsis* NPL1: a phototropin homolog controlling the chloroplast high-light avoidance response. *Science*, 291, 2138–2141.

Kagawa, T. and Wada, M. (2000) Blue light-induced chloroplast relocation in *Arabidopsis thaliana* as analyzed by microbeam irradiation. *Plant Cell Physiol.*, 41, 84–93.

Kagawa, T. and Wada, M. (2002) Blue light-induced chloroplast relocation. *Plant Cell Physiol.*, 43, 367–371.

Kandasamy, M.K. and Meagher, R.B. (1999) Actin–organelle interaction: association with chloroplast in *Arabidopsis* leaf mesophyll cells. *Cell Motil. Cytoskeleton*, 44, 110–118.

Kasahara, M., Kagawa, T., Oikawa, K., Suetsugu, N., Miyao, M. and Wada, M. (2002a) Chloroplast avoidance movement reduces photodamage in plants. *Nature*, 420, 829–832.

Kasahara, M., Kagawa, T., Sato, Y., Kiyosue, T. and Wada, M. (in press) Phototropins mediate blue and red light-induced chloroplast movements in *Physcomitrella patens*. *Plant Physiol.*

Kasahara, M., Swartz, T.E., Olney, M.A. *et al.* (2002b) Photochemical properties of the flavin mononucleotide-binding domains of the phototropins from *Arabidopsis*, rice, and *Chlamydomonas reinhardtii*. *Plant Physiol.*, 129, 762–773.

Kawai, H., Kanegae, T., Christensen, S. *et al.* (2003) Responses of ferns to red light are mediated by an unconventional photoreceptor. *Nature*, 421, 287–290.

Lechowski, Z. (1974) Chloroplast arrangement as a factor of photosynthesis in multilayered leaves. *Acta Soc. Bot. Pol.*, 43, 531–540.

Malec, P., Rinaldi, R.A. and Gabrys, H. (1996) Light-induced chloroplast movements in *Lemna trisulca*: identification of the motile system. *Plant Sci.*, 120, 127–137.

Niyogi, K.K. (1999) Photoprotection revisited: genetic and molecular approaches. *Annu. Rev. Plant Physiol. Plant Mol. Biol.*, 50, 333–359.

Nozue, K., Kanegae, T., Imaizumi, T. *et al.* (1998) A phytochrome from the fern *Adiantum* with features of the putative photoreceptor NPH1. *Proc. Natl. Acad. Sci. U.S.A.*, 95, 15826–15830.

Oikawa, K., Kasahara, M., Kiyosue, T. *et al.* (2003) CHLOROPLAST UNUSUAL POSITIONING1 is essential for proper chloroplast positioning. *Plant Cell.*, 15, 2805–2815.

Park, Y.I., Chow, W.S. and Anderson, J.M. (1996) Chloroplast movement in the shade plant *Tradescantia albiflora* helps protect photosystem II against light stress. *Plant Physiol.*, 111, 867–875.

Sakai, T., Kagawa, T., Kasahara, M. *et al.* (2001) *Arabidopsis* nph1 and npl1: blue light receptors that mediate both phototropism and chloroplast relocation. *Proc. Natl. Acad. Sci. U.S.A.*, 98, 6969–6974.

Sakamoto, K. and Briggs, W.R. (2002) Cellular and subcellular localization of phototropin 1. *Plant Cell*, 14, 1723–1735.

Salomon, M., Christie, J.M., Knieb, E., Lempert, U. and Briggs, W.R. (2000) Photochemical and mutational analysis of the FMN-binding domains of the plant blue light receptor, phototropin. *Biochemistry*, 39, 9401–9410.

Sato, Y., Kadota, A. and Wada, M. (2003) Chloroplast movement: dissection of events downstream of photo- and mechano-perception. *J. Plant Res.*, 116, 1–5.

Sato, Y., Wada, M. and Kadota, A. (2001a) Choice of tracks, microtubules and/or actin filaments for chloroplast photo-movement is differentially controlled by phytochrome and a blue light receptor. *J. Cell Sci.*, 114, 269–279.

Sato, Y., Wada, M. and Kadota, A. (2001b) External Ca^{2+} is essential for chloroplast movement induced by mechanical stimulation but not by light stimulation. *Plant Physiol.*, 127, 497–504.

Schönbohm, E. (1980) Phytochrome and non-phytochrome dependent blue light effects on intracellular movements in fresh-water algae. In *The Blue Light Syndrome* (ed. H. Senger), Springer-Verlag, Berlin, pp. 69–96.

Senn, G. (1908) *Die Gestalts- und Lageveranderung der Pflanzen-Chromatophoren*, Engelmann, Stuttgart.

Stoelzle, S., Kagawa, T., Wada, M., Hedrich, R. and Dietrich, P. (2003) Blue light activates calcium-permeable channels in *Arabidopsis* mesophyll cells via the phototropin signaling pathway. *Proc. Natl. Acad. Sci. U.S.A.*, 100, 1456–1461.

Swartz, T.E., Corchnoy, S.B., Christie, J.M. *et al.* (2001) The photocycle of a flavin-binding domain of the blue light photoreceptor phototropin. *J. Biol. Chem.*, 276, 36493–36500.

Taylor, B.L. and Zhulin, I.B. (1999) PAS domains: internal sensors of oxygen, redox potential, and light. *Microbiol. Mol. Biol. Rev.*, 63, 479–506.

Terashima, I. and Hikosaka, K. (1995) Comparative ecophysiology of leaf and canopy photosynthesis. *Plant Cell Environ.*, 18, 1111–1128.

Tlalka, M. and Fricker, M. (1999) The role of calcium in blue-light-dependent chloroplast movement in *Lemna trisulca* L. *Plant J.*, 20, 461–473.

Tlalka, M. and Gabrys, H. (1993) Influence of calcium on blue-light-induced chloroplast movement in *Lemna trisulca* L. *Planta*, 189, 491–498.

Trojan, A. and Gabrys, H. (1996) Chloroplast distribution in *Arabidopsis thaliana* (L.) depends on light conditions during growth. *Plant Physiol.*, 111, 419–425.

Wada, M., Kadota, A. and Furuya, M. (1983) Intracellular localization and dichronic orientation of phytochrome in plasma membrane and/or ectoplasm of a centrifuged protonema of fern *Adiantum capillus-veneris* L. *Plant Cell Physiol.*, 24, 1441–1447.

Wada, M. and Kagawa, T. (2001) Light-controlled chloroplast movement. In *Photomovement* (eds D.-H. Häder and M. Lebert), Elsevier, Amsterdam, pp. 897–924.

Wada, M., Kagawa, T. and Sato, Y. (2003) Chloroplast movement. *Annu. Rev. Plant Biol.*, 54, 455–468.

Wagner, G., Haupt, W. and Laux, A. (1972) Reversible inhibition of chloroplast movement by cytochalasin B in the green alga *Mougeotia*. *Science*, 176, 808–809.

Yatsuhashi, H., Kadota, A. and Wada, M. (1985) Blue- and red-light action in photoorientation of chloroplasts in *Adiantum* protonemata. *Planta*, 165, 43–50.

Zurzycki, J. (1955) Chloroplasts arrangement as a factor in photosynthesis. *Acta Soc. Bot. Pol.*, 24, 27–63.

Zurzycki, J. (1957) The destructive effect of intense light on the photosynthetic apparatus. *Acta Soc. Bot. Pol.*, 26, 157–175.

Zurzycki, J. (1980) Blue light-induced intracellular movements. In *The Blue Light Syndrome* (ed. H. Senger), Springer-Verlag, Berlin, pp. 50–68.

10 Chloroplast genetic engineering for enhanced agronomic traits and expression of proteins for medical/industrial applications

Andrew L. Devine and Henry Daniell

10.1 Introduction

The chloroplast genome is a circular molecule that self-replicates and varies in size from 120 to 220 kb among different plant species. The chloroplast genome exists predominantly as a single, monomeric, circular molecule, but recent cytogenomic analysis has revealed that a small percentage may exist in the form of circular dimers, trimers, tetramers, or even as linear multimeric forms (Lilly *et al.*, 2001). The typical plant cell contains approximately 100 chloroplasts and each chloroplast contains about 100 copies of the identical plastid genomes. A single gene is therefore represented at least 10,000 times within a plant cell. Most plant species contain two inverted repeat regions within their plastid genomes, and therefore the copy number of the genes encoded by this region approaches ~20,000. These facts, taken together, make it quite appealing to introduce transgenes into the plastid genome in order to achieve high levels of expression, taking advantage of the extremely high copy number of the plastid genome.

Site-directed insertion of transgenes into the plastid genome differs from random integration that occurs during nuclear transformation. Transgenes are integrated via homologous recombination within the plastid genome (Figure 10.1). Plastid transformation vectors are therefore designed to contain homologous flanking sequences on either side of transgenes and are introduced into the plastids of plant cells via particle bombardment (Sanford *et al.*, 1993) or into protoplasts by polyethylene glycol (PEG) treatment (Golds *et al.*, 1993). Following bombardment, integration of transgenes occurs into a few plastid genomes, followed by approximately 15–20 cell divisions under selective pressure (in two to three rounds of selection), resulting in a homogeneous population of plastid genomes. When genes of interest are introduced into the inverted repeat region, integration into one inverted repeat region is followed by the phenomenon of copy correction that recruits the introduced transgenes into another inverted repeat region (Figure 10.1). Several sites of integration, including some in the inverted repeat region, have been used for integration of foreign genes into the plastid genome (Tables 10.1 and 10.2). Plastid transformation vectors may also carry a plastid origin of replication that facilitates replication of the introduced plastid vector within the chloroplast, thereby increasing the templates to be presented for homologous recombination, and consequently, this enhances the probability of transgene integration: chloroplast vectors with such origins of replication have been shown to achieve homoplasmy (replacement of all native plastid

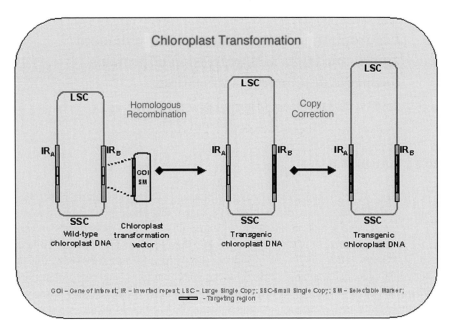

Figure 10.1   Schematic representation of chloroplast transformation showing the phenomenon of homologous recombination and copy correction.

genomes with transformed genomes) even in the first round of selection (Daniell *et al.*, 1990, 1998; Guda *et al.*, 2000).

10.2   Historical aspects

The concept of chloroplast genetic engineering originated in the 1980s, when it was possible to introduce isolated intact chloroplasts into protoplasts and regenerate plants (Daniell *et al.*, 2002). Therefore, early investigations of chloroplast transformation in vascular plants focused on the development of chloroplast systems capable of efficient, prolonged protein synthesis and expression of foreign genes (Daniell and McFadden, 1987). With the invention of the Gene Gun by John Sanford and colleagues (reviewed in 1993), it became possible to transform plastids without the need to isolate them. Chloroplast genetic engineering in higher plants was initiated through the use of autonomously replicating chloroplast vectors in dicot plastids (Daniell *et al.*, 1990) and transient expression in monocot plastids (Daniell *et al.*, 1991).

In 1988, the first successful chloroplast genome complementation was reported for *Chlamydomonas reinhardtii*, a unicellular green alga that has a single chloroplast that occupies ~60% of the cell volume (Boyton *et al.*, 1988). The strategy utilized photosynthetically incompetent mutants with a deletion in the *atpB* gene. These mutants lacked chloroplast ATP synthase activity. The wild-type *atpB* gene

Table 10.1 List of agronomic traits engineered via the chloroplast genome.[1,2]

Agronomic traits	Gene	Site of Integration	Promoter	5'/3' regulatory elements	Reference
Insect resistance	Cry1A(c)	trnV/rps12/7	Prrn	rbcL/Trps16	McBride et al., 1995
Herbicide resistance	aroA (petunia)	rbcL/accD	Prrn	ggagg/TpsbA	Daniell et al., 1998
Insect resistance	Cry2Aa2	rbcL/accD	Prrn	ggagg (native)/TpsbA	Kota et al., 1999
Herbicide resistance	Bar	rbcL/accD	Prrn	rbcL/TpsbA	Iamtham and Day, 2000
Insect resistance	Cry2Aa2 operon	trnI/trnA	Prrn	Native 5'UTRs/TpsbA	DeCosa et al., 2001
Disease resistance	MSI-99	trnI/trnA	Prrn	ggagg/TpsbA	DeGray et al., 2001
Drought tolerance	Tps	trnI/trnA	Prrn	ggagg/TpsbA	Lee et al., 2003
Phytoremediation	merA[a]/merB[b]	trnI/trnA	Prrn-F	Ggagg[a,b]/TpsbA	Ruiz et al., 2003
Salt tolerance	Badh	trnI/trnA	Prrn-F	ggagg/rps16	Kumar et al., submitted-a

[1] Modified from Daniell et al. (2004).
[2] Only first reports of genes or gene products are included.

Table 10.2 List of biopharmaceutical proteins and vaccine antigens expressed via the chloroplast genome.[1]

Therapeutic proteins	Gene	Site of integration	Promoter	5'/3' regulatory elements	% tsp expression	Reference
Elastin-derived polymer	EG121	trnI/trnA	Prrm	T7gene10/TpsbA	ND	Guda et al., 2000
Human somatotropin	hST	trnV/rps12/7	Prrm[a], PpsbA[b]	T7gene10[a] or psbA[b]/Trps16	7.0%[a] and 1.0%[b]	Staub et al., 2000
Cholera toxin	CtxB	trnI/trnA	Prrm	Ggagg/TpsbA	4%	Daniell et al., 2001a
Antimicrobial peptide	MSI-99	trnI/trnA	Prrm	Ggagg/TpsbA	Not tested	DeGray et al., 2001
Insulin-like growth factor	IGF-1	trnI/trnA	Prrm	PpsbA/TpsbA	33%	Ruiz, 2002
Interferon alpha 5	INFα5	trnI/trnA	Prrm	PpsbA/TpsbA	ND	Torres, 2002
Interferon alpha 2b	INFα2B	trnI/trnA	Prrm	PpsbA/TpsbA	19%	Falconer, 2002
Human serum albumin	Hsa	trnI/trnA	Prrm[c], PpsbA[d]	ggagg[c], psbA[d]/TpsbA	0.02%[c], 11.1%[d]	Fernandez et al., 2003
Interferon gamma	IFN-g	rbcL/accD	PpsbA	PpsbA/TpsbA	6%	Leelavathi & Reddy, 2003
Monoclonal antibodies	Guy's 13	trnI/trnA	Prrm	Ggagg/TpsbA	ND	Daniell, 2004
Anthrax protective antigen	Pag	trnI/trnA	Prrm	PpsbA/TpsbA	4–5%	Daniell et al., in press
Plague vaccine	CaF1~LcrV	trnI/trnA	Prrm	PpsbA/TpsbA	4.6%	Singleton, 2003
CPV VP2	CTB-2L21[e] GFP2L21[f]	TrnI/trnA	Prrm	psbA/TpsbA	31.1%[e] 22.6%[f]	Molina et al., in press
Tetanus toxin	TetC	Trnv/rps12/7	Prrm	T7gene10[g] atpB[h]/TrbcL	25%[g] 10%[h]	Tregoning et al., 2003

[1]Modified from Daniell et al. (2004).

was deposited on tungsten microprojectiles and propelled into cells spread on agar plates, utilizing a gunpowder charge (Klein *et al.*, 1987). The single large chloroplast provided an ideal target for DNA delivery. The wild-type *atpB* gene introduced into cells was able to correct the deletion mutant phenotype. Upon selection in the light, it was demonstrated that photoautotrophic growth was restored and the wild-type gene carried by the vector was successfully integrated into the *C. reinhardtii* chloroplast genome via homologous recombination, replacing the deleted *atpB* gene. Plastids have inherited from their cyanobacterial ancestors an efficient recA-type system to facilitate homologous recombination (Cerutti *et al.*, 1992). It was then demonstrated that the foreign DNA flanked by chloroplast DNA sequences is incorporated and stably maintained in the *C. reinhardtii* chloroplast genome; although the introduced *uid*A gene was transcribed, translated product could not be detected (Blowers *et al.*, 1989). Daniell *et al.* (1990) demonstrated the first foreign gene expression in plastids of cultured tobacco cells upon biolistic delivery of autonomously replicating chloroplast vectors and in wheat plants (leaves, calli, and somatic embryos; Daniell *et al.*, 1991). Previously, foreign genes were introduced and expressed only in isolated but intact plastids (Daniell and McFadden, 1987). Goldschmidt-Clermont (1991) then transformed the chloroplast genome of *C. reinhardtii* with the *aadA* selectable marker gene to confer spectinomycin/streptomycin resistance (Goldschmidt-Clermont, 1991); this was an important contribution because the majority of higher plants genetically transformed via the chloroplast genome now use this selectable marker. This work was then extended to demonstrate stable integration of the *aadA* gene into the tobacco chloroplast genome (Svab and Maliga, 1993). This field has advanced quite rapidly in the past few years, and there are now biotechnology companies based on this technology. Chloroplast genetic engineering technology is currently applied to other useful crops such as potato, tomato, carrot, cotton and soybean (Sidorov *et al.*, 1999; Ruf *et al.*, 2001; Kumar *et al.*, submitted-a, -b). Foreign proteins expressed via the plastid genome to confer useful agronomic traits and therapeutic proteins have been listed with appropriate citations in Tables 10.1 and 10.2 respectively.

10.3 Unique features of chloroplast genetic engineering

The world of plant genetic engineering research was revolutionized with the accumulation of *Bacillus thuringiensis* (Bt) Cry2Aa2 at 46.1% total soluble protein in transgenic tobacco chloroplasts (DeCosa *et al.*, 2001). It was not only the highest accumulation of protein in transgenic plants but also for the first time, a complete bacterial operon was successfully expressed, resulting in the formation of stable Cry2Aa2 crystals. This was achieved because native bacterial genes were expressed several hundred fold higher in the chloroplast compartment than in nuclear transgenic plants (Daniell *et al.*, 2001a; DeCosa *et al.*, 2001; Molina *et al.*, 2004). Chloroplast transformation provides not only higher protein expression capability than does nuclear genetic engineering, but also several other advantages. Even though transgenic chloroplasts may be present in the pollen, the foreign gene will not escape to other crops because chloroplast DNA in not passed on to the egg cell. In most crop plants, plastid genes are inherited uniparentally in a strictly maternal fashion. Although pollen from plants shown to exhibit maternal plastid inheritance contains metabolically active plastids, the plastid DNA itself is lost during the process of pollen maturation and hence is not transmitted to the next generation (Nagata *et al.*, 1999). This is described in greater detail below.

The chloroplast could be a good place to accumulate certain proteins or their biosynthetic products that would be harmful if they were in the cytoplasm and perhaps not affect the physiology of the transgenic plant (Bogorad, 2000). This has been demonstrated by the nontoxic effect of cholera toxin B (CTB) subunit, a candidate oral subunit vaccine for cholera, when it was accumulated in large quantities within transgenic plastids (Daniell *et al.*, 2001a; Molina *et al.*, 2004); however, even very small quantities of LTB (protein toxin with similar structure, function, and immunochemistry as CTB) were toxic when expressed in the cytoplasm (Mason *et al.*, 1998). Similarly, trehalose, (Lee *et al.*, 2003) is used in the pharmaceutical industry as a preservative and xylanase (Leelavathi *et al.*, 2003) an industrial enzyme, were very toxic when they were accumulated in the cytosol but nontoxic when compartmentalized within plastids.

Transgenes are integrated into the spacer regions of the chloroplast genome by homologous recombination of chloroplast DNA flanking sequences (Figure 10.1). This allows site-specific integration and eliminates concerns of position effect, frequently observed in nuclear transgenic plants. As a result, it is not necessary to screen large numbers of putative transgenic lines to choose for high-level expression of transgenes. In contrast, all chloroplast transgenic lines express the same level of foreign protein, within the range of physiological variations (Daniell *et al.*, 2001a). In addition, site-specific integration by homologous recombination eliminates introduction of vector sequences, which is often a concern in nuclear transformation by nonhomologous recombination. Yet another advantage is the lack of transgene silencing in chloroplast transgenic plants, which is also a serious concern in nuclear transformation. It has been shown that there is no gene silencing in chloroplast transgenic lines at the transcriptional level in spite of accumulation of transcripts 169-fold higher than in nuclear transgenic plants (Lee *et al.*, 2003). Similarly, there is no transgene silencing at the translational level in spite of accumulation of foreign protein up to 46.1% of the total plant protein in chloroplast transgenic lines (DeCosa *et al.*, 2001).

10.4   Maternal inheritance and gene containment

In the vast majority of angiosperm plant species, plastids are maternally inherited (Hagemann, 1992; Birky, 1995; Mogensen, 1996, Hagemann, in press). This minimizes the concern of outcross of transgenes to related weeds or crops (Daniell *et al.*, 1998; Scott and Wilkenson, 1999; Daniell, 2002) and reduces the potential toxicity of transgenic pollen to nontarget insects (DeCosa *et al.*, 2001). During meiosis, haploid egg and sperm cells are formed. The synergid cell attracts the pollen tube by secretion of calcium, carbohydrates, and proteins. Fertilization begins when the pollen tube enters the synergid cell. Once inside the cytoplasm of the synergid cell, the pollen tube ruptures, releasing its contents. The paternal chloroplasts are disintegrated and only the sperm nucleus enters the egg cell and fuses with the egg to form a zygote. The zygote contains only maternal plastids because the paternal plastids disintegrate in the synergid cell. Thus, maternal inheritance of transgenes offers containment because of lack of gene flow through pollen (Daniell, 2002; Daniell and Parkinson, 2003). This has been repeatedly demonstrated in tobacco by pollination of wild-type plants with pollen from chloroplast transgenic lines and lack of growth of resultant seedlings on an antibiotic selection medium. The pollinated plants still remain antibiotic sensitive (Figure 10.2). Conversely, one could pollinate a transgenic plant with pollen from wild-type plants and the progeny will all be antibiotic resistant. Chloroplast genetic engineering provides transgene containment, which ensures a much more ecologically safe transformation system, compared to nuclear transformation in which transgene containment is a potential concern.

10.5   Crop species stably transformed via the plastid genome

The concept of the use of a "universal vector" that utilized the plastid DNA flanking sequences from one plant species to transform another species of unknown genome sequence was proposed several years ago (Daniell *et al.*, 1998). Utilizing this concept, both potato and tomato plastid genomes have thus far been transformed using flanking sequences from tobacco. However, the transformation efficiency may be lower if the homology of flanking sequences is not very high. In general, tobacco plastid transformation is highly efficient when tobacco endogenous flanking sequences (100% homology) are used. However, when Petunia flanking sequences were used to transform the tobacco plastid genome, transformation efficiency was

Figure 10.2   Maternal inheritance of transgenes. (A) Wild-type or (B) chloroplast transgenic lines were germinated on MSO medium (Daniell *et al.*, 2004, Kumar and Daniell, 2004.) supplemented with 500 mg/L spectinomycin.

lower than that was observed with endogenous flanking sequences (DeGray et al., 2001).

10.5.1 Tobacco

Tobacco has by far been the most extensively utilized plastid transformation system because of its ease of genetic manipulation. More transgenes have been expressed in tobacco through the nuclear and chloroplast genomes than in all other plant species combined. Tobacco is an ideal system for scale up. A single tobacco plant is capable of producing a million seeds and 1 acre of tobacco produces more than 40 metric tons of leaves in multiple harvests per year (Cramer et al., 1999). Moreover, the machinery for harvesting and processing tobacco leaves is already in place for commercialization. Furthermore, an alternate use for this hazardous crop is highly desirable. It has been estimated that the cost of production of recombinant proteins in tobacco leaves will be 50-fold lower than that of *Escherichia coli* fermentation systems (Kusnadi, 1997). Most importantly, tobacco is a nonfood/feed crop. Therefore, it is ideal for producing therapeutic proteins. Harvesting leaves before appearance of reproductive structures offers complete containment of transgene. For a recent list of foreign genes expressed via the tobacco plastid genome, please refer to Tables 10.1 and 10.2.

10.5.2 Potato

Sidorov et al., (1999) reported a potato plastid transformation protocol by optimization of microprojectile bombardment parameters and the selection process. These researchers utilized a vector that was originally designed for tobacco plastid transformation. The flanking sequence between tobacco and potato was ~98% homologous. The vector contained the aadA selectable marker and GFP (green fluorescent protein) screenable marker. Upon particle bombardment and regeneration of shoots, quantification of GFP in the green tissues showed 5% total soluble protein (tsp), but the microtubers expressed only 0.5% tsp. The authors explain this to be due to the lower plastid DNA copy number coupled with generally lower transcriptional/translational activities in plastids of this tissue type (Sidorov et al., 1999). The plastid transformation of potato was a positive step in the right direction for applying this technology to other crop plants. However, since 1999, no useful traits have been introduced via the plastid genome in potato. This may be partly due to very low efficiency of plastid transformation. More than a hundred bombardments resulted in two transgenic calli and one transgenic plant. Therefore, the efficiency of plastid transformation in potato needs to be greatly improved.

10.5.3 Tomato

Researchers used the tomato cultivar (*Lycopersicon esculentum* var. IAC-Santa Clara), a red-fruited tomato variety that is native to South America to demonstrate plastid transformation using the *add*A gene as a selectable marker that confers resistance to spectinomycin. It took the research group two years from the date of bombardment of leaf material until the harvest of the first ripe plastid transformants. Transformed tomato plants produced fruits with viable seeds, which transmitted the transgene in a uniparentally maternal fashion as expected for a plastid-encoded trait (Ruf *et al.*, 2001). Plastid transformation of tomato opens the door for many possibilities, including the ability to engineer a crop suitable for human consumption. However, since the publication of this article, no useful traits have been engineered via the plastid genome. Again, this may be due to the extremely low efficiency of plastid transformation in tomato; among the 540 plates examined, only six transgenic calli were recovered.

10.5.4   Carrot, cotton, and monocots

In all of the aforementioned examples, chloroplast genomes have been transformed using fully mature chloroplasts as recipients of foreign DNA and regeneration was achieved via direct organogenesis. However, in order to transform more economically important crops, it is essential to transform nongreen cells that contain proplastids, regenerate plants via somatic embryogenesis and achieve homoplasmy without subsequent rounds of regeneration. This requires knowledge of regulatory sequences in nongreen plastids and the ability to achieve homoplasmy in the absence of repetitive regeneration (which is routinely possible via organogenesis). Therefore, using carrot and cotton as two model systems, transformation of non-green plastids and regeneration of transgenic plants via somatic embryogenesis have been recently demonstrated (Kumar *et al.*, submitted-a, -b). These findings have paved the way for plastid transformation of several monocots including corn, wheat, barley and sugarcane. Yet another limitation in achieving plastid transformation of crop plants is that their plastid genomes have not yet been sequenced. Once this is accomplished, more crop plants will be transformed via their plastid genomes.

Carrot (*Daucus carota* L.) is a vegetable that is one of the world's most important crop plants for human and animal consumption. This crop plant is an excellent source of vitamins A and C, fiber, and carbohydrates. The carrot is a biennial plant that completes its life cycle in 2 years. In the first year the edible fleshy taproot appears, and then in the second year, in response to cold weather, the plant flowers. Therefore, carrot is an environmentally safe crop and is doubly protected against transgene flow via pollen and seeds to achieve zero-contamination of food crops by pharmaceutical crops, expected by various regulatory agencies. Kumar *et al.* (submitted-a) reported stable and highly efficient plastid transformation in carrot and cotton utilizing nongreen tissues as explants, regenerated via somatic embryogenesis. Carrot cultures were bombarded with two separate vectors, one containing aadA/GFP and the second aadA/BADH targeted to the 16S/trnI-trnA/23S plastid genome inverted repeat region. The flanking sequences were doubled (2 kb) on either side to increase the efficiency of homologous recombination. The expression of the betaine aldehyde dehydrogenase (BADH) enzyme not only conferred the highest level of salt tolerance (up to 400 mM), but also greatly facilitated the visual selection of the transgenic green cells from the nontransformed yellow cells. Carrot plastid transformation using non-green cells containing proplastids was as efficient as transformation of tobacco green cells, containing mature chloroplasts. Maximum transformation efficiency (13.3%) was observed in nongreen cell cultures when compared with cotyledons (10%) and stem segments (6.25%). This was the first plant species in which stable plastid transformation was achieved using various explants and stable transgene expression in proplastids. High levels of foreign gene expression in chromoplasts (up to 75% of expression observed in leaves) should facilitate oral delivery of therapeutic proteins.

Cotton cultures were bombarded with *aph*A-6/*npt*II vectors driven by regulatory elements that function in green and nongreen plastids; this provided the ability to overcome selection pressure day and night, without any interruption, in all tissue types. The flanking sequences were doubled (2 kb) on either side to increase the efficiency of homologous recombination and one of the flanks contained the complete chloroplast origin of replication. Cotton chloroplast transgenic plants regenerated via somatic embryogenesis were fertile, flowered and set bolls/seeds as control untransformed plants. Transgenes stably integrated into cotton plastid genomes were maternally inherited and were not passed through transgenic pollen when tested by cross-pollination with untransformed control plants (Kumar *et al.*, submitted-b). This highly efficient and reproducible process for plastid transformation through somatic embryogenesis using various selectable markers should pave the way to engineer the plastid genome of several major crops in which regeneration is mediated through somatic embryogenesis.

10.6   Agronomic traits conferred via the plastid genome

Several useful genes conferring valuable traits via chloroplast genetic engineering have been demonstrated recently. For example, plants resistant to *Bt*-sensitive insects were obtained by integrating the *cryIAc* gene into the tobacco chloroplast genome (McBride *et al.*, 1995). Plants resistant to *Bt*-resistant insects (up to 40,000-fold) were obtained by hyperexpression of the *cry2A* gene within the tobacco chloroplast genome (Kota *et al.*, 1999). Plants have also been genetically engineered via the chloroplast genome to confer herbicide resistance and the introduced foreign genes were maternally inherited, overcoming the problem of outcross with weeds (Daniell *et al.*, 1998). More recently, plants exhibiting tolerance to bacterial and fungal diseases (DeGray *et al.*, 2001), drought (Lee *et al.*, 2003), salt (Kumar *et al.*, submitted-a), capable of phytoremediation (Ruiz *et al.*, 2003), have been reported. It is also possible to engineer multiple genes, novel pathways, or bacterial operons via the chloroplast genome (DeCosa *et al.*, 2001; Daniell and Dhingra, 2002; Ruiz *et al.*, 2003). A few examples of such useful agronomic traits are discussed below.

10.6.1   Herbicide resistance

Expression of herbicide resistance genes within the plastid of higher plants is very logical considering the high expression levels seen in transgenic lines and maternal inheritance of transgenes. In addition, the target proteins or enzymes of most herbicides are compartmentalized within plastids. The most commonly used herbicide glyphosate is a broad spectrum, nonselective systemic herbicide that is used to control annual and perennial plants such as grasses, sedges, broad-leaved weeds, and woody plants. It can be used on non-cropland as well as on a great variety of crops. Glyphosate when applied to crops is absorbed by leaves and rapidly moves through the plant. The herbicide inhibits the enzyme 5-enolpyruvylshikimate-3-phosphate synthase (EPSPS), a nuclear-encoded gene in plants that is imported into plastids where it functions in biosynthesis of aromatic amino acids. EPSP is synthesized from shikimate-3-phosphate and phosphenolpyruvate catalyzed by the enzyme EPSPS. Unfortunately, glyphosate does not possess the ability to distinguish between crops and weeds. Plants engineered for herbicide resistance via the nuclear genome were quite effective in protecting crops but one drawback that was encountered was the outcross of transgenes with other plants that might confer herbicide resistance to unwanted weeds (Daniell, 1999, 2000). By transforming the plastid genome of higher plants, this problem could be minimized because of the maternal inheritance of the plastid genome.

Transgenic plants resistant to glyphosate have been engineered either by overexpression of EPSPS or expression of a gene such as *aroA* that is glyphosate resistant. Since the target of glyphosate resides within the chloroplast, the next logical step was expression directly in the chloroplast. Daniell *et al.* (1998) expressed a wild-type petunia EPSPS enzyme by integration of the *aroA* gene into the rbcL/orf512 spacer region of the tobacco chloroplast genome. Transgenic plants were confirmed to be homoplasmic by Southern blot analysis. Eighteen-week-old wild-type and control plants were sprayed with 0.5–5 mM glyphosate; the wild type died within 7 days. On the other hand, the chloroplast transgenics were able to survive concentrations as high as 5 mM (Figure 10.3), which is ~10 times the lethal concentration (Daniell *et al.*, 1998). This was the first report of a eukaryotic nuclear gene expressed within the prokaryotic chloroplast compartment and the transit peptide was properly cleaved within chloroplasts. These results were surprising to investigators since the petunia EPSPS gene used is known to have a very low tolerance to glyphosate. The EPSPS gene allowed the tobacco plant to be resistant to the herbicide glyphosate because of the overexpression of EPSPS enzyme.

Recently, the *Agrobacterium* strain CP4 EPSPS gene was expressed in tobacco plastids and resulted in 250-fold higher levels of the glyphosate-resistant EPSPS enzyme than the levels achieved via nuclear transformation (Ye *et al.*, 2001). Although EPSPS (strain CP4) expression in plastids was enhanced more than nuclear expression levels, field tolerance to glyphosate remained the same, showing that higher levels of expression do not always proportionately increase herbicide tolerance. In two different studies, transgenic tobacco plants expressing *bar* genes via the chloroplast genome exhibited field-level tolerance to phosphinothricin (PPT) (Iamtham and Day, 2000; Lutz *et al.*, 2001). In this case, even plants with the lowest levels of *bar* expression were resistant to the highest levels of PPT tested, and no pollen transmission of the transgene was detected. However, high-level expression of

Figure 10.3 Herbicide resistance assay. Eighteen-week-old chloroplast transgenic and wild-type plants were sprayed with 5 mM glyphosate solution (10 times the lethal concentration). (A) Chloroplast transgenic line; (B) wild-type control.

bar genes in the chloroplast was not sufficient to allow direct selection of chloroplast transformants on medium containing PPT (Lutz *et al.*, 2001).

10.6.2   *Insect resistance*

Cry2Aa2 is an insecticidal protein produced by the bacterium *B. thuringensis*. Plants that have been genetically modified to express insecticidal proteins have shown a significantly increased resistance against target insects. The Cry2Aa2 protein is encoded within an operon and has been expressed in the chloroplast genome both as a single gene (Kota *et al.*, 1999) and as an operon (DeCosa *et al.*, 2001; Daniell and Dhingra, 2002). The Cry2Aa2 operon consists of the *cry2Aa2* gene, *orf* 1 and *orf* 2. The function of orf 1 is unknown; orf 2 functions as a chaperone that has the ability to fold the Cry2Aa2 protein into cuboidal crystals. DeCosa *et al.* (2001) demonstrated the presence of cuboidal crystals upon expression of the cry2Aa2 operon, using transmission electron microscopy (Figure 10.4). In addition to the ORF 2 protein, crystal formation was also facilitated by hyperexpression of the insecticidal protein through chloroplast genetic engineering. Upon expression of the cry2Aa2 operon in chloroplast transgenic plants, the Cry2Aa2 protein had accumulated up to 46.1% tsp and this is by far the highest expressed foreign protein in transgenic plants to date. The chloroplast transgenic plants expressing the single *cry2Aa2* gene or the complete operon showed very high insecticidal activity when compared with the untransformed wild-type plants. The wild-type control plant leaf material was fed to the tobacco budworm (*Heliothis virescens*), cotton bollworm (*Helicoverpa zea*), and beet armyworm (*Spodoptera exigua*), and the leaves were

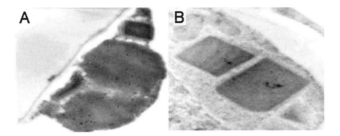

Figure 10.4 Transmission electron micrographs. (A) Detection of Cry2A protein by immunogold-labeling using Cry2A antibody. (B) Accumulation of folded Cry2A protein as cuboidal crystals in transgenic chloroplasts.

totally devoured within 24 h (Figures 10.5A, 10.5D, and 10.5G). When tobacco budworm was fed with chloroplast transgenic leaves expressing the single gene, all the insects died after 5 days (Figure 10.5B), whereas the insects fed with the leaf material expressing the Bt operon died in 3 days (Figure 10.5C). When the same assays were applied to cotton bollworm (Figure 10.5D–F) and the beet armyworm (Figure 10.5G–I), similar results were obtained. These results show that the hyper-expression of the Cry2Aa2 protein in the chloroplasts of transgenic tobacco plants conferred 100% resistance to insects that fed on the chloroplast transgenic plants. These results also showed that even old bleached senescent leaves contained high levels of the insecticidal protein, in spite of the high protease activity associated with senescence. This may be due to chaperone-assisted cuboidal crystal formation that prevented proteolytic degradation of the protein and this should have contributed to the high-level accumulation of the insecticidal protein. Most importantly, chloroplast transgenic plants killed insects that were 40,000-fold resistant to insecticidal proteins, even at lower levels of expression (Kota *et al.*, 1999).

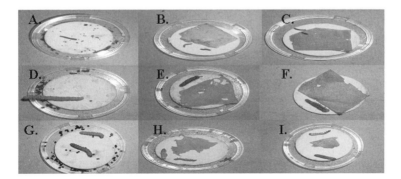

Figure 10.5 Insect bioassays. (A, D, G) Untransformed tobacco leaves; (B, E, H) single gene-derived cry2Aa2 transformed leaves; (C, F, I) operon-derived cry2Aa2 transformed leaves. (A–C) Bioassays with *Heliothis virescens* (tobacco budworm); (D–F) bioassays with *Helicoverpa zea* (cotton bollworm); (G–I) bioassays with *Spodoptera exigua* (beet armyworm). For each replicate samples from the same leaf were used.

10.6.3 Pathogen resistance

Many protective agents are expressed against pathogens in many organisms, and some are helical structured antimicrobial peptides (AMP). MSI-99 is an analog of maganin 2; this peptide confers protection against prokaryotic organisms because of its high specificity for negatively charged phospholipids, which are found mostly in bacteria and less abundantly in eukaryotic organisms. An *in planta* assay was performed on these chloroplast transgenic plants expressing MSI-99 (DeGray *et al.*, 2001). The leaves were inoculated with a syringe containing the phytopathogen *Pseudomonas syringae pv tabaci*, and then observed for the absence of necrosis around the site of inoculation; this observation led to the conclusion that the chloroplast transgenic plant had developed increased resistance to phytopathogen colonization and infection. No necrotic tissue could be observed in transgenic plants even when 8×10^5 cells were inoculated into the transgenic plant. In the wild-type plants inoculated with 8×10^3 cells of the similar phytopathogen (much lower number of cells than used in transgenics), a large necrotic area could be seen (Figures 10.6C and 10.6D). These data suggest that the high levels of AMP expressed in the chloroplast of the transgenic plant are released from the chloroplast during infection, by lysis of chloroplasts during cell death; thus, the phytopathogen comes in contact with the AMP. Further studies of the bacterial populations at the site of inoculation 4 days following inoculation showed that the wild-type nontransgenic plants had a cell population of $13{,}750 \pm 750$ CFU compared to the lower count in transgenic plants, 4650 ± 125 CFU. When the same bioassays were performed with the plant fungal pathogen *Colletotrichum destructivum* in nontransgenic lines, the plant developed anthracnose lesions whereas the chloroplast transgenic plants expressing MSI-99 did not incur any lesions whatsoever (Figures 10.6A and 10.6B). The minimum inhibitory concentration of MSI-99 was investigated based on total inhibition of 1000 *P. syringae* cells; MSI-99 was most effective against *P. syringae*, requiring only 1 µg/1000 bacteria (Figure 10.6; DeGray *et al.*, 2001). Because the lytic activity of antimicrobial peptides is concentration dependent, the amount of AMP required to kill bacteria was used to estimate the level of expression in transgenic plants. Based on the minimum inhibitory concentration, it was estimated that chloroplast transgenic tobacco plants expressed MSI-99 at 21.5–43% of the tsp (Daniell, 2004). These results clearly show that chloroplast genetically engineered plants expressing AMP can confer high levels of resistance to phytopathogenic organisms.

10.6.4 Drought tolerance

Several environmental stress factors such as drought, salinity, and freezing are extremely hazardous to plants because of their sessile way of life. Plants, yeast, and other organisms produce osmoprotectants that confer resistance to several factors including drought. Trehalose phosphate synthase (encoded by *TPS1* gene) catalyzes the reaction to form an osmoprotectant trehalose. Nuclear genetic engineering of plants expressing trehalose phosphate synthase has thus far been unsuccessful,

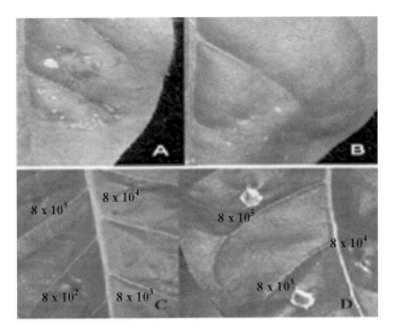

Figure 10.6 *In planta* bioassays for disease resistance. (A, B) Fungal disease resistance. Leaves were inoculated on the adaxial surface with eight drops of 10 μL each of the culture containing 1×10^6 spores/mL of the fungal pathogen *Colletotrichum destructivum*. (A) Wild-type leaf; (B) transgenic leaf. (C, D) Bacterial disease resistance: 8×10^5, 8×10^4, 8×10^3, and 8×10^2 cell cultures of bacterial pathogen *Pseudomonas syringae pv tabaci* were added to a 7-mm-scraped area in transgenic and nontransgenic tobacco lines. Photographs were taken 5 days after inoculation.

because of adverse pleotrophic affects seen in the plants, even at very low levels of trehalose accumulation (Holmstrom *et al.*, 1996). We have reported the hyperexpression of trehalose phosphate synthase and the increased accumulation of trehalose in the chloroplasts of transgenic plants (Lee *et al.*, 2003). Upon high expression of the *TPS1* gene in chloroplast transgenic plants, no pleotrophic effects were observed and the phenotype of transgenic plants was similar to that of untransformed control plants. Drought tolerance bioassays in which transgenic and wild-type seeds were germinated in MS medium containing concentrations of 3–6% PEG showed that the chloroplast transgenic plants producing high levels of trehalose germinated, grew, maintained green color, and remained healthy (Plate 4). Untransformed wild-type seeds germinated under similar conditions showed severe dehydration, loss of chlorophyll (chlorosis), and retarded growth that resulted in the death of the seedlings. Loss of chlorophyll in the untransformed plants suggests that drought destabilizes the thylakoid membrane, but accumulation of trehalose in transgenic chloroplasts conferred thylakoid membrane stability. In a separate assay, when seedlings from chloroplast transgenic and untransformed tobacco plants were dried for 7 h, they showed symptoms of dehydration, but when the seedlings were rehydrated in MS medium for 48 h, all chloroplast transgenic plants that accumulated trehalose recovered and grew normally. The untransformed controls became bleached out and died. Additionally, when potted chloroplast transgenics and untransformed plants were not watered for 24 days and were then rehydrated for 24 h, the chloroplast transgenic plants recovered while the untransformed control plants did not recover. These results clearly indicate that expression of the enzyme trehalose phosphate synthase in the chloroplast of transgenic plants confers drought tolerance.

10.6.5   Phytoremediation

Organomercurial compounds are the most toxic form of mercury; they present a serious hazard to our environment and ecosystems. Current methods of chemical and physical remediation as well as bacterial bioremediation methods have thus far proven to be ineffective owing to the high cost and environmental concerns. An alternative method, phytoremediation, has been proposed as a safe and cost-efficient system for the remediation of toxic chemicals in the environment. Chloroplast is the primary target for mercury and organomercurials in plants. For this reason, it is advantageous to utilize chloroplast genetic engineering to increase resistance to mercury and organomercurials and at the same time detoxify the highly toxic organomercurials and metal mercury forms present in the contaminated environment. In order to achieve this, two bacterial enzymes that confer resistance to different forms of mercury, mercuric ion reductase (*mer*A) and organomercurial lyase (*mer*B), were expressed by Ruiz *et al.* (2003) as an operon in chloroplasts of transgenic tobacco plants. Upon creation of chloroplast transgenic plants expressing the operon containing mercuric ion reductase and organomercurial lyase, the transgenic lines were tested by a bioassay using the extremely toxic organomercurial phenyl mercuric acetate (PMA). The chloroplast transgenic plants were shown to be substantially more resistant than untransformed wild-type tobacco plants grown under similar conditions. The 16-day-old tobacco plants (seedlings) were capable of growth in soil containing PMA concentrations of 50 and 100 μM, and were even able to survive concentrations as high as 200 μM (Figure 10.7), even though they absorbed several hundred fold more PMA than did untransformed controls plants, which struggled to survive at concentration of 50 μM (Figure 10.7). When nuclear transgenic seedlings containing *mer*A and *mer*B genes were germinated in a medium containing PMA, they were capable of resistance to concentrations of only 5 μM (Bizily *et al.*, 2000). In sharp contrast, when chloroplast transgenic tobacco plants were treated with 100, 200, 300, and 400 μM PMA, they showed an increase in the total dry weight when compared with the untransformed wild-type plants grown at similar concentrations; the total dry weight of the untransformed plants progressively decreased with each increase in PMA concentration from 0 to 400 μM. The chlorophyll of the leaf is an indicator of the chloroplast structural and functional integrity. When 15-mm-diameter leaf discs from the untransformed wild-type and chloroplast transgenic

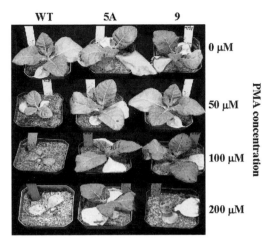

Figure 10.7 Effect of PMA concentration on the growth of wild-type or transgenic tobacco lines. Plants were treated with 200 mL Hoagland's nutrient solution supplemented with 0, 50, 100, and 200 µM PMA. Photographs were taken 14 days after treatment. WT: negative control Petit Havana, 5A: pLDR-MerAB transgenic line, 9: pLDR-MerAB-3'UTR transgenic line.

plants were grown for 10 days in 10 µM PMA, the chlorophyll concentration of the transgenic plants increased, while it decreased in untransformed plants. These bioassays show the efficiency and activity of the chloroplast-expressed *mer*A and *mer*B operon enzymes and establish that plastid genetic engineering can be used successfully for phytoremediation. This is the first report of chloroplast genetic engineering for the purpose of phytoremediation.

10.7   Transgenic plastids as bioreactors

There are several reasons for utilizing plastid genetic engineering for the purpose of expressing foreign proteins and using them as bioreactors. Plant systems are more economical than industrial facilities using fermentation systems; technology is available for harvesting and processing plants and plant products on a large scale. Elimination of the expensive purification requirement when the plant tissue containing the recombinant protein is used as a food (for oral delivery of therapeutic proteins) greatly reduces the cost of vaccines and biopharmaceuticals. When foreign genes are expressed directly in chloroplasts, the amount of recombinant protein produced approaches industrial levels; health risks due to contamination with potential human pathogens/toxins are minimized in chloroplast derived proteins.

A remarkable feature of chloroplast genetic engineering is the observation of exceptionally large accumulation of bacterial proteins in transgenic plants, as much as 46% of CRY protein in tsp, even in bleached old leaves (DeCosa *et al.*, 2001). Stable expression of a pharmaceutical protein in chloroplasts was first reported for GVGVP, a protein-based polymer with varied medical applications such as the prevention of postsurgical adhesions and scars, wound coverings, artificial pericardia, tissue reconstruction, and programmed drug delivery (Guda *et al.*, 2000). Subsequently, expression of the human somatotropin (hST) via the tobacco chloroplast genome (Staub *et al.*, 2000) to high levels (7% of tsp) was observed; chloroplast-derived hST was shown to be properly disulfide bonded and fully functional. Peptides as small as 20 amino acids (magainin; DeGray *et al.*, 2001) or as large as 83 kDa (anthrax protective antigen; Daniell *et al.*, 2004) have been expressed in transgenic chloroplasts. Vaccine antigens that require oligomeric proteins with stable disulfide bridges (Daniell *et al.*, 2001a) or large fusion proteins (for plague; Singleton, 2003) or viral epitopes fused with transmucosal carriers (for canine parvovirus; Molina *et al.*, 2004) have been expressed in transgenic chloroplasts.

Both the viral canine parvovirus 2L21 peptide (Molina *et al.*, 2004) and bacterial fragment C of tetanus toxin (TetC; Tregoning *et al.*, 2003) vaccine antigens have been expressed to very high levels (up to 31.1 and 25% tsp, respectively) via chloroplast genetic engineering. Chloroplast-derived TetC intranasal-immunized mice were both immunogenic and immunoprotective against pathogen challenge, and the chloroplast-derived 2L21 peptide was shown to be immunogenic. Various therapeutic proteins expressed in transgenic chloroplasts are summarized in Table 10.2 and a few examples are described below.

10.7.1 Human somatotropin

Human somatotropin (hST) is utilized in the treatment of hypopituitary dwarfism in children, Turner syndrome, chronic renal failure, and HIV wasting syndrome. It is produced in the pituitary gland and contains a pair of disulfide bonds. Staub *et al.* (2000) expressed hST in transgenic tobacco chloroplasts by site-specific integration of transgenes between the *trn*V gene and *rps*7/3′–*rps*12 operon located in the inverted repeat region of the plastid genome and achieved a maximum of 7% hST in the total soluble protein. Upon purification of hST, it was demonstrated that proper disulfide bond formation had occurred and the plastid-derived hST was identical to native human hST. The results of these experiments have demonstrated that plastids possess the proper machinery to fold eukaryotic proteins and add disulfide bonds, possibly utilizing the chloroplast enzyme protein disulfide isomerase. This report demonstrates the feasibility of using plastid transformation for hyperexpression of human blood proteins.

10.7.2 Human serum albumin

Human serum albumin (HSA) accounts for approximately 60% of proteins in blood serum. It is the most widely distributed intravenous protein and is prescribed in multigram quantities to replace blood volume in trauma and in various other clinical situations. HSA is a monomeric 66.5-kDa protein which contains 17 disulfide bonds. The current need for HSA worldwide exceeds 500 tons, with an approximate market value of $1.5 billion. Current systems for producing HSA to date are not

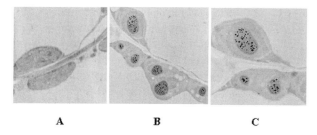

A   B   C

Figure 10.8 HSA accumulation in transgenic chloroplasts. Electron micrographs of immunogold-labeled tissues from untransformed (A) or transformed (B, C) mature leaves using the chloroplast vector pLDApsbAHSA. Magnifications are (A) ×10,000, (B) ×5000, and (C) ×6300.

yet commercially feasible. Initial attempts at expressing HSA in nuclear transgenic plants have achieved disappointingly low levels of HSA (0.02% tsp). Current estimates by industry suggest that the cost-effective yield for pharmaceutical production is 0.1 mg of HSA per gram of fresh weight (Farran *et al.*, 2002). Fernandez-San Millan *et al.* (2003) utilizing chloroplast transformation technology, expressed HSA in transgenic chloroplasts using two different 5′-regulatory sequences; in both cases, transgenes were integrated between the *trn*I/*trn*A genes within the inverted repeat region of the plastid genome. The Shine Dalgarno (SD) ribosome binding site (ggagg) construct was regulated by the *Prrn* promoter (transcribed as a polycistron) and the light-regulated psbA 5′ untranslated region (UTR) (translation enhancer) was driven by its own promoter (transcribed as a monocistron). The chloroplast transgenic plants showed a 360-fold difference in expression levels, with the SD construct at 0.02% total protein (tp), and the 5′UTR construct at 7.2% tp. Since the psbA 5′UTR is light regulated, chloroplast transgenic plants were subjected to continuous illumination. The maximum HSA levels were observed when the transgenic plants were exposed to 50 h of continuous illumination in which expression levels reached up to 11.2% tp in the mature green leaves. The phenotypes of the chloroplast transgenic plants expressing HSA were identical to those of untransformed wild-type plants. High levels of HSA expression resulted in the formation of inclusion bodies (Figure 10.8) and the transgenic chloroplasts increased in size to accommodate the foreign protein. These inclusion bodies not only protected HSA from proteolytic degradation but also facilitated single-step purification by centrifugation. The HSA molecule has a chemical and structural function rather than an enzymatic activity; therefore, complex studies are necessary to fully demonstrate the functionality of HSA (Watanabe *et al.*, 2001).

10.7.3 Antimicrobial peptide

There are many human pathogenic bacteria that are resistant to known drugs or have acquired that trait over time. Therefore the need to explore alternate ways to combat such bacteria is essential. Magainin and its analogs have been investigated as a broad-spectrum topical agent, a systemic antibiotic, a wound-healing stimulant, and an anticancer agent. A magainin analog MSI-99, a synthetic lytic peptide, has been expressed via the tobacco chloroplast genome (DeGray *et al.*, 2001). This AMP is an amphipathic alpha helix molecule that possess an affinity for negatively charged phospholipids present in the outer membrane of all bacteria. The probability that bacteria can adapt to the lytic activity of this synthetic peptide is very low. It was observed that this lytic peptide accumulated in transgenic chloroplasts at extremely high levels (~21.5% tsp). *Pseudomonas aeruginosa*, a multidrug-resistant Gram-negative bacteria, which acts as an opportunistic pathogen in plants, animals, and humans, was used for *in vitro* assays to determine the effectiveness of the lytic peptide expressed in tobacco chloroplasts (Figure 10.9). Cell extracts that were prepared from T_1 generation plants resulted in 96% inhibition of growth of this opportunistic pathogen. The results observed are highly encouraging for exploration treatments against drug-resistant bacteria in general and particularly to cystic fibrosis patients because of their high susceptibility to *P. aeruginosa*. Pharmaceutical companies are exploring the use of lytic peptides as broad-spectrum topical antibiotics and systemic antibiotics. Reports have demonstrated that the outer leaflet of melanoma and colon carcinoma cells express three- to sevenfold more phosphatidylserine than their noncancerous counterparts. Previous studies have shown that analogs of magainin 2 were quite effective against hematopoietic, melanoma, sarcoma, and ovarian teratoma cell lines. The preference of this lytic peptide for negatively charged phospholipids makes MSI-99 an ideal candidate for an anticancer agent.

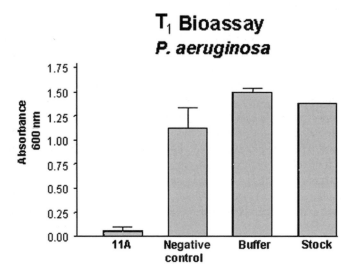

Figure 10.9 *In vitro* bioassay for T1 generation (11A) against *Pseudomonas aeruginosa*. Bacterial cells from an overnight culture were diluted to A_{600} 0.1–0.3 and incubated for 2 h at 25°C with 100 μg of total protein extract. One milliliter of Luria-Bertani Broth was added to each sample and incubated overnight at 26°C. Absorbance at 600 nm was recorded. Data were analyzed using GraphPad Prism.

10.7.4 Human interferon alpha

Human cytokines of the immune system termed interferon alphas (IFNαs) are known to interfere with viral replication and cell proliferation. They are also known as potent enhancers of the immune response and have many uses in clinical treatments. A specific subtype, IFNα2b, was first approved in 1986 by the Food and Drug Administration to treat patients with hairy cell leukemia. It has also shown efficacy in a growing number of treatments for various viral and malignant diseases. However, the recombinant IFNα2b, now on the market, is produced by an *E. coli* expression system, and because of the necessary *in vitro* processing and purification processes, the average cost for treatment is $26,000 per year. Recently, recombinant IFNα2b is being used to treat patients suffering from West Nile virus and the cost for a 2-week treatment is $2500 per patient. IFNα2b is administered by injection and severe side effects are quite common. Also, up to 20% of patients produce anti-IFNα antibodies when IFNα2b aggregates with HSA in blood. These antibodies are not desirable because they lessen the effectiveness of the treatment. The negative side effects have been linked to the route of administration and dosage parameters. It has been demonstrated that oral administration of natural human IFNα is therapeutically useful for the treatment of various infectious diseases.

Therefore, IFNα2b expressed via the tobacco chloroplast genome for oral delivery may eliminate some of the side effects. A recombinant IFNα2b containing a polyhistidine tag (for single-step purification) and a thrombin cleavage site was generated and introduced into the tobacco chloroplast genome of petit Havana and into a low nicotine variety of tobacco, LAMD-609 (Falconer, 2002). It is well known that chloroplasts possess the ability to correctly process and fold human proteins as well as form disulfide bonds. Also, bioencapsulation by plant cells can protect recombinant proteins from degradation in the gastrointestinal tract, and plant-derived expression systems are free of human pathogens. Western blotting detected both monomers and multimers of IFNα2b in both the tobacco varieties, using interferon alpha monoclonal antibody. Southern blot analysis both confirmed stable site-specific integration of transgenes into the chloroplast genome and determined homoplasmy. In the Petit Havana chloroplast transgenic lines, homoplasmy of chloroplast genomes occurred in the first generation and this corresponds to the highest level of IFNα2b expression. Quantification of IFNα2b recombinant protein by ELISA showed 18.8% tsp in Petit Havana and up to 12.5% in LAMD-609 in T_0 transgenic plants. The next generation (T_1) of the Petit Havana chloroplast transgenic lines had accumulated even higher levels, clearly observable in a commassie-stained SDS-PAGE (Figure 10.10). In contrast to these observations, interferon gamma could be barely observed (0.1% tsp) in transgenic chloroplasts unless they were fused with β-glucuronidase (GUS, see below; Leelavathi and Reddy, 2003). IFNα2b functionality was determined by the ability of IFNα2b to protect HeLa cells against the cytopathic effect of encephalomyocarditis (EMC) virus and RT-PCR. Chloroplast-derived IFNα2b was shown to be just as active as commercially produced Intron A. The mRNA levels of two genes induced by IFNα2b, $2'$–$5'$ oligoadenylate synthase and STAT-2, were tested by RT-PCR using primers specific

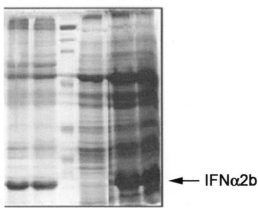

1 2 M PH 3 4

◄— IFNα2b

Figure 10.10 Commassie-stained SDS-PAGE. Untransformed or transgenic lines expressing high levels of IFNα2b. Lanes 1 and 2: tsp(total soluble protein); Lane M: protein marker; Lanes PH: untransformed Petit Havana; Lanes 3 and 4: tp (total protein).

for each gene. Chloroplast-derived IFNα2b induced expression of both genes the same way as commercially available IFNα2b. The expression levels observed and proper functionality are ideal for purification and further use in oral IFNα2b delivery or preclinical trials (Daniell, 2004).

10.7.5 Human interferon gamma

Leelavathi and Reddy (2003) chose the human interferon gamma (INF-γ) for their expression and purification studies because of important medical applications. INF-γ is a major cytokine of the immune system that interferes with viral replication, prevents proliferation, and has several other roles in immunoregulatory actions in response to pathogenic bacteria and viruses. Researchers generated nuclear or chloroplast transgenic lines, expressing INF-γ, either independently or as a GUS fusion protein. Eight tobacco nuclear transgenic lines expressing INF-γ were screened and the highest level of expression obtained was 0.001% tsp. In order to compare GUS protein accumulation along with INF-γ, the two genes were expressed in separate tobacco chloroplast transgenic plants. The estimated levels of expression of GUS protein and INF-γ were 3 and 0.1% (100-fold greater than nuclear transgenic lines), respectively. Pulse labeling experiments demonstrated that GUS had a ∼48-h half-life whereas INF-γ had a relatively short half-life of ∼4–6 h.

A recombinant fusion protein GUS/INF-γ was then created and expressed in tobacco chloroplast transgenic lines. Based on western blot analysis, it was estimated that the GUS/INF-γ fusion protein had accumulated to 6% tsp. Pulse labeling experiments demonstrated that the GUS/INF-γ had a half-life of ∼48 h, which is similar to that of GUS alone. GUS/INF-γ fusion protein was purified to homogeneity by a two-step His-tag based chromatography purification scheme. Utilizing this procedure, researchers were able to obtain ~360 μg of fusion protein from 1 g of leaf tissue with a >75% recovery. Upon cleavage of the fusion with factor Xa, GUS estimated recovery was ~210 μg/g fresh weight tissue and INF-γ had an estimated recovery of ~40 μg/g fresh weight tissue. It was demonstrated that the INF-γ was highly pure, as observed by silver stain and the protein cross-reacted with anti-INF-γ antibody. A bioassay was performed with the purified INF-γ, and it was shown to offer complete protection to human lung carcinomas against infection with the EMC virus. This demonstrated that INF-γ produced in the chloroplast of tobacco plants was just as biologically active as native human INF-γ.

10.7.6   Insulin-like growth factor 1

Human insulin-like growth factor 1 (IGF-1) acts as a potent multifunctional anabolic hormone that is produced in the liver. IGF-1 is a 7.6-kDa protein composed of 70 amino acids and has three disulfide bonds. IGF-1 functions in the regulation of cell proliferation and differentiation in a variety of human cell and tissue types and also plays an important role in tissue renewal and repair. One cirrhotic patient requires 600 mg of IGF-1 per year and the cost of IGF-1 is $30,000 per mg (Nilsson et al., 1991). IGF-1 is currently produced in E. coli but the protein is not produced in a mature form, because E. coli is unable to form disulfide bonds in the cytoplasm. In order to increase expression levels, a synthetic IGF-1 gene that contains chloroplast-preferred codons was synthesized. Integration of the IGF-1 gene into the tobacco chloroplast genome was confirmed by PCR and Southern blot analyses. Western blotting analysis confirmed the presence of IGF-1 protein in large quantities within the tobacco chloroplast transgenic plants. Quantification by ELISA of tobacco chloroplast transformants showed that the IGF-1 protein had accumulated up to 32% tsp, in both the human native gene and the synthetic gene (Figure 10.11; Ruiz, 2002). However, determination of expression levels was complicated by the interaction of IGF-1 antibody with the zz-tag used for purification. If confirmed, these observations demonstrate that the chloroplast translation machinery is quite flexible, unlike the bacterial translation machinery that translated only the synthetic chloroplast codon-optimized IGF-1 gene (Daniell, 2004).

10.7.7   Guy's 13 – monoclonal antibody against dental cavities

Monoclonal antibodies possess remarkable specificity and a therapeutic nature for defined targets; therefore, monoclonal antibodies are emerging as therapeutic drugs at a fast rate. The chloroplast genome was chosen for transformation with antibody genes because of (i) tremendously high levels of foreign protein expression, (ii) ability to fold, process, assemble foreign proteins with disulfide bridges, and (iii) transgene containment via maternal inheritance (Daniell, 2004). In order to enhance translation of a monoclonal antibody, the Guy's 13 gene was codon optimized and placed under the control of specific 5′UTR. Guy's 13 monoclonal antibody targets

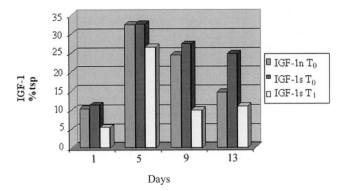

Figure 10.11 Expression of IGF-1 in transgenic chloroplasts under continuous light exposure (13 days). Plant tissue was collected at different times and IGF-1 expression is shown as a percentage of total soluble protein. IGF-1n is the native gene and IGF-1s is the chloroplast codon optimized gene. T_0 is the first generation and T_1 is the second generation.

the surface antigen of *Streptococcus mutans*, a bacterium that causes dental cavities. IgA-G, a humanized, chimeric monoclonal antibody (Guy's 13), has been successfully synthesized in transgenic tobacco chloroplasts with proper disulfide bridges (Daniell and Wycoff, 2001). This study confirmed integration of the Guy's 13 gene into the tobacco chloroplast genome by PCR and Southern blot analyses. Western blot analysis revealed the expression of heavy and light chains individually on a reducing gel, as well as the fully assembled antibody on a nonreducing gel (Figure 10.12). However, the levels of expression should be further enhanced to meet

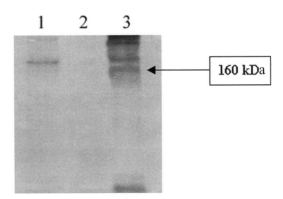

Figure 10.12 Western blot analysis of transgenic lines showing the assembled antibody in transgenic chloroplasts. Lane 1: extract from a transgenic line; Lane 2: negative control-extract from an untransformed plant; Lane 3: positive control-human IgA. The gel was run under nonreducing conditions. The blot was developed with AP-conjugated goat anti-human kappa antibody.

commercial feasibility. These results may suggest the presence of chaperones for proper protein folding and enzymes for formation of disulfide bonds within tobacco chloroplasts (Daniell, 2004).

10.7.8 Vaccines

Subunit vaccines produced in the plastid of transgenic plants offer an attractive alternative to current vaccines that are injected. Besides oral delivery, the low costs associated with its purification should make plastid-derived subunit vaccines the wave of the future. For developing countries that cannot afford vaccines because of the high cost of production, storage, and transportation, elimination of the "cold chain," a practice that must keep the vaccines refrigerated during long-time delivery/ storage, is a major advantage. It is well known that subunit vaccines expressed in plants are capable of inducing a mucosal immune response when given parenterally or orally in experimental animals, and these animals are also able to withstand a pathogen challenge. One major concern about plant-derived oral vaccines is the degradation of the proteins in the stomach and gut before an immune response can be elicited. The advantage in using plant cells is that they contain cellulose and sugars, which provide protection to the subunit vaccine in the stomach and is released gradually into the gut; this is also known as *bioencapsulation* (Mor *et al.*, 1998).

Arntzen *et al.* (1997) has demonstrated that nuclear-transformed tomato and potato plants were capable of synthesizing the subunit vaccines for Norwalk virus, enterotoxingenic *E. coli*, *Vibrio cholerae*, and the hepatitis B surface antigen (Langridge, 2000). The hepatitis B surface antigen and Norwalk virus capsid protein expressed via the nuclear genome in potato have been shown to elicit a serum immunoglobulin response when fed to mice and humans (Mason *et al.*, 1996; Kapusta *et al.*, 1999; Richter *et al.*, 2000; Tacket *et al.*, 2000; Kong *et al.*, 2001). The expression levels seen in nuclear transformants are very low and an increased dosage is required for an effective immune response; therefore, the most logical step is to hyperexpress subunit vaccines in the plastids of transgenic plants.

In order for the oral delivery of vaccines to be successful, two requirements must be fulfilled. First, large quantities of vaccine antigens should be expressed in edible parts of plants; the following examples demonstrate that it is possible to express large quantities of vaccine antigens. Also, as described above, it is possible to transform chromoplasts of carrot and tomato. The second requirement is the need to transform plastids without the use of antibiotic resistance genes or delete selectable markers. Several such examples are described below.

10.7.9 Antibiotic free selection using BADH

A major concern of the public is that genetically engineered plants containing antibiotic resistance genes may inactivate oral doses of the antibiotic (Daniell *et al.*, 2001c). Another major concern is the transfer of antibiotic resistance genes to pathogenic microbes in the gastrointestinal tract or soil creating antibiotic resistance pathogens

(Daniell *et al.*, 2001b). The BADH gene from spinach encodes an enzyme that converts toxic betaine aldehyde (BA) to nontoxic glycine betaine, which also serves as an osmoprotectant (drought/salt tolerance; Rathinasabapathy *et al.*, 1994). It has been demonstrated that the chloroplast transformation efficiency was 25-fold higher in BA than spectinomycin; also, rapid regeneration of transgenic shoots was observed (Daniell *et al.*, 2001b). The public concern regarding antibiotic resistance genes should be eased with the use of BADH as a selectable marker because the gene for BADH is naturally present in plant species that are routinely consumed (e.g. spinach).

10.7.10 Selectable marker excision

Another approach to develop marker-free transgenic plants is to eliminate the antibiotic resistance gene after transformation. Strategies have been developed to remove genes using endogenous chloroplast recombinases that delete the marker genes via engineered direct repeats (Iamtham and Day, 2000). Early experiments with *C. reinhardtii* showed that it is possible to exploit these recombination events to eliminate introduced selectable marker genes. Homologous recombination between two direct repeats, engineered to flank a selectable marker, enabled marker removal under nonselective growth conditions (Fischer *et al.*, 1996). A similar approach applied in tobacco was effective in generating aadA-free T_0 transplastomic lines while leaving a third unflanked transgene, *bar*, in the genome to confer herbicide resistance (Iamtham and Day, 2000). Recently, another strategy to eliminate selectable marker genes has been developed, using the P1 bacteriophage CRE/lox site-specific recombination system. In two separate studies, a marker gene flanked by lox sites was introduced into the tobacco chloroplast genome. Removal of marker genes was subsequently induced by expression of the CRE protein via the nuclear genome (Corneille *et al.*, 2001; Hajdukiewicz *et al.*, 2001). These reports show that efficient removal of selectable marker(s) from chloroplast genomes is feasible.

10.7.11 Cholera vaccine

The cholera toxin B (CTB) subunit encoded in the genome of *V. cholerae* is an A–B holotoxin. The B subunit forms a pentamer that binds to the GM_1-ganglioside receptor, which is present on all nucleated mammalian cells and is predominantly abundant in the intestinal epithelial cells. The B subunit contributes to toxicity to the host when the A subunit binds to it, and upon entry into the cell it leads to ADP-ribosylation of G protein α which inactivates GTPase, thus activating adenylate cyclase which affects signal transduction. The heat-labile toxic B subunit of cholera toxin has been shown to be a very powerful adjuvant and a potential candidate for a subunit vaccine. Site-directed integration of the native CTB gene lacking the leader sequence was expressed in the chloroplasts of transgenic tobacco plants; CTB antigen accumulated up to 4.1% tsp (Daniell *et al.*, 2001a). The results not only showed the accumulation of CTB but also enabled the pentameric protein to be properly assembled within the chloroplast; this is essential because of the critical role of quaternary structure necessary for the proper folding of the subunit vaccine. The functionality of the protein was demonstrated by binding aggregates of assembled pentamers in plant extracts similar to purified bacterial antigen, and binding assays confirmed that chloroplast synthesized and bacterial CTB both bind to the intestinal membrane GM_1-ganglioside receptor (Figure 10.13). These results confirm that the tobacco chloroplast can correctly fold vaccine antigens and form disulfide bonds. Subsequent studies using translational enhancer elements resulted in hyperexpression (up to 31% of tsp) of CTB in transgenic chloroplasts and proper assembly of pentamers (Molina *et al.*, 2004).

10.7.12   Anthrax vaccine

Unfortunately, inhalation anthrax makes for an ideal biological warfare agent for several reasons: it is almost always fatal unless treated immediately; weapons grade spores can be produced and stored for decades; and the spores can be spread by rockets, missiles, aerial bombs, and even through the mail. *Bacillus anthracis* is the causative agent of anthrax. It is a spore-forming, nonmotile, aerobic bacterium. The mode of action in humans is as follows: Protective antigen (PA) binds to anthrax toxin receptor followed by the formation of a heptamer, which then accommodates either the edema factor (EF) or the lethal factor (LF). When the EF binds PA, this toxin increases cyclic AMP levels in the cell, which upsets the water homeostasis, resulting in the accumulation of fluid, termed edema. When the LF binds to PA, this toxin stimulates macrophages to release interleukin 1β, tumor necrosis factor α, and other cytokines, which leads to shock and sudden death. The only current vaccine licensed for human use in the United States is the Anthrax Vaccine Absorbed (AVA) produced by BioPort Corporation in Lansing, Michigan. The current problem that

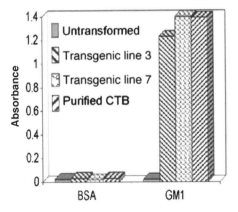

Figure 10.13   CTB–GM_1-ganglioside binding ELISA assay. Plates, coated first with GM_1-ganglioside and BSA, respectively, were plated with total soluble plant protein from chloroplast transgenic lines (3 and 7) and untransformed plant total soluble protein and 300 ng of purified bacterial CTB. The absorbance of the GM1-ganglioside–CTBantibody complex in each case was measured at 405 nm.

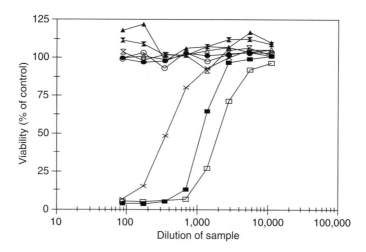

Figure 10.14 Macrophage cytotoxic assays for extracts from untransformed and chloroplast transgenic plants. Supernatant samples from T_1 pLD-JW1 tested (proteins extracted in buffer containing no detergent and MTT added after 5 h). ■ pLD-JW1 (extract stored 2 days); □ pLD-JW1 (extract stored 7 days); × PA 5 μg/ml; ● Control wild type (extract stored 2 days); ○ Control wild type (extract stored 7 days); ▲ Control wild type no LF (extract stored 2 days); △ Control wild type no LF (extract stored 7 days); ✕ Control pLD-JW1 no LF (extract stored 2 days); ⊠ Control pLD-JW1 no LF (extract stored 7 days).

exists with this vaccine is that it is produced by a cell-free filtrate of a toxigenic, nonencapsulated strain of *B. anthracis* (Baillie, 2001). The immunogenic portion termed PA is bound to an adjuvant, but trace amounts of the EF and LF are present in the filtrate, which are toxic.

Taken altogether, it is imperative to devise a clean and safer vaccine for anthrax. Therefore, the PA gene (*pag*) has been introduced into the chloroplasts of transgenic tobacco plants (Daniell *et al.*, 2004). Chloroplast transgenic plants contained up to 2.5 mg PA/g fresh weight. The functionality of PA was confirmed by macrophage lysis assays. Chloroplast-derived PA was capable of efficiently binding to anthrax toxin receptor (formed heptamers), underwent proper cleavage, and bound to LF, followed by internalization of LF; completion of a sequence of these events resulted in macrophage lysis (Figure 10.14). Quantitative analysis of these events yielded up to 25 μg of functional PA per mL in transgenic leaves. With observed expression levels, 400 million doses of vaccine (free of EF and LF) could be produced per acre of transgenic tobacco using an experimental cultivar in a green house, which could be further enhanced 18-fold utilizing a commercial cultivar in the field to combat bioterrorism.

10.7.13   Plague vaccine

The Gram-negative bacterium *Yersinia pestis* is the causative agent of the plague and causes three different forms of plague: Bubonic plague, the most common, is carried by an infected flea; upon entry into the lymph system this pathogen then drains into the lymph nodes, causing swelling termed *bubos*. Septacemic plague occurs when bacteremia without bubos is present, and pneumonic plague occurs when the pathogen colonizes the alveolar spaces within the lung. The current vaccine for the plague is a killed whole vaccine, which is only moderately effective against the bubonic form and ineffective against the pneumonic and septicemic plague. Several subunit vaccines have been produced for the plague and the most effective are CaF1 and LcrV. The F1 capsular protein is located on the surface of the pathogen and may have antiphagocytic properties. The V antigen is part of the organism's Type III secretion system and may form an injectosome. It has been demonstrated that both the F1 and V antigens elicit a protective immune response individually, but the additive effect of both F1 and V has been shown to be more protective when mice were immunized. An F1/V fusion protein has proven to be a very effective subunit vaccine against both subcutaneous and aerosolized challenges and has the potential to induce protective immunity against pneumonic as well as bubonic plague. Singleton (2003) has expressed the F1/V fusion protein in chloroplasts of transgenic tobacco plants. The maximum levels of expression that were attained of the F1/V antigen was 14.8% tsp. Future studies are needed to determine if the plastid-derived F1/V fusion protein can be both immunogenic and immunoprotective against pathogen challenge.

10.7.14   Canine parvovirus anti-viral animal vaccine

Canine parvovirus (CPV) infects dogs and other Canidae family members such as wolves, coyotes, South American dogs, and Asiatic raccoon dogs. Infection leads to hemorrhagic gastroenteritis and myocarditis. CPV-1 was the first strain of parvovirus discovered, initially reported in 1967. It only posed a medical threat to newborn puppies. However in 1978, a new strain, CPV-2, appeared in the United States. The new CPV-2 strain appeared to represent a mutated form of the feline parvovirus, most commonly known as *feline distemper virus*. Infected animals shed CPV-2 in large numbers, which is resilient in the environment. Therefore, worldwide distribution of the virus has occurred rapidly. Attempts to protect a puppy from exposure to this virus are completely ineffective. Another mutant form termed CPV-2a was identified in 1979 and determined to be even more aggressive. The vaccine needed for prevention was in short supply and many veterinarians substituted the distemper vaccine because it was the most similar vaccine available. Nowadays, the current form of the virus is termed CPV-2b and young animals are commonly vaccinated with an attenuated viral vaccine against the virus.

Recent advances in molecular biology have lead to the identification of CPV antigens that are capable of eliciting a protective immune response. In this case, a linear antigenic peptide (2L21) from the VP2 capsid protein (amino acids 1–23) of CPV was selected and its DNA sequence was successfully utilized to transform the chloroplasts of tobacco plants (Molina *et al.*, 2004). The 2L21 synthetic peptide, chemically coupled to a KLH carrier protein, has been extensively researched and has been demonstrated to effectively protect dogs and minks against parvovirus infection (Langeveld *et al.*, 1994, 1995). This peptide has been expressed in nuclear transgenic plants as an N-terminal fusion with GUS (Gil *et al.*, 2001). The 2L21

peptide was fused to either the CTB subunit or GFP and has been shown to be efficiently expressed and accumulated in transgenic chloroplasts at levels up to 10-fold
greater than those previously reported in nuclear transformation. The expression
levels were dependent on the age of the plant. Both young and senescent chloroplast transgenic plants accumulated lower amount of recombinant protein than was
observed in mature plants, demonstrating that the time of harvest is important when
attempting to scale up the process of recombinant protein production. The maximum level of expression of CTB-2L21 was 7.49 mg/g fresh weight, which equates to
31.1% tsp; the GFP-2L21 transgenic plants expressed 5.96 mg/g fresh weight, which
equates to 22.6% tsp. The chloroplast expressed 2L21 epitope was detected with
a CPV-neutralizing monoclonal antibody, which indicated that the epitope is correctly positioned at the C-terminus of the fusion proteins. The tobacco-chloroplast
derived CTB-2L21 fusion protein retained the ability to form pentamers and bind
GM_1-ganglioside receptor, which are characteristics of native CTB. Antibodies that
were induced recognized the VP2 protein from CPV in intraperitoneal immunized
mice (Figure 10.15). These results demonstrate that chloroplast-derived CTB-2L21
fusion protein is immunogenic when administered intraperitoneally, confirmed by
its ability to induce a humoral response that can cross-react with native VP2 protein. Additional experiments are underway to test the ability of this fusion protein
to induce specific immune responses following mucosal delivery. This is the first
report of an animal vaccine epitope expressed in transgenic tobacco chloroplasts.

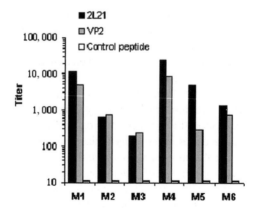

Figure 10.15   Titers of antibodies at day 50 induced by chloroplast-derived CTB-2L21 recombinant protein. Balb/c mice were intraperitoneally immunized with leaf extract from CTB-2L21
transgenic plants. Animals were boosted at days 21 and 35. Each mouse received 20 μg/mouse
of CTB-2L21 recombinant protein. Individual mice sera were titrated against 2L21 synthetic
peptide, VP2 protein and control peptide (amino acids 122–135 of hepatitis B virus surface
antigen) Titers were expressed as the highest serum dilution to yield two times the absorbance
mean of preimmune sera. M1–M6: mouse 1 to 6. 2L21: epitope from the VP2 protein of the
canine parvovirus; CTB: cholera toxin B; VP2: protein of the canine parvovirus that includes
the 2L21 epitope.

10.8   Biomaterials, enzymes, and amino acids

10.8.1   Chorismate pyruvate lyase

Chorismate pyruvate lyase (CPL) is an enzyme encoded by the *ubi*C gene present in *E. coli*. CPL catalyzes the direct conversion of chorismate to pyruvate and *p*-hydroxybenzoic acid (pHBA). pHBA is the principal monomer found in all commercial thermotropic liquid crystal polymers. Chorismate, the substrate for CPL, is an intermediate of the shikimate pathway, which is located in chloroplasts. CPL is not normally present in plants; instead, the entire reaction leading from chorismate to pHBA involves up to 10 successive enzymatic reaction steps as opposed to the direct conversion with CPL. In order to accumulate high levels of pHBA, tobacco plants were transformed via the chloroplast genome to express the CPL enzyme. The plastid transgenic plants achieved homoplasmy and were morphologically indistinguishable from untransformed and did not show any pleitrophic effects, in spite of the high accumulation of pHBA (Devine *et al.*, submitted).

The pHBA in the T_0 line, in normal light/dark cycle (16 h light/8 h dark), continued to steadily increase up to 100 days in soil, reaching a dry weight of $\sim$15% pHBA glucose conjugates in mature leaves. Upon switch to continuous illumination, after 22 days this further increased pHBA accumulation to about 25%. This shows that the 5′UTR in continuous light increases translation of the *ubi*C gene. A few transgenic plants were sacrificed to determine the yield of pHBA in the total plant leaf material (young, mature, and old) or the total stalk. In T_0 transgenic lines, the pHBA levels varied between 10.87 and 15.22 in total leaves and between 1.65 and 2.87 in total stalk. These plants were morphologically identical to untransformed plants.

Seeds from T_0 lines were then germinated on spectinomycin (500 μg/mL), transferred to jars for another round of selection, and then transferred to soil. Once these plants reached optimum size of about a meter (73 days in soil), grown in a normal light/dark cycle, they were transferred to continuous illumination. After 27 days in continuous light, pHBA accounted for 17.44% dry weight in old leaves (100 days in soil). By day 93 in continuous light, the pHBA content had reached about 25% dry weight. The T_1 plants were morphologically indistinguishable from untransformed tobacco plants even after 93 days in continuous light (167 days in soil).

The highest CPL activity observed in chloroplast transgenic T_1 lines in the total leaf material was 50,783 pkat/mg of protein and this equates to approximately 30% tsp. Such high levels of expression are not surprising because foreign proteins up to 46.1% of tp have been reported in transgenic chloroplasts (DeCosa *et al.*, 2001). In sharp contrast, the highest level observed in nuclear transgenic plants was only 208 pkat/mg of protein. The pHBA accumulation in leaf sample is 50-fold higher in transgenic chloroplasts when compared to nuclear expression whereas the CPL enzyme activity in leaves is about 240-fold higher than in nuclear transgenic plants. The 5-fold difference between CPL enzyme activity and pHBA accumulation indicates that chorismate may be the limiting factor; thus there is an upper limit to pHBA production in transgenic plants but observed levels are well within commercial feasibility. The CPL activity in the T_1 generation whole stalk was analyzed and the results showed 8378 pkat/mg of protein ($\sim$5% tsp); the CPL activity in the whole stalk is 40-fold higher than any other data published and this is 250-fold higher in the leaf material.

10.8.2   Polyhydroxybutyrate

Lossl et al. (2003) expressed the pathway for the synthesis of polyhydroxybutyrate (PHB), a polyester used as biodegradable plastics and elastomers, in the chloroplasts of transgenic tobacco plants. There are three bacterial enzymes encoded in the polycistronic *phb* operon in which transcription of the three genes is divided into two transcripts, with *phb*B being transcribed separately from *phb*C and *phb*A. First, condensation of two molecules of acetyl-CoA is catalyzed by β-ketothiolase to form acetoacetyl-CoA; acetyl-CoA reductase then reduces acetoacetyl-CoA to β-hydroxybutyryl-CoA, which is then polymerized by PHB synthase to PHB. The chloroplast transgenic lines reached up to 1.7% dry weight PHB. Unfortunately, even in the mature plants expressing low levels of PHB, male sterility and stunted growth were observed.

10.8.3   Xylanase

Leelavathi et al. (2003) expressed the alkali-thermostable xylanase gene from *Bacillus* sp. strain NG-6 within the chloroplast of tobacco plants. Alkali-thermostable xylanases are of considerable interest because of their uses in paper, fiber, baking, brewing, and animal feed industries (Biely, 1985). In spite of the several uses for xylanases, they are not routinely utilized because of its high production cost. Previous nuclear transgenic plants expressing xylanases encountered the problem of degradation of cell wall component, thereby affecting their growth. Therefore, xylanases were targeted to the apoplast (Herbers et al., 1995), seed oil body (Liu et al., 1997), or by secretion of enzyme through roots into the culture medium (Borisjuk et al., 1999). In all the above nuclear transgenic approaches, the expression levels were too low to make it economically feasible.

Therefore researchers transformed the *xyn*A gene into the chloroplast genome of tobacco plants. A zymography assay demonstrated direct evidence for the presence of xylanase in its biologically active form in leaf tissues. It was estimated that the expression of xylanase was $\sim$6% tsp, and the estimated activity was 140,755 U/kg of fresh leaf tissue. Nuclear expression of a thermostable xylanase from *Clostridium thermocellum* accumulated only 0.1–0.3% tsp. Since chloroplast-derived xylanase was thermostable, researchers devised a purification technique that utilized heat in the first step in order to possibly increase yield by reducing proteolytic degradation by plant proteases. Researchers first heated the crude protein (60–70°C for 30 min) extract and followed this with anion-exchange chromatography. There was a strong correlation observed between enzyme activity and the intensity of the protein band on an SDS-PAGE gel. Several generations of transgenic lines were tested and all of them appeared to be no different from untransformed control in height, chloroplhyll content, flowering time, and biomass. Drying out of the transgenic leaves in the sun or at 42°C showed that greater than 85% activity could be recovered. The same results were seen in leaves undergoing senescence (85% recovery). Characterization of chloroplast-derived xylanase showed that the enzyme is biologically as active as the bacterially produced enzyme in the pH range of 6–11, with peak activity at pH 8.4 (Leelavathi *et al.*, 2003). Also, the chloroplast-derived xylanase retained substrate specificity. This study demonstrates that chloroplasts are quite capable of expressing industrially important cellulolytic enzymes that cannot be expressed at high levels via the nuclear genome because of adverse affects.

10.8.4 Amino acid biosynthesis: ASA2 – anthranilate synthase alpha subunit

It is well known that plastids possess the ability to synthesize some of the essential amino acids in higher plants. The majority of enzymes involved in amino acid biosynthesis are encoded in the nucleus, synthesized in the plant cytosol, and then transported to plastids. Tryptophan (Trp) biosynthesis branches off the shikimate pathway at chorismate, which is the last common precursor of many aromatic compounds. Anthranilate synthase (AS) catalyzes the first committed reaction for Trp biosynthesis, converting chorismate to anthranilate, and undergoes feedback inhibition by the end product (Haslam, 1993). Plant AS holoenzyme has been shown to be made up of tetramers of two α and two β subunits that are encoded in the nucleus and synthesized in the cytosol and targeted to plastids via transit peptide. Upon entering the plastid, the transit peptide is cleaved and the holoenzyme tetramer is assembled. The β subunit acts as an aminotransferase that cleaves Gln to make ammonia, which the α subunit utilizes to convert chorismate to anthranilate. The α subunit binds to Trp and acts as a feedback inhibitor, and can utilize free ammonia as an alternative substrate *in vitro* (Bohlmann *et al.*, 1995). Biosynthesis of Trp and other related secondary products in plants is tightly controlled, not only through Trp feedback inhibition of the AS enzyme, but also by regulation of the abundance of AS mRNA (Radwanski and Last, 1995). It has been shown that a tobacco suspension culture selected for its resistance to the Trp analog termed 5-methyl-Trp (5MT) contains Trp feedback insensitive AS enzyme activity and high levels of free Trp (Brotherton *et al.*, 1986). Therefore the cDNA encoding the AS α subunit (ASA2) was cloned.

Zhang *et al.* (2001) expressed the *ASA2* gene within the plastid genome of transgenic tobacco plants. Their theory was that the regulation of ASA2 transcription within the nucleus could be overcome by expression within plastids. Researchers also wanted to determine what effects a plastid-expressed ASA2 gene would have on the expression of nuclear-encoded AS genes and the β subunit genes, and to determine if Trp biosynthesis can be increased in plastids. The chloroplast transgenic plants showed a high level of accumulation of ASA2 mRNA, an increased expression of the AS α subunit protein, and demonstrated a 4-fold increase in AS enzyme activity that was less sensitive to feedback inhibition by Trp. Researchers also observed a 10-fold increase in free Trp in chloroplast transgenic tobacco leaves. Upon immunological probing the researchers detected a much higher level of the α subunit compared to wild type, indicating that the abundance of the α subunit encoded by ASA2 may stabilize the β subunit, which may explain the increase in the holoenzyme (Zhang *et al.*, 2001).

10.9   Conclusions

Plastid transformation has opened the door to engineer valuable agronomic traits, production of industrially valuable biomaterials and therapeutic proteins such as antibodies, biopharmaceuticals, or vaccine antigens. The high-level expression in transgenic chloroplasts holds the promise of unlimited quantities of therapeutic proteins to people around the world, produced at lower cost. However, purification methods such as chromatography are still very expensive. Oral delivery of therapeutic proteins should completely eliminate the need for expensive purification. Recent advancements augur well for production of therapeutic proteins in transgenic plastids as well as an environmentally friendly approach for introduction of agronomic traits in various crops.

Acknowledgements

Results of investigations from the Daniell laboratory were supported in part by the United States Department of Agriculture (Grant 3611-21000-017-00D) and the National Institutes of Health (Grant R01 GM 63879).

References

Arntzen, C.J., Mason, H.S., Tariq, H.A. and Clements JD. (1997) Oral immunization with transgenic plants. US Patent WO96/12801.

Baillie, L. (2001) The development of new vaccines against *Bacillus anthracis*. *J. Appl. Micro.*, 91, 609–613.

Biely, P. (1985) Microbial xylanolytic enzymes. *Trends Biotechnol.*, 3, 286–290.

Birky, C.W. (1995) Uniparental inheritance of mitochondrial and chloroplast genes: mechanisms and evolution. *Proc. Natl. Acad. Sci. U.S.A.*, 92, 11331–11338.

Bizily, S.P., Rugh, C.L. and Meagher, R.B. (2000) Phytodetoxification of hazardous organomercurials by genetically engineered plants. *Nat. Biotechnol.*, 18, 213–217.

Blowers, A.D., Bogorad, L., Shark, K.B. and Sanford, J.C. (1989) Studies on *Chlamydomonas* chloroplast transformation: foreign DNA can be stably maintained in the chromosome. *Plant Cell*, 1, 123–132.

Bogorad, L. (2000) Engineering chloroplasts: an alternative site for foreign genes, proteins, reactions and products. *Trends Biotechnol.*, 18, 257–263.

Bohlmann, J., De Luca, V., Eilert, U. and Martin, W. (1995) Purification and cDNA cloning of anthranilate synthase from *Ruta graveolens*: modes of expression and properties of native and recombinant enzymes. *Plant J.*, 7, 491–501.

Borisjuk, N.V., Borisjuk, L.J., Logendra, S., Petersen, F., Gleba, Y. and Raskin, I. (1999) Production of recombinant proteins in plant root exudates. *Nat. Biotechnol.*, 17, 466–469.

Boyton, J.E., Gillham, N.W., Harris, E.H. *et al.* (1988) Chloroplast transformation in *Chlamydomonas* with high velocity microprojectiles. *Science*, 240, 1534–1538.

Brotherton, J.E., Hauptmann, R.M. and Widholm, J.M. (1986) Anthranilate synthase forms in plants and cultured cells of *Nicotiana tabacum* L. *Planta*, 168, 214–221.

Cerutti, H., Osman, M., Grandoni, P. and Jagendorf, A.T. (1992) A homolog of *Escherichia coli* RecA protein in plastids of higher plants. *Proc. Natl. Acad. Sci. U.S.A.*, 89, 8068–8072.

Corneille, S., Lutz, K., Svab, Z. and Maliga, P. (2001) Efficient elimination of selectable marker genes from the plastid genome by the CRE-lox site-specific recombination system. *Plant J.*, 27, 171–178.

Cramer, C.L., Boothe, J.G., and Oishi, K.K. (1999) Transgenic plants for therapeutic proteins: linking upstream and downstream strategies. *Curr. Top. Microbiol. Immunol.*, 240, 95–118.

Daniell, H. (1999) Environmentally friendly approaches to genetic engineering. *In Vitro Cell Dev. Biol. Plant*, 35, 361–368.

Daniell, H. (2000) Genetically modified crops: current concerns and solutions for next generation crops. *Biotechnol. Gen. Engr. Rev.*, 17, 327–352.

Daniell, H. (2002) Molecular strategies for gene containment in transgenic crops. *Nat. Biotechnol.*, 20, 581–586.

Daniell, H. (2004) Medical molecular pharming: therapeutic recombinant antibodies, biopharmaceuticals, and edible vaccines in transgenic plants engineered via the chloroplast genome. In *Encyclopedia of Plant and Crop Science* (ed. R.M. Goodman), Marcel Decker, New York, pp. 704–710.

Daniell, H., Datta, R., Varma, S., Gray, S. and Lee, S.B. (1998) Containment of herbicide resistance through genetic engineering of the chloroplast genome. *Nat. Biotechnol.*, 16, 345–348.

Daniell, H. and Dhingra, A. (2002) Multigene engineering: dawn of an exciting new era in biotechnology. *Curr. Opin. Biotechnol.*, 13, 136–141.

Daniell, H., Kahn, M. and Allison, L. (2002) Milestones in chloroplast genetic engineering: an environmentally friendly era in biotechnology. *Trends Plant Sci.*, 7 (2), 84–91.

Daniell, H., Krishnan, M. and McFadden, B.F. (1991) Expression of β-glucoronidase gene in different cellular compartments following biolistic delivery of foreign DNA into wheat leaves and calli. *Plant Cell Rep.*, 9, 615–619.

Daniell, H., Lee, S.B., Panchal, T. and Wiebe, P.O. (2001a) Expression of cholera toxin B subunit gene and assembly as functional oligomers in transgenic tobacco chloroplasts. *J. Mol. Biol.*, 311, 1001–1009.

Daniell, H. and McFadden, B.A. (1987) Uptake and expression of bacterial and cyanobacterial genes by isolated cucumber etioplasts. *Proc. Natl. Acad. Sci. U.S.A.*, 84, 6349–6353.

Daniell, H., Muthukumar, B. and Lee, S.B. (2001b) Marker free transgenic plants: engineering the chloroplast genome without the use of antibiotic selection. *Curr. Genet.*, 39, 109–116.

Daniell, H. and Parkinson, C.L. (2003) Jumping genes and containment. *Nat. Biotechnol.*, 21, 374–375.

Daniell, H., Ruiz, O.N. and Dhingra, A. (2004) Chloroplast genetic engineering to improve argonomic traits. *Methods Mol. Biol.*, 286, 111–137.

Daniell, H., Vivekananda, J., Nielsen, B.L., Ye, G.N., Tewari, K.K. and Sanford, J.C. (1990) Transient foreign gene expression in chloroplasts of cultured tobacco cells after biolistic delivery of chloroplast vectors. *Proc. Natl. Acad. Sci. U.S.A.*, 87, 88–92.

Daniell, H., Watson, J., Koya, V. and Leppla, SH. (in press) Expression of *Bacillus anthracis* protective antigen in transgenic tobacco chloroplasts: development of an improved anthrax vaccine in a non-food/feed crop. *Vaccine*.

Daniell, H., Wiebe, P. and Fernandez-San Millan, A. (2001c) Antibiotic-free chloroplast genetic engineering – an environmentally friendly approach. *Trends Plant Sci.*, 6, 237–239.

Daniell, H. and Wycoff, K. (2001) Production of antibodies in transgenic plastids. WO 01/64929.

DeCosa, B., Moar, W., Lee, S.B., Miller, M. and Daniell, H. (2001) Overexpression of the *Bt* cry2Aa2 operon in chloroplasts leads to formation of insecticidal crystals. *Nat. Biotechnol.*, 19, 71–74.

DeGray, G., Rajasekaran, K., Smith, F., Sanford, J. and Daniell, H. (2001) Expression of an antimicrobial peptide via the chloroplast genome to control phytopathogenic bacteria and fungi. *Plant Physiol.*, 127, 852–862.

Devine, A.L., Kahn, S., Deuel, D.L., Van Dyk, D.E., Viitanen, P.V. and Daniell, H. (submitted) Metabolic engineering via the chloroplast genome to produce 4-hydroxybenzoic acid a principle onomer of liquid crystal polymers.

Falconer, R. (2002) *Expression of Interferon alpha 2b in transgenic chloroplasts of a low-nicotine tobacco.* M.S. thesis, University of Central Florida, Orlando, FL.

Farran, I., Sanchez-Serrano, J.J., Medina, J.F., Prieto, J. and Mingo-Castel, AM. (2002) Targeted expression of human serum albumin to potato tubers. *Transgenic Res.*, 11, 337–346.

Fernandez-San Millan, A., Mingeo-Castel, A.M., Miller, M. and Daniell, H. (2003) A chloroplast transgenic approach to hyper-express and purify human serum albumin, a protein highly susceptible to proteolytic degradation. *Plant Biotechnol. J.*, 1, 71–79.

Fischer, N., Stampacchia, O., Redding, K. and Rochaix, J.D. (1996) Selectable marker recycling in the chloroplast. *Mol. Gen. Genet.*, 251, 373–380.

Gil, F., Brun, A., Wigdorovitz, A. *et al.* (2001) High yield expression of a viral peptide vaccine in transgenic plants . *FEBS Lett.*, 488, 13–17.

Golds, T., Maliga, P. and Koop, HU. (1993) Stable plastid transformation in PEG-treated proto-plasts of *Nicotiana tabacum. Bio/Technology*, 11, 95–97.

Goldschmidt-Clermont, M. (1991) Transgenic expression of aminoglycoside adenine transferase in the chloroplast: a selectable marker for site-directed transformation of *Chlamydomonas. Nucl. Acids Res.*, 19, 4083–4089.

Guda, C., Lee, S.B. and Daniell, H. (2000) Stable expression of biodegradable protein based polymer in tobacco chloroplasts. *Plant Cell Rep.*, 19, 257–262.

Hagemann, R. (1992) Plastid genetics in higher plants. In *Cell Organelles* (ed. R.G. Herrmann), Springer-Verlag, Berlin, pp. 65–69.

Hagemann, R. (in press) The sexual inheritance of plant organelles. In *Molecular Biology and Biotechnology of Plant Organelles*, (eds. H. Daniel and C. Chase), Kluwer Academic Pub-lishers, Dordrecht.

Hajdukiewicz, P.T., Gilbertson, L. and Staub, J.M. (2001) Multiple pathways for Cre/lox-mediated recombination in plastids. *Plant J.*, 27, 161–170.

Haslam, E. (1993) *Shikimic Acid: Metabolism and Metabolites*, John Wiley & Sons, Chichester, UK.

Herbers, K., Wilke, I. and Sonnewald, U.A. (1995) Thermostable xylanase from *Clostridium ther-mocellum* expressed at high levels in the apoplast of transgenic tobacco has no detrimental effects and is easily purified. *Bio/Technology*, 13, 63–66.

Holmstrom, K.O., Mantyla, E., Welin, B. *et al.* (1996) Drought tolerance in tobacco. *Nature*, 379, 683–684.

Iamtham, S. and Day, A. (2000) Removal of antibiotic resistance genes from transgenic tobacco plastids. *Nat. Biotechnol.*, 18, 1172–1176.

Kapusta, J., Modelska, A., Figlerowicz, M. *et al.* (1999) A plant-derived edible vaccine against hepatitis B virus. *FASEB J.*, 13, 1796–1799.

Klein, T.M., Wolf, E.D. and Sanford, J.C. (1987) High-velocity microprojectiles for delivering nucleic acids into living cells. *Nature*, 327, 70–73.

Kong, Q., Richter, L., Yang, Y.F., Arntzen, C., Mason, H. and Thanavala, Y. (2001) Oral immu-nization with hepatitis B surface antigen expressed in transgenic plants. *Proc. Natl. Acad. Sci. U.S.A.*, 98, 11539–11544.

Kota, M., Daniell, H., Varma, S., Garczynski, S.F., Gould, F. and William, M.J. (1999) Over-expression of the *Bacillus thuringiensis* (Bt) Cry2Aa2 protein in chloroplasts confers resistance to plants against susceptible and *Bt*-resistant insects. *Proc. Natl. Acad. Sci. U.S.A.*, 96, 1840–1845.

Kumar, S. and Daniell, H. (2004) Engineering the chloroplast genome for hyperexpression of human therapeutic proteins and vaccine antigens. *Methods Mol. Biol.*, 267, 365–385.

Kumar, S., Dhingra, A. and Daniell, H. (submitted-a) Highly efficient stable genetic transformation of carrot plastid via somatic embryogenesis to confer salt tolerance. *Plant Physiol.*

Kumar, S., Dhingra, A. and Daniell, H. (submitted-b) Manipulation of gene expression facilitates cotton plastid transformation via somatic embryogenesis and maternal inheritance of transgenes. *Plant Mol. Bio.*

Kusnadi, A., Nikolov, G. and Howard, J. (1997) Production of recombinant proteins in plants: practical considerations. *Biotechnol. Bioeng.*, 56, 473–484.

Langeveld, J.P., Casal, J.I., Osterhaus, A.D. *et al.* (1994) First peptide vaccine providing protection against viral infection in the target animal: studies of canine parvovirus in dogs. *J. Virol.*, 68, 4506–4513.

Langeveld, J.P., Kamstrup, S., Uttenthal, A. *et al.* (1995) Full protection in mink against enteritis virus with new generation canine parvovirus vaccines based on synthetic peptide or recombinant protein. *Vaccine*, 13, 1033–1037.

Langridge, W. (2000) Edible vaccines. *Sci. Am.*, 283, 66–71.

Lee, S.B., Kwon, H.B., Kwon, S.J. *et al.* (2003) Accumulation of trehalose within transgenic chloroplasts confers drought tolerance. *Mol. Breed.*, 11, 1–13.

Leelavathi, S., Gupta, N., Maiti, S., Ghosh, A. and Reddy, V.S. (2003) Overproduction of an alkali- and thermo-stable xylanase in tobacco chloroplasts and efficient recovery of the enzyme. *Mol. Breed.*, 11, 59–67.

Leelavathi, S. and Reddy, V.S. (2003) Chloroplast expression of His-tagged GUS-fusions: a general strategy to overproduce and purify foreign proteins using transplastomic plants as bioreactors. *Mol. Breed.*, 11, 49–58.

Lilly, J.W., Havey, M.J., Jackson, S.A. and Jiang, J.M. (2001) Cytogenomic analyses reveal the structural plasticity of the chloroplast genome in higher plants. *Plant Cell*, 13, 245–254.

Liu, J.H., Selinger, L.B., Cheng, K.J., Beauchemin, K.A. and Moloney, M.M. (1997) Plant seed oil-bodies as an immobilization matrix for a recombinant xylanase from the rumen fungus *neocallimastix patriciarum*. *Mol. Breed.*, 3, 463–470.

Lossl, A., Eibl, C., Harloff, H.J., Jung, C. and Koop, H.U. (2003) Polyester synthesis in transplastomic tobacco (*Nicotiana tabacum* L.): significant contents of polyhydroxybutyrate are associated with growth reduction. *Plant Cell Rep.*, 21, 891–899.

Lutz, K.A., Knapp, J.E. and Maliga, P. (2001) Expression of bar in the plastid genome confers herbicide resistance. *Plant Physiol.*, 125, 1585–1590.

Mason, H.S., Ball, J.M., Shi, J.J., Jiang, X., Estes, M.K. and Arntzen, C.J. (1996) Expression of Norwalk virus capsid protein in transgenic tobacco and potato and its oral immunogenecity in mice. *Proc. Natl. Acad. Sci. U.S.A.*, 93, 5335–5340.

Mason, H.S., Haq, T.A., Clements, J.D. and Arntzen C.J. (1998) Edible vaccine protects mice against *Escherichia coli* heat-labile enterotoxin (LT): potatoes expressing a synthetic LT-B gene. *Vaccine*, 16, 1336–1343.

McBride, K.E., Svab, Z., Schaaf, D.J., Hogan, P.S., Stalker, D.M. and Maliga, P. (1995) Amplification of a chimeric *Bacillus* gene in chloroplasts leads to an extraordinary level of an insecticidal protein in tobacco. *Bio/Technology*, 13, 362–365.

Mogensen, H.L. (1996) The hows and whys of cytoplasmic inheritance in seed plants. *Am. J. Bot.*, 83, 383–404.

Molina, A., Herva-Stubbs, S., Daniell, H., Mingo-Castel, A.M. and Veramendi, J. (2004) High yield expression of a viral peptide animal vaccine in transgenic tobacco chloroplasts. *Plant Biotechnol. J.*, 2, 141–153.

Mor, T.S., Gomez-Lim, M.A. and Palmer, K.E. (1998) Perspective: edible vaccines – a concept coming of age. *Trends Microbiol.*, 6, 449–453.

Nagata, N., Saito, C., Sakai, A., Kuroiwa, H. and Kuroiwa, T. (1999) The selective increase or decrease of organellar DNA in generative cells just after pollen mitosis one controls cytoplasmic inheritance. *Planta*, 209, 53–65.

Nilsson, B., Forsberg, G. and Hartmanis, M. (1991) Expression and purification of recombinant insulin-like growth factors from *E. coli*. *Methods Enzymol.*, 198, 3–16.

Radwanski, E.R. and Last, R.L. (1995) Tryptophan biosynthesis and metabolism: biochemical and molecular genetics. *Plant Cell*, 7, 921–934.

Rathinasabapathy, B., McCue, K.F., Gage, D.A. and Hanson, A.D. (1994) Metabolic engineering of glycine betaine synthesis: plant betaine aldehyde dehydrogenasese lacking a typical transit peptides are targeted to tobacco chloroplasts where they confer aldehyde resistance. *Planta*, 193, 155–162.

Richter, L.J., Thanavala, Y., Arntzen, C.J. and Mason, H.S. (2000) Production of hepatitis B surface antigen in transgenic plants for oral immunization. *Nat. Biotechnol.*, 18, 1167–1171.

Ruf, S., Hermann, M., Berger, I., Carrer, H. and Bock, R. (2001) Stable genetic transformation of tomato plastids and expression of a foreign protein in fruit. *Nat. Biotechnol.*, 19, 870–875.

Ruiz, G. (2002) *Optimization of Codon Composition and Regulatory Elements for Expression of the Hhuman IGF-1 in Transgenic Chloroplasts*. M.S. thesis, University of Central Florida, Orlando, FL.

Ruiz, O.N., Hussein, H., Terry, N. and Daniell, H. (2003) Phytoremediation of organomercurial compounds via chloroplast genetic engineering. *Plant Physiol.*, 132, 1–9.

Sanford, J.C., Smith, F.D. and Russell, J.A. (1993) Optimizing the biolistic process for different biological applications. *Methods Enzymol.*, 217, 483–509.

Scott, S.E. and Wilkenson, M.J. (1999) Low probability of chloroplast movement from oilseed rape (*Brassica napus*) into wild *Brassica rapa*. *Nat. Biotechnol.*, 17, 390–392.

Sidorov, V.A., Kasten, D., Pang, S.Z., Hajdukiewicz, P.T., Staub, J.M. and Nehra, N.S. (1999) Technical advance: stable chloroplast transformation in potato: use of green fluorescent protein as a plastid marker. *Plant J.*, 19, 209–216.

Singleton, M.L. (2003) *Expression of CaF1 and LcrV as a Fusion Protein for a Vaccine Against Yersinia pestis via Chloroplast Genetic Engineering*. M.S. thesis, University of Central Florida, Orlando, FL.

Staub, J.M., Garcia, B., Graves, J. *et al.* (2000) High-yield production of a human therapeutic protein in tobacco chloroplasts. *Nat. Biotechnol.*, 18, 333–338.

Svab, Z. and Maliga, P. (1993) High-frequency plastid transformation in tobacco by selection for a chimeric *aadA* gene. *Proc. Natl. Acad. Sci. U.S.A.*, 90, 913–917.

Tacket, C.O., Mason, H.S., Losonsky, G., Estes, M.K., Levine, M.M. and Arntzen, C.J. (2000) Human immune responses to a novel norwalk virus vaccine delivered in transgenic potatoes. *J. Infect. Dis.*, 182, 302–305.

Torres, M. (2002) *Expression of Interferon α5 in Transgenic Chloroplasts of Tobacco*, M.S. thesis, University of Central Florida, USA.

Tregoning, J.S., Nixon, P., Kuroda, H. *et al.* (2003) Expression of tetanus toxin Fragment C in tobacco chloroplasts. *Nucleic Acids Res.*, 31, 1174–1179.

Watanabe, H., Yamasaki, K., Kragh-Hansen, U. *et al.* (2001) *In vitro* and *in vivo* properties of recombinant human serum albumin from *Pichia pastoris* purified by a method of short processing time. *Pharm. Res.*, 18, 1775–1781.

Ye, G.N., Hajdukiewicz, P.T.J., Broyles, D. *et al.* (2001) Plastid-expressed 5-enolpyruvylshikimate-3-phosphate synthase genes provide high level glyphosate tolerance in tobacco. *Plant J.*, 25, 261–270.

Zhang, X.H., Brotherton, J.E., Widholm, J.M. and Portis, A.R., Jr. (2001) Targeting a nuclear anthranilate synthase subunit gene to the tobacco plastid genome results in enhanced tryptophan biosynthesis. Return of a gene to its pre-endosymbiotic origin. *Plant Physiol.*, 127, 131–141.

Index

DATE D